THERMODYNAMIC AND TRANSPORT PROPERTIES

THERMODYNAMIC AND TRANSPORT PROPERTIES

CLAUS BORGNAKKE

RICHARD E. SONNTAG

University of Michigan

John Wiley & Sons, Inc.
New York Chichester Brisbane Toronto Singapore Weinheim

Acquisitions Editor Cliff Robichaud
Marketing Manager Harper Mooy
Senior Production Editor Nancy Prinz
Text Designer Jack Hilton Cunningham
Cover Designer Dawn Stanley
Illustration Editor Sigmund Malinowski
Manufacturing Manager Dorothy Sinclair

Recognizing the importance of preserving what has been written, it is a policy of John Wiley & Sons, Inc. to have books of enduring value published in the United States printed on acid-free paper, and we exert our best efforts to that end.

ISBN: 0-471-12170-3 (paper)

Printed in the United States of America

10 9 8 7 6 5 4 3 2

This book of *Tables of Thermodynamic and Transport Properties* consists of an expanded set of tables from the textbook *Fundamentals of Classical Thermodynamics, Fourth Edition,* by Gordon J. Van Wylen, Richard E. Sonntag and Claus Borgnakke, John Wiley & Sons, 1994. We have added tables for many substances not included in the textbook appendix, and also added tables and charts for several transport properties, in order to make this book more useful in calculations involving fluid mechanics and heat transfer applications.

All the tables and property values are listed in SI metric units, and selected tables are also printed in standard English units. The computer disk that accompanies this book includes both sets, as well as a number of alternative choices (as, for example, a choice of temperature in Celsius or Kelvin in the SI-unit tables). There are also numerous choices of sets of independent properties by which the disk thermodynamic tables can be entered, for the convenience of the user.

It is our hope that these tables will be useful to the practicing engineer working with thermal-fluid systems, and also to students and faculty in the educational environment. We would certainly appreciate any comments or suggestions for inclusion of other properties or substances in future editions of this book.

CLAUS BORGNAKKE
RICHARD E. SONNTAG
Ann Arbor, Michigan
January 1997

A	Area m^2
B	Second virial coefficient m^3/kmol
c	velocity of sound m/s
c	mass fraction
C_p	constant-pressure specific heat
C_v	constant-volume specific heat
e, E	specific energy and total energy
f	fugacity kPa or MPa
g	acceleration due to gravity
g	specific Gibbs function
g_c	a constant that relates force, mass, length, and time
h	specific enthalpy
h	heat transfer coefficient
k	conductivity
K	equilibrium constant
L	length
m	mass
M	molecular mass
M	Mach number
n	number of moles
P	pressure
P_r	relative pressure as used in gas tables
q, Q	heat transfer per unit mass and total heat transfer
R	gas constant
$\bar{R}$	universal gas constant
s	specific entropy
t	time
T	temperature
u	specific internal energy
v, V	specific volume and total volume
v_r	relative specific volume as used in gas tables
$\mathbf{V}$	velocity
x	quality
Z	compressiblity factor

GREEK LETTERS

α	diffusivity
β	volume expansivity
β_S	adiabatic compressibility
γ	specific heat ratio: C_p/C_v
μ	viscositgy, dynamic
μ	viscosity, hinematic
ν	stoichiometric coefficient
ρ	density
Φ	equivalence ratio
ϕ	relative humidity
ω	acentric factor
ω	humidity ratio or specific humidity
σ	surface tension

SUBSCRIPTS

c	property at the critical point
f	formation
f	property of saturated liquid
fg	difference in property for saturated vapor and saturated liquid
g	property of saturated vapor
i	property of saturated solid
ig	difference in property for saturated vapor and saturated solid
r	reduced property
0	stagnation property

SUPERSCRIPTS

$\bar{}$	bar over symbol denotes property on a molal basis
$^\circ$	property at standard-state condition
*	ideal gas
*	property at the throat of a nozzle

Fundamental Physical Constants

Avogadro	$N_0 = 6.022\ 136 \times 10^{23}$ l/mol
Boltzmann	$k = 1.380\ 658 \times 10^{-23}$ J/K
Planck	$\underline{h} = 6.626\ 076 \times 10^{-34}$ Js
Gas Constant	$\overline{R} = N_0 k = 8.314\ 51$J/mol K
Atomic Mass Unit	$m_0 = 1.660\ 540 \times 10^{-27}$ kg
Velocity of Light	$c = 2.997\ 925 \times 10^{8}$ m/s
Electron Charge	$e = 1.602\ 177 \times 10^{-19}$ C
Electron Mass	$m_e = 9.109\ 389 \times 10^{-31}$ kg
Proton Mass	$m_p = 1.672\ 623 \times 10^{-27}$ kg
Neutron Mass	$m_n = 1.674\ 929 \times 10^{-27}$ kg
Gravitation (Std.)	$g = 9.806\ 65$ m/s^2
Stefan-Boltzman	$\sigma = 5.670\ 51 \times 10^{-8}$ W/m^{2}K^4

Prefixes

10^{-1}	deci	d
10^{-2}	centi	c
10^{-3}	milli	m
10^{-6}	micro	mu
10^{-9}	nano	n
10^{-12}	pico	p
10^{-15}	femto	f
10^{1}	deka	da
10^{2}	hecto	h
10^{3}	kilo	k
10^{6}	mega	M
10^{9}	giga	G
10^{12}	tera	T
10^{15}	peta	P

Concentration

10^{-6}	parts per million ppm

SECTION A

SINGLE STATE PROPERTIES

TABLE A.1 *Conversion Factors*

Area

1 mm^2 = 1.0×10^{-6} m^2	1 ft^2 = 144 in.^{2}
1 cm^2 = 1.0×10^{-4} m^2 = 0.1550 in.^{2}	1 in.^{2} = 6.4516 cm^2 = 6.4516×10^{-4} m^2
1 m^2 = 10.7639 ft^2	1 ft^2 = 0.092 903 m^2

Conductivity

1 W/m-K = 1 J/s-m-K	
= 0.577 789 Btu/h-ft-R	1 Btu/h-ft-R = 1.730 735 W/m-K

Diffusivity

1 m^2/s = 10.7639 ft^2/s	1 ft^2/s = 0.092903 m^2/s

Density

1 kg/m^3 = 0.06242797 lbm/ft^3	1 lbm/ft^3 = 16.018 46 kg/m^3

Energy

1 J = 1 N-m = 1 kg-m^2/s^2	
1 J = 0.737 562 lbf-ft	1 lbf-ft = 1.355 818 J
1 cal (Int.) = 4.1868 J	= 1.28507×10^{-3} Btu
	1 Btu (Int.) = 1.055 056 kJ
1 erg = 1.0×10^{-7} J	= 778.1693 lbf-ft
1 eV = $1.602\,177\,33 \times 10^{-19}$ J	

Force

1 N = 0.224809 lbf	1 lbf = 4.448 222 N

Gravitation

g = 9.80665 m/s^2	g = 32.17405 ft/s^2

Heat capacity, specific entropy

1 kJ/kg-K = 0.238 846 Btu/lbm-R	1 Btu/lbm-R = 4.1868 kJ/kg-K

Heat flux (per unit area)

1 W/m^2 = 0.316 998 Btu/h-ft^2	1 Btu/h-ft^2 = 3.15459 W/m^2

Heat transfer coefficient

1 W/m^2-K = 0.176 11 Btu/h-ft^2-R	1 Btu/h-ft^2-R = 5.67826 W/m^2-K

Length

1 mm = 0.001 m = 0.1 cm	1 ft = 12 in.
1 cm = 0.01 m = 10 mm = 0.3970 in.	1 in. = 2.54 cm = 0.0254 m
1 m = 3.28084 ft = 39.370 in.	1 ft = 0.3048 m
1 km = 0.621 371 mi	1 mi = 1.609344 km
1 mi = 1609.3 m (US statute)	1 yd = 0.9144 m

(Continued)

TABLE A.1 (Continued) ***Conversion Factors***

Mass

1 kg = 2.204 623 lbm	1 lbm = 0.453 592 kg
1 ton = 1000 kg	1 slug = 14.5939 kg
1 grain = 6.47989 10^{-5} kg	

Moment (torque)

1 N-m = 0.737 562 lbf-ft	1 lbf-ft = 1.355 818 N-m

Momentum (mV)

1 kg-m/s = 7.232 94 lbm-ft/s	1 lbm-ft/s = 0.138 256 kg-m/s
= 0.224809 lbf-s	

Power

1 W	= 1 J/s = 1 N-m/s	1 lbf-ft/s	= 1.355 818 W
	= 0.737 562 lbf-ft/s		= 4.626 24 Btu/h
1 kW	= 3412.14 Btu/h	1 Btu/s	= 1.055 056 kW
1 hp (metric)	= 0.735 499 kW	1 hp (UK)	= 0.7457 kW
			= 550 lbf-ft/s
			= 2544.43 Btu/h
1 ton of refrigeration	= 3.516 85 kW	1 ton of refrigeration	= 12 000 Btu/h

Pressure

1 Pa	= 1 N/m^2 = 1 kg/m-s^2	1 lbf/in.^{2}	= 6.894 757 kPa
1 bar	= 1.0×10^5 Pa = 100 kPa		
1 atm	= 101.325 kPa	1 atm	= 14.695 94 lbf/in.^{2}
	= 1.01325 bar		= 29.921 in. Hg [32 F]
	= 760 mm Hg [0°C]		= 33.899 5 ft H_2O [4°C]
	= 10.332 56 m H_2O [4°C]		
1 torr	= 1 mm Hg [0°C]		
1 mm Hg [0°C]	= 0.133 322 kPa	1 in. Hg [0°C]	= 0.49115 lbf/in.^{2}
1 m H_2O [4°C]	= 9.806 38 kPa	1 in. H_2O [4°C]	= 0.036126 lbf/in.^{2}

Specific energy

1 kJ/kg	= 0.42992 Btu/lbm	1 Btu/lbm	= 2.326 kJ/kg
	= 334.55 lbf-ft/lbm	1 lbf-ft/lbm	= 2.98907×10^{-3} kJ/kg
			= 1.28507×10^{-3} Btu/lbm

Specific kinetic energy (V^2)

1 m^2/s^2	= 0.001 kJ/kg	1 ft^2/s^2	= 3.9941×10^{-5} Btu/lbm
1 kJ/kg	= 1000 m^2/s^2	1 Btu/lbm	= 25037 ft^2/s^2

Specific potential energy (Zg)

1 m-g_{std}	= 9.80665×10^{-3} kJ/kg	1 ft-g_{std}	= 1.0 lbf-ft/lbm
	= 4.21607×10^{-3} Btu/lbm		= 0.001285 Btu/lbm
			= 0.002989 kJ/kg

(Continued)

TABLE A.1 (Continued) *Conversion Factors*

Specific volume

$1\ cm^3/g = 0.001\ m^3/kg$

$1\ m^3/kg = 16.018\ 46\ ft^3/lbm$ | $1\ ft^3/lbm = 0.062\ 428\ m^3/kg$

Temperature

1 K = 1 °C = 1.8 R = 1.8 F | 1 R = (5/9) K

TC = TK − 273.15 = (TF − 32)/1.8 | TF = TR − 459.67 = 1.8 TC + 32

TK = TR/1.8 | TR = 1.8 TK

Thermal resistance

1 K/W = 0.52753 R-h/Btu | 1 R-h/Btu = 1.89563 K/W

Universal Gas Constant

$R = N_0\, k$ = 8.31451 kJ/kmol-K
= 1.98589 kcal/kmol-K
= 82.0578 atm-L/kmol-K

R = 1.98589 Btu/lbmol-R
= 1545.36 lbf-ft/lbmol-R
= $0.73024\ atm\text{-}ft^3/lbmol\text{-}R$
= $10.7317\ (lbf/in.^2)\text{-}ft^3/lbmol\text{-}R$

Velocity

1 m/s = 3.6 km/h
= 3.28084 ft/s
= 2.23694 mi/h

1 km/h = 0.27778 m/s
= 0.91134 ft/s
= 0.62137 mi/h

1 ft/s = 0.681818 mi/h
= 0.3048 m/s
= 1.09728 km/h

1 mi/h = 1.46667 ft/s
= 0.44704 m/s
= 1.609344 km/h

Viscosity (absolute)

1 kg/s-m = 1000 cP = 1 Pa-s
= 0.6719695 lbm/s-ft

1 P (Poise) = 1 g/s-cm = 0.1 kg/s-m

1 lbm/s-ft = 1.488163 kg/s-m

Viscosity (kinematic)

Same units as diffusivity

$1\ m^2/s = 1 \times 10^4$ Stokes | $1\ Stoke = 1\ cm^2/s$

Volume

$1\ m^3 = 35.3147\ ft^3$

$1\ L = 1\ dm^3 = 0.001\ m^3$

1 Gal (US) = 3.785 412 L
= $3.785\ 412 \times 10^{-3}\ m^3$

$1\ ft^3 = 2.831\ 685 \times 10^{-2}\ m^3$

$1\ in.^3 = 1.6387 \times 10^{-5}\ m^3$

1 Gal (UK) = 4.546 090 L

$1\ Gal\ (US) = 231.00\ in.^3$

TABLE A.2 ***Dimensionless Numbers***

Biot number	$Bi = hL/k_{solid}$	Fluid convection/solid conduction
Boussinesq	$Bo = g\beta\Delta TL^3/\alpha^2$	Bouyancy/diffusivity
Eckert number	$Ec = \mathbf{V}^2/C_p\Delta T$	Kinetic energy/enthalpy
Euler number	$Eu = \Delta p/\rho\mathbf{V}^2$	Pressure force/inertia
Fourier number	$Fo = \alpha t/L^2$	Dimensionless time
Froude number	$Fr = \mathbf{V}/(gL)^{1/2}$	Inertia/gravity force
Graetz number	$Gz = L^2\mathbf{V}/\alpha x$	Regrouping as Gz = Pe L/x
Grashof number	$Gr = g\beta\Delta TL^3/\nu^2$	Buoyancy/viscous force
Lewis number	$Le = \alpha/D = Sc/Pr$	Thermal/species diffusivity
Mach number	$Ma = \mathbf{V}/c$	Dimensionless velocity
Nusselt number	$Nu = hL/k$	Convection/conduction
Peclet number	$Pe = \mathbf{V}L/\alpha = Re\ Pr$	Inertia/conductivity
Prandtl number	$Pr = \nu/\alpha$	Friction/conductivity
Rayleigh number	$Ra = g\beta\Delta TL^3/\alpha\nu$	Regrouping as Ra = Gr Pr
Reynolds number	$Re = \mathbf{V}L/\nu$	Inertia/ friction force
Schmidt number	$Sc = \nu/D = Le\ Pr$	Friction/species diffusivity
Sherwood number	$Sh = hL/D$	Heat transfer/diffusivity
Stanton number	$St = h/\rho C_p\mathbf{V}$	Heat transfer/convection energy
Stefan number	$Ste = C_{p,liq}\Delta T/(h_f - h_s)$	Superheat/latent heat of melting
Strouhal number	$St = \omega L/\mathbf{V}$	Dimensionless frequency
Weber number	$We = \rho\mathbf{V}^2L/\sigma$	Inertia force/surface tension

Symbols used:

$\alpha = k/\rho C_p$	Diffusivity $[m^2/s]$
$\beta = (\partial v/\partial T)_p/v$	Volume expansivity [1/K]
c	Speed of sound [m/s]
C_p	Heat capacity [J/kg K] = [kJ/kg K]/1000
D	Diffusion coefficient $[m^2/s]$
g	Gravitation $[m/s^2]$
h	Heat transfer coefficient $[W/m^2\ K]$
h_f, h_s	Enthalpy of liquid, solid [J/kg] = [kJ/kg]/1000
k	Conductivity [W/m K]
L	Length, characteristic [m]
μ	Viscosity, dynamic [kg/m s]
ω	Frequency [1/s]
ρ	Density $[kg/m^3]$
σ	Surface tension $[kg/s^2]$
$\mathbf{V}$	Velocity [m/s]
$\nu = \mu/\rho$	Viscosity, kinematic $[m^2/s]$

TABLE A.3 ***Critical Constants (SI Units)***

Substance	Formula	Molec. Mass	Temp. K	Press. MPa	Vol. m^3/kmol	Acentric Factor
Ammonia	NH_3	17.031	405.5	11.35	0.0725	0.250
Argon	Ar	39.948	150.8	4.87	0.0749	0.001
Bromine	Br_2	159.808	588	10.30	0.1272	0.108
Carbon dioxide	CO_2	44.01	304.1	7.38	0.0939	0.239
Carbon monoxide	CO	28.01	132.9	3.50	0.0932	0.066
Chlorine	Cl_2	70.906	416.9	7.98	0.1238	0.090
Deuterium (normal)	D_2	4.032	38.4	1.66	—	−0.160
Fluorine	F_2	37.997	144.3	5.22	0.0663	0.054
Helium	He	4.003	5.19	0.227	0.0574	−0.365
Helium[3]	He	3.017	3.31	0.114	0.0729	−0.473
Hydrogen (normal)	H_2	2.016	33.2	1.30	0.0651	−0.218
Krypton	Kr	83.80	209.4	5.50	0.0912	0.005
Neon	Ne	20.183	44.4	2.76	0.0416	−0.029
Nitric oxide	NO	30.006	180	6.48	0.0577	0.588
Nitrogen	N_2	28.013	126.2	3.39	0.0898	0.039
Nitrogen dioxide	NO_2	46.006	431	10.1	0.1678	0.834
Nitrous oxide	N_2O	44.013	309.6	7.24	0.0974	0.165
Oxygen	O_2	31.999	154.6	5.04	0.0734	0.025
Sulfur dioxide	SO_2	64.063	430.8	7.88	0.1222	0.256
Water	H_2O	18.015	647.3	22.12	0.0571	0.344
Xenon	Xe	131.30	289.7	5.84	0.1184	0.008
Acetylene	C_2H_2	26.038	308.3	6.14	0.1127	0.190
Benzene	C_6H_6	78.114	562.2	4.89	0.2590	0.212
n-Butane	C_4H_{10}	58.124	425.2	3.80	0.2550	0.199
Carbon tetrachloride	CCL_4	153.823	556.4	4.56	0.2759	0.193
Chlorodifluoroethane[a] (142b)	CH_3CCLF_2	100.495	410.3	4.25	0.2310	0.250
Chlorodifluoromethane (22)	$CHCLF_2$	86.469	369.3	4.97	0.1656	0.221
Chloroform	$CHCL_3$	119.378	536.4	5.37	0.2389	0.218
Dichlorodifluoromethane (12)	CCL_2F_2	120.914	385.0	4.14	0.2167	0.204
Dichlorofluoroethane[a] (141)	CH_3CCL_2F	116.95	481.5	4.54	0.2520	0.215
Dichlorofluoromethane (21)	$CHCL_2F$	102.923	451.6	5.18	0.1964	0.210
Dichlorotrifluoroethane[a] (123)	$CHCL_2CF_3$	152.93	456.9	3.67	0.2781	0.282
Difluoroethane[a] (152a)	CHF_2CH_3	66.05	386.4	4.52	0.1795	0.275
Ethane	C_2H_6	30.070	305.4	4.88	0.1483	0.099
Ethyl alcohol	C_2H_5OH	46.069	513.9	6.14	0.1671	0.644
Ethylene	C_2H_4	28.054	282.4	5.04	0.1304	0.089
n-Heptane	C_7H_{16}	100.205	540.3	2.74	0.4320	0.349
n-Hexane	C_6H_{14}	86.178	507.5	3.01	0.3700	0.299
Methane	CH_4	16.043	190.4	4.60	0.0992	0.011
Methyl alcohol	CH_3OH	32.042	512.6	8.09	0.1180	0.556

(Continued)

TABLE A.3 (Continued) ***Critical Constants (SI Units)***

Substance	Formula	Molec. Mass	Temp. K	Press. MPa	Vol. m^3/kmol	Acentric Factor
Methyl chloride	CH_3CL	50.488	416.3	6.70	0.1389	0.153
n-Octane	C_8H_{18}	114.232	568.8	2.49	0.4920	0.398
n-Pentane	C_5H_{16}	72.151	469.7	3.37	0.3040	0.251
Propane	C_3H_8	44.094	369.8	4.25	0.2030	0.153
Propene	C_3H_6	42.081	364.9	4.60	0.1810	0.144
Propyne	C_3H_4	40.065	402.4	5.63	0.1640	0.215
Tetrafluoroethane[a] (134a)	CF_3CH_2F	102.03	374.2	4.06	0.1980	0.327

Source: R. C. Reid, J. M. Prausnitz, and B. E. Poling, *The Properties of Gases and Liquids*, fourth edition, McGraw-Hill Book Company, New York, 1987.

[a]Data from M. O. McLinden, NIST Thermophysics Division, 1989.

TABLE A.4 ***The Triple Point Constants***

Substance	Formula	T K	P kPa	ρ_f kg/m^3
Ammonia	NH_3	195.5	6.06	735.3
Argon	Ar	83.804	68.95	1377
Carbon dioxide	CO_2	216.6	517.96	1179
Copper	Cu	1356	0.79E-4	7940
Ethane	C_2H_6	90.352	1.13E-3	651.5
Ethylene	C_2H_4	103.986	0.1225	655
Helium	He	2.17	5.07	146.3
Hydrogen	H_2	13.84	7.09	77
Isobutane	C_4H_{10}	113.55	1.95E-5	741.4
n-butane	C_4H_{10}	134.86	6.74E-4	735.3
Mercury	Hg	234.2	1.30E-7	13690
Methane	CH_4	90.685	11.696	451.5
Methanol	CH_3OH	175.59	1.835E-4	904.4
Neon	Ne	24.56	43.38	1250
Nitrogen	N_2	63.148	12.52	869.7
Oxygen	O_2	54.361	0.14633	1306
Propane	C_3H_8	85.47	1.685E-7	733.5
Silver	Ag	1234	0.01	9330
Water	H_2O	273.16	0.6113	1000
Zinc	Zn	692	5.066	6640

The gas phase is well approximated as an ideal gas for all substances

TABLE A.5 ***Properties of Selected Solids at 25° C***

Substance	ρ kg/m³	C_p kJ/kg-K	k W/m-K	$\alpha \times 10^6$ m²/s
Asphalt	2120	0.92	0.70	0.358
Bark	340	1.26	0.074	0.173
Brick, common	1800	0.84	0.7	0.463
Carbon, diamond	3250	0.51	1350	814.5
Carbon, graphite	2000-2500	0.61	155	113
Clay	1450	0.88	1.28	1.00
Coal	1200-1500	1.26	0.26	0.153
Concrete	2200	0.88	1.28	0.661
Cork	150	1.88	0.042	0.149
Earth, coarse	2040	1.84	0.59	0.157
Glass, plate	2500	0.80	1.40	0.70
Glass, wool	200	0.66	0.037	0.280
Granite	2750	0.89	2.9	1.185
Ice (0 C)	917	2.04	2.25	1.203
Leather	860	1.5	0.12-0.15	0.105
Paper	700	1.2	0.12	0.143
Plaster	1690	0.8	0.79	0.584
Plexiglas	1180	1.44	0.184	0.108
Polystyrene	920	2.3	0.35	0.165
Polyvinyl chloride	1380	0.96	0.15	0.113
Quartz	2100-2500	0.78	1.40	0.780
Rubber, foam	500	1.67	0.09	0.108
Rubber, hard	1150	2.01	0.16	0.069
Rubber, soft	1100	1.67	0.13	0.0708
Salt, rock	2100-2500	0.92	7	3.308
Sand, dry	1500	0.8	0.3	0.25
Sandstone	2150-2300	0.71	1.6-2.1	1.171
Silicon	2330	0.70	153	93.81
Snow, firm	560	2.1	0.46	0.391
Teflon	2200	1.04	0.23	0.101
Wood, hard (oak)	720	1.26	0.16	0.176
Wood, soft (pine)	510	1.38	0.12	0.171
Wool	100	1.72	0.036	0.209

TABLE A.6 *Properties of Metals at 25° C*				
Substance	**ρ kg/m³**	**C_p kJ/kg-K**	**k W/m-K**	**α × 10⁶ m²/s**
Aluminum, duralumin	2787	0.883	164	66.64
Copper, commercial	8300	0.419	372	107
Brass, 60-40	8400	0.376	113	35.78
Gold	19300	0.129	315	126.5
Iron, cast	7272	0.42	52	17.03
Iron, steel 0.5% C	7833	0.465	54	14.83
Iron. 304 St Steel	7817	0.46	13.8	3.838
Lead	11340	0.13	34.8	23.61
Magnesium, 2% Mn	1778	1.00	114	64.12
Nickel, 10% Cr	8666	0.444	17	4.418
Silver, 99.9% Ag	10524	0.236	411	165.5
Sodium	971	1.206	133	113.6
Tin	7304	0.220	67	41.7
Tungsten	19300	0.134	179	69.2
Zinc	7144	0.388	121	43.65

TABLE A.7 *Properties of Some Liquids at 25° C*								
Substance	**ρ kg/m³**	**C_p kJ/kg-K**	**k W/m-K**	**α × 10⁶ m²/s**	**μ × 10³ kg/s-m**	**ν × 10⁶ m²/s**	**β × 10³ 1/K**	**Pr**
Ammonia	604	4.84	0.514	0.176	0.214	0.354	2.5	2.01
Benzene	879	1.72	0.141	0.093	0.6	0.682	--	7.32
Butane	556	2.469	0.118	0.086	0.164	0.295	1.6	3.43
CCL_4	1584	0.826	0.10	0.076	0.91	0.574	--	7.51
CO_2	680	2.9	0.081	0.041	0.071	0.104	14.0	2.54
Engine oil	885	1.91	0.144	0.085	490	554	0.7	6500
Ethanol	783	2.456	0.168	0.087	1.04	1.33	1.08	15.2
Gasoline	750	2.08	0.116	0.074	0.52	0.693	--	9.32
Glycerine	1260	2.42	0.286	0.094	800	635	0.48	6770
Kerosine	815	2.0	0.116	0.071	1.5	1.84	--	24.3
Mercury	13580	0.139	8.54	4.52	1.52	0.112	0.181	0.025
Methanol	787	2.55	0.190	0.095	0.55	0.699	1.2	7.38
n-octane	692	2.23	0.128	0.083	0.508	0.734	1.04	8.85
Propane	510	2.54	0.095	0.073	0.091	0.178	5.3	2.43
R-12	1310	0.97	0.070	0.055	0.25	0.191	2.75	3.46
R-22	1190	1.255	0.088	0.059	0.20	0.168	3.25	2.85
R-134a	1206	1.43	0.081	0.047	0.20	0.166	3.1	3.53
Water	997	4.179	0.60	0.144	0.89	0.893	0.26	6.21

TABLE A.8 *Properties of Liquid Metals*

Substance	ρ kg/m^3	C_p kJ/kg-K	k W/m-K	$\alpha \times 10^6$ m^2/s	$\nu \times 10^6$ m^2/s	Pr	T_m C	h_{sf} kJ/kg
Bismuth, Bi	10040	0.14	16.5	11.7	0.17	0.014	271	54
Lead, Pb	10660	0.16	16.3	9.6	0.24	0.025	327	23
Mercury, Hg	13690	0.14	8.5	4.4	0.11	0.025	-38.8	11.4
Potassium, K	828	0.81	46	68.6	0.52	0.0076	63.4	59.3
Sodium, Na	929	1.38	86	67.1	0.71	0.011	97.7	113
Tin, Sn	6950	0.24	32.7	19.6	0.26	0.013	232	59.2
Zinc, Zn	6570	0.50	58.8	16.6	0.45	0.027	419.5	112
NaK (56/44)	887	1.13	25	27	0.66	0.026	19	--

TABLE A.9 *Surface Tension at 25° C, Interphase to Air, Normal Boiling Point and Heat of Evaporation for Some Liquids*

Substance	Formula	$\sigma \times 10^3$ N/m	T_b C	h_{fg} kJ/kg
Water	H_2O	71.99	100	2257
Benzene	C_6H_6	28.22	80	393
Carbon tetrachloride	CCL_4	27.0	77	194
Ethanol	C_2H_5OH	21.97	78	837
Glycerin	$C_2H_5(OH)_3$	63.0	290	974
Hexane	C_6H_{14}	18.4	69	335
Mercury	Hg	485.48	357	295
Methanol	CH_3OH	22.07	65	1099
Octane	C_8H_{18}	21.14	126	301
Pentane	C_5H_{12}	15.49	36	357

TABLE A.10 ***Properties of Various Ideal Gases at 25° C, 100 kPa* (SI Units)***

Gas	Chemical Formula	Molecular Mass	R kJ/kg-K	ρ kg/m^3	C_{po} kJ/kg-K	C_{vo} kJ/kg-K	γ C_{po}/C_{vo}
Steam	H_2O	18.015	0.4615	0.0231	1.872	1.410	1.327
Acetylene	C_2H_2	26.038	0.3193	1.05	1.699	1.380	1.231
Air	--	28.97	0.287	1.169	1.004	0.717	1.400
Ammonia	NH_3	17.031	0.4882	0.694	2.130	1.642	1.297
Argon	Ar	39.948	0.2081	1.613	0.520	0.312	1.667
Butane	C_4H_{10}	58.124	0.1430	2.407	1.716	1.573	1.091
Carbon monoxide	CO	28.01	0.2968	1.13	1.041	0.744	1.399
Carbon dioxide	CO_2	44.01	0.1889	1.775	0.842	0.653	1.289
Ethane	C_2H_6	30.07	0.2765	1.222	1.766	1.490	1.186
Ethanol	C_2H_5OH	46.069	0.1805	1.883	1.427	1.246	1.145
Ethylene	C_2H_4	28.054	0.2964	1.138	1.548	1.252	1.237
Helium	He	4.003	2.0771	0.1615	5.193	3.116	1.667
Hydrogen	H_2	2.016	4.1243	0.0813	14.209	1.008	1.409
Methane	CH_4	16.043	0.5183	0.648	2.254	1.736	1.299
Methanol	CH_3OH	32.042	0.2595	1.31	1.405	1.146	1.227
Neon	Ne	20.183	0.4120	0.814	1.03	0.618	1.667
Nitric oxide	NO	30.006	0.2771	1.21	0.993	0.716	1.387
Nitrogen	N_2	28.013	0.2968	1.13	1.042	0.745	1.400
Nitrous oxide	N_2O	44.013	0.1889	1.775	0.879	0.690	1.274
n-octane	C_8H_{18}	114.23	0.07279	0.092	1.711	1.638	1.044
Oxygen	O_2	31.999	0.2598	1.292	0.922	0.662	1.393
Propane	C_3H_8	44.094	0.1886	1.808	1.679	1.490	1.126
R-12	CCL_2F_2	120.914	0.06876	4.98	0.616	0.547	1.126
R-22	$CHCLF_2$	86.469	0.09616	3.54	0.658	0.562	1.171
R-134a	CF_3CH_2F	102.03	0.08149	4.20	0.852	0.771	1.106
Sulfur dioxide	SO_2	64.059	0.1298	2.618	0.624	0.494	1.263
Sulfur trioxide	SO_3	80.053	0.10386	3.272	0.635	0.531	1.196

*Or saturation pressure if it is less than 100 kPa.

TABLE A.11 ***Transport Properties of Various Ideal Gases at 25° C, 100 kPa* (SI Units)***

Gas	ρ kg/m^3	$k \times 10^3$ W/m-K	$\alpha \times 10^6$ m^2/s	$\mu \times 10^6$ kg/s-m	$\nu \times 10^6$ m^2/s	Pr	$10^{-6} g\beta/\alpha\nu$ 1/K-m^3
Steam	0.0231	19.6	453	9.05	392	0.865	0.185
Acetylene	1.05	21.4	12	10.4	9.9	0.825	277
Air	1.169	26.3	22.4	18.46	15.79	0.704	93
Ammonia	0.694	25	16.9	10.1	14.55	0.86	134
Argon	1.613	17.9	21.33	22.9	14.2	0.667	108
Butane	2.407	16.4	3.97	7.5	3.12	0.785	2655
CO	1.13	25	21.25	17.8	15.75	0.741	98.3
CO_2	1.775	16.6	11.11	15	8.45	0.761	350
Ethane	1.222	20.9	9.68	9.33	7.64	0.788	445
Ethanol	1.883	14.4	5.36	8.6	4.57	0.852	1343
Ethylene	1.138	20.5	11.63	10.4	9.14	0.785	309
Helium	0.1615	155	184.8	19.9	123.2	0.667	1.44
Hydrogen	0.0813	187	161.9	8.9	109.5	0.676	1.85
Methane	0.648	34.1	23.31	11.1	17.1	0.734	82.5
Methanol	1.31	25	13.58	11	8.4	0.618	288
Neon	0.814	49.5	59.05	31.9	39.2	0.664	14.2
Nitric oxide	1.21	25.7	21.39	19.1	15.8	0.738	97.3
Nitrogen	1.13	25.9	21.99	17.8	15.74	0.716	95
Nitrous oxide	1.775	17.8	11.41	14.9	8.39	0.735	344
n-octane	0.092	19	120.7	7.3	79.3	0.657	3.44
Oxygen	1.292	26.6	22.34	20.7	16.0	0.717	92
Propane	1.808	17.6	5.8	8.3	4.59	0.792	1235
R-12	4.98	9.7	3.16	12.6	2.53	0.80	4114
R-22	3.54	11	4.72	13	3.67	0.778	1900
R-134a	4.20	13.4	3.74	~12.6	~3.0	~0.8	~2900
Sulfur dioxide	2.618	9.6	5.88	12.8	4.89	0.832	1144
Sulfur trioxide	3.272	13	6.26	~16.4	~5.0	~0.8	~1050

*Or saturation pressure if it is less than 100 kPa. ~ indicates estimated value

TABLE A.12 ***Constant-Pressure Specific Heats of Various Ideal Gases (SI Units)***

Gas	C_{p0} = kJ/kmol K $\theta = T$(Kelvin)/100	Range *K*	Max Error %
N_2	$\bar{C}_{p0} = 39.060 - 512.79\ \theta^{-1.5} + 1072.7\ \theta^{-2} - 820.40\ \theta^{-3}$	300–3500	0.43
O_2	$\bar{C}_{p0} = 37.432 + 0.020\ 102\ \theta^{1.5}\ 2\ 178.57\ \theta^{-1.5} + 236.88\ \theta^{-2}$	300–3500	0.30
H_2	$\bar{C}_{p0} = 56.505 - 702.74\ \theta^{-0.75} + 1165.0\ \theta^{-1} - 560.70\ \theta^{-1.5}$	300–3500	0.60
CO	$\bar{C}_{p0} = 69.145 - 0.704\ 63\ \theta^{0.75}\ 2\ 200.77\ \theta^{20.5} + 176.76\ \theta^{20.75}$	300–3500	0.42
OH	$\bar{C}_{p0} = 81.546 - 59.350\ \theta^{0.25} + 17.329\ \theta^{0.75} - 4.2660\ \theta$	300–3500	0.43
NO	$\bar{C}_{p0} = 59.283 - 1.7096\ \theta^{0.5}\ 2\ 70.613\ \theta^{20.5} + 74.889\ \theta^{21.5}$	300–3500	0.34
H_2O	$\bar{C}_{p0} = 143.05 - 183.54\ \theta^{0.25} + 82.751\ \theta^{0.5} - 3.6989\ \theta$	300–3500	0.43
CO_2	$\bar{C}_{p0} = 23.7357 + 30.529\ \theta^{0.5} - 4.1034\ \theta + 0.024\ 198\ \theta^{2}$	300–3500	0.19
NO_2	$\bar{C}_{p0} = 46.045 + 216.10\ \theta^{20.5} - 363.66\ \theta^{20.75} + 232.550\ \theta^{22}$	300–3500	0.26
CH_4	$\bar{C}_{p0} = -672.87 + 439.74\ \theta^{0.25} - 24.875\ \theta^{0.75} + 323.88\ \theta^{20.5}$	300-2000	0.15
C_2H_4	$\bar{C}_{p0} = -95.395 + 123.15\ \theta^{0.5} - 35.641\ \theta^{0.75} + 182.77\ \theta^{23}$	300–2000	0.07
C_2H_6	$\bar{C}_{p0} = 6.895 + 17.26\ \theta - 0.6402\ \theta^{2} + 0.007\ 28\ \theta^{3}$	300–1500	0.83
C_3H_8	$\bar{C}_{p0} = -4.042 + 30.46\ \theta - 1.571\ \theta^{2} + 0.031\ 71\ \theta^{3}$	300–1500	0.40
C_4H_{10}	$\bar{C}_{p0} = 3.954 + 37.12\ \theta - 1.833\ \theta^{2} + 0.034\ 98\ \theta^{3}$	300–1500	0.54

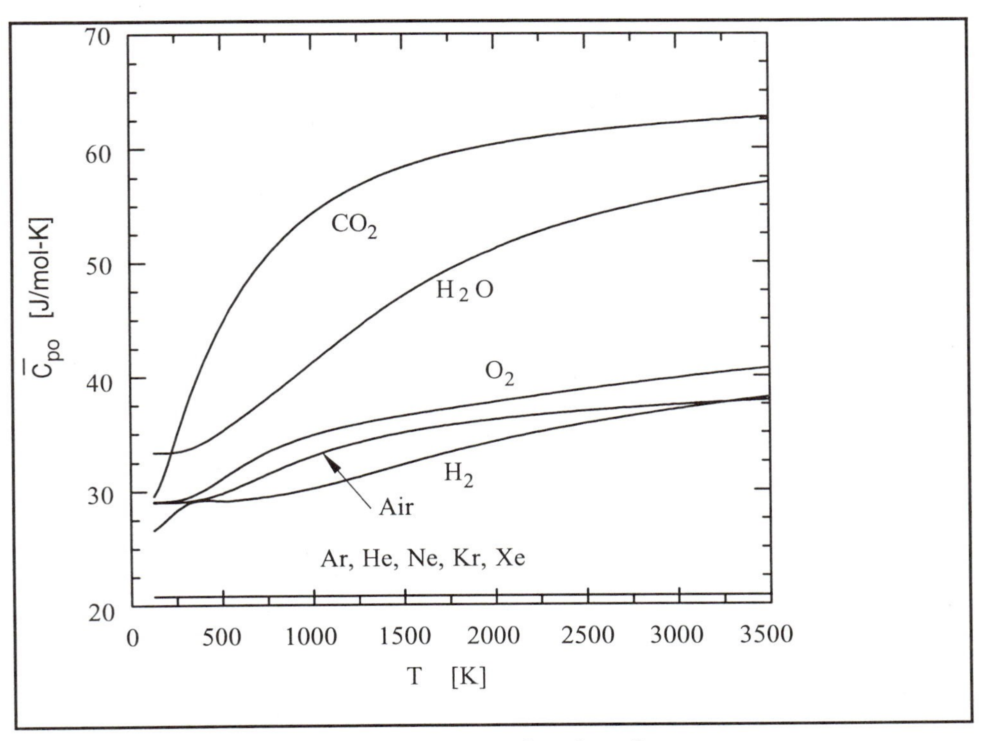

FIGURE A.1 Heat capacity for some gases as function of temperature.

TABLE A.13 *Atomic Masses, and Thermodynamic Properties of Selected Elements Based on the Assigned Relative Atomic Mass of $^{12}C = 12$*

Name	Symbol	Atomic Mass	Melting Point, °C	Boiling Point, °C	C_p kJ/kg K	k W/m K	Δh_{fus} kJ/kg	Δh_{evap} kJ/kg
Aluminum	Al	26.9815	660.4	2467	0.897	237	397	10896
Antimony	Sb	121.75	630.7	1750	0.207	24.4	163	---
Argon	Ar	39.948	−189.2	−185.7	0.521	0.0177	28	161
Arsenic	As	74.9216	817	613	0.329	50	326	1703
Barium	Ba	137.33	725	1640	0.204	18.4	51.8	1019
Beryllium	Be	9.012 18	1278	2970	1.825	201	877	---
Bismuth	Bi	208.980	271.3	1560	0.122	7.92	54.1	723
Boron	B	10.81	2079	2550	1.026	27.4	4644	44403
Bromine liq.	Br	79.904	−7.2	58.78	0.474	0.122	66.2	188
Bromine gas	Br_2	159.808			0.226	0.0048		
Cadmium	Cd	112.41	320.9	765	0.232	96.9	55.1	888
Calcium	Ca	40.08	839	1484	0.647	201	213	3.8
Carbon	C	12.011	3652	t	0.709	1.59	8709	---
Cesium	Cs	132.9054	28.40	669.3	0.242	35.9	15.8	394
Cesium liq.	Cs					19.7		---
Chlorine	Cl	35.453	−100.98	−34.6	0.479	0.0089	90.3	288
Chlorine liq.	Cl	35.453				0.134		
Chromium	Cr	51.996	1857	2672	0.449	93.9	404	6.6
Cobalt	Co	51.9332	1495	2870	0.421	100	312	6390
Copper	Cu	63.546	1083.4	2567	0.385	401	209	4700
Fluorine	F	18.9984	−219.6	−188.1	0.824	0.0279	13.4	174
Gold	Au	196.967	1064.4	3080	0.129	318	63.7	1645
Helium	He	4.002 6	−272.2$^{26\ atm}$	−268.9	5.193	0.152	2.1	20.7
Hydrogen	H	1.007 94	−259.1	−252.9	14.304	0.1815	59.5	446
Indium	In	114.82	156.6	2080	0.233	81.8	28.6	2019
Iodine	I	126.905	113.5	184.4	0.145	0.449	61	164
Iodine liq.	I					0.116		
Iridium	Ir	192.22	2410	4130	0.131	147	214	3185
Iron	Fe	55.847	1535	2750	0.449	80.4	247	6260
Krypton	Kr	83.80	−156.6	−152.3	0.248	0.0095	16.3	108
Lanthanum	La	138.906	920	3454	0.195	13.4	44.6	2980
Lead	Pb	207.2	327.5	1740	0.129	35.3	23.0	866
Lithium	Li	6.941	180.5	1342	3.582	84.8	432	21340

(Continued)

The values apply to elements as they exist in materials of terrestrial origin and to certain artificial elements.
Source: *Handbook of Chemistry and Physics*, sixty-seventh edition, 1986–1987, CRC Press, Boca Raton, FL, 1986.

TABLE A.13 (Continued) *Atomic Masses, and Thermodynamic Properties of Selected Elements Based on the Assigned Relative Atomic Mass of $^{12}C = 12$*

Name	Symbol	Atomic Mass	Melting Point, °C	Boiling Point, °C	C_p kJ/kg K	k W/m K	Δh_{fus} kJ/kg	Δh_{evap} kJ/kg
Magnesium	Mg	24.305	648.8	1090	1.023	156	349	5240
Manganese	Mn	54.9380	1244	1962	0.479	7.8	235	4110
Mercury	Hg	200.59	−38.87	356.6	0.140	8.3	11.4	295
Molybdenum	Mo	95.94	2617	4612	0.251	138	391	---
Neon	Ne	20.1179	−248.7	−246.0	1.03	0.0493	16.9	85
Nickel	Ni	58.69	1453	2732	0.444	90.9	298	6310
Nitrogen	N	14.0067	−209.9	−195.8	1.04	0.026	25.4	199
Oxygen	O	15.9994	−218.4	−182.96	0.918	0.0267	13.8	213
Phosphorus, white	P	30.9738	44.1	280	0.769	0.236	21.3	400
Platinum	Pt	195.08	1772	3827	0.133	71.6	114	2610
Plutonium	Pu	(244)	641	3232	--	6.7	11.6	1410
Potassium, kalium	K	39.0983	63.25	759.9	0.757	102.5	59.3	2050
Radon	Rn	(222)	−71	−61.8	0.094	0.00364	12.3	83
Rubidium	Rb	85.4678	38.9	686	0.363	58.2	25.6	810
Selenium	Se	78.96	217	684.9	0.321	0.52	84.7	1209
Silicon	Si	28.0855	1410	2355	0.705	149	1788	14050
Silver	Ag	107.868	961.9	2212	0.235	429	105	2323
Sodium, natrium	Na	22.9898	97.8	882.9	1.228	142	113	4260
Strontium	Sr	87.62	769	1384	0.301	35.4	84.8	---
Sulfur	S	32.06	112.8	444.7	0.71	0.2	53.6	1404
Thallium	Tl	204.383	303.5	1457	0.129	46.1	20.3	806
Tin, stannum	Sn	118.71	232.0	2270	0.228	66.8	59.2	2500
Titanium	Ti	47.88	1660	3287	0.523	21.9	296	8790
Tungsten, wolfram	W	183.85	3410	5660	0.132	173	285	4500
Uranium	U	238.029	1132	3818	0.116	27.5	38.4	1950
Vanadium	V	50.9415	1890	3380	0.489	30.7	422	8870
Xenon	Xe	131.29	−111.9	−107.1	0.158	0.00569	13.8	96
Zinc	Zn	65.39	419.6	907	0.388	116	112	1770
Zirconium	Zr	91.224	1852	4377	0.278	22.7	230	6400

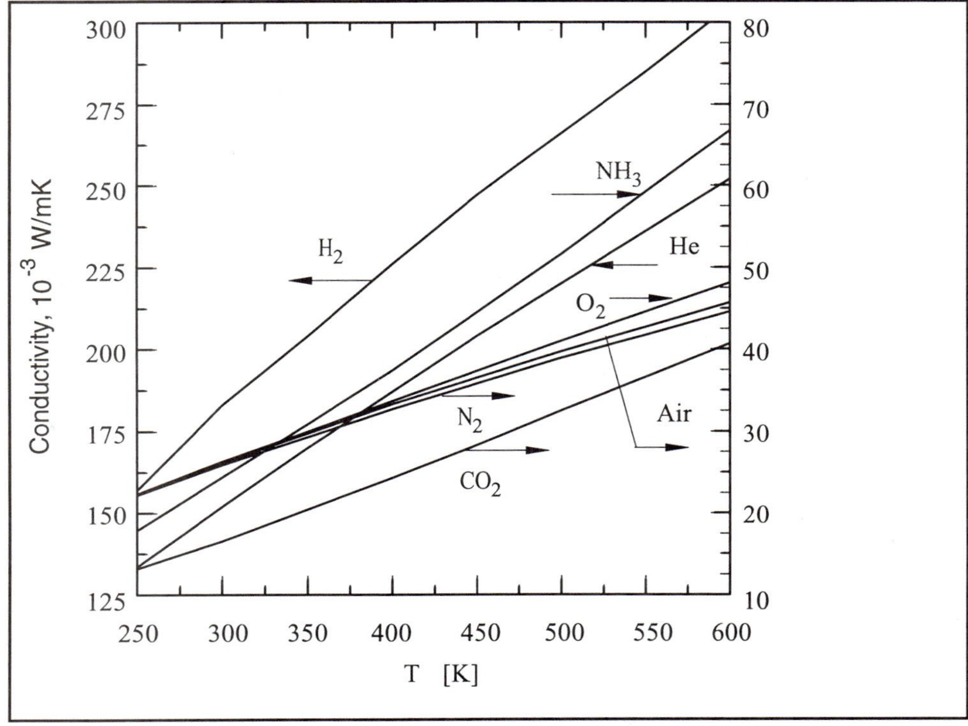

FIGURE A.2 Conductivity of gases as function of temperature.

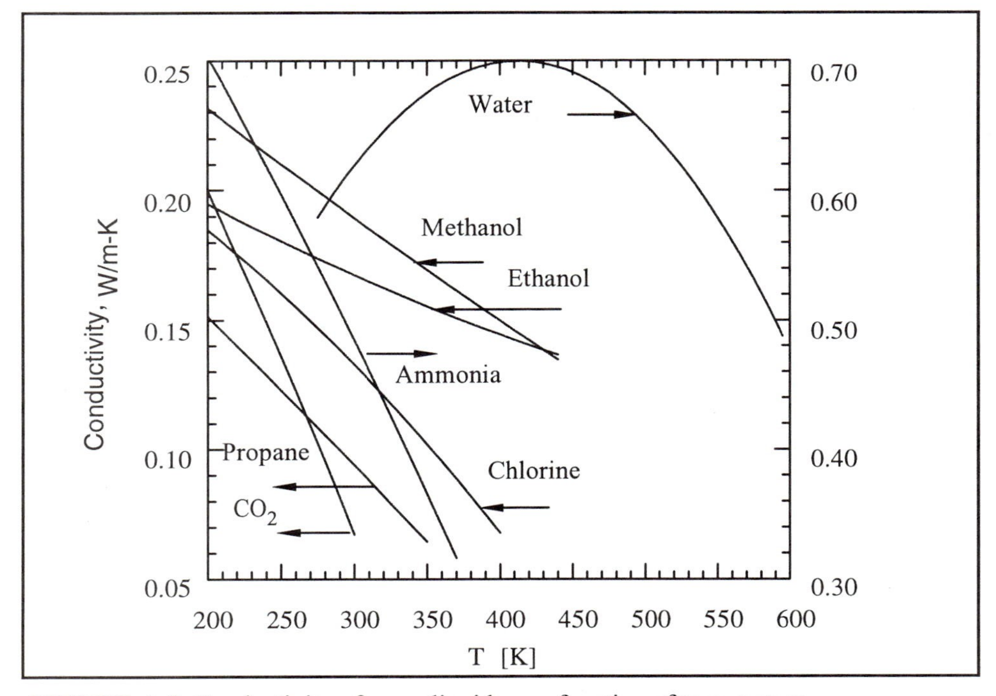

FIGURE A.3 Conductivity of some liquids as a function of temperature.

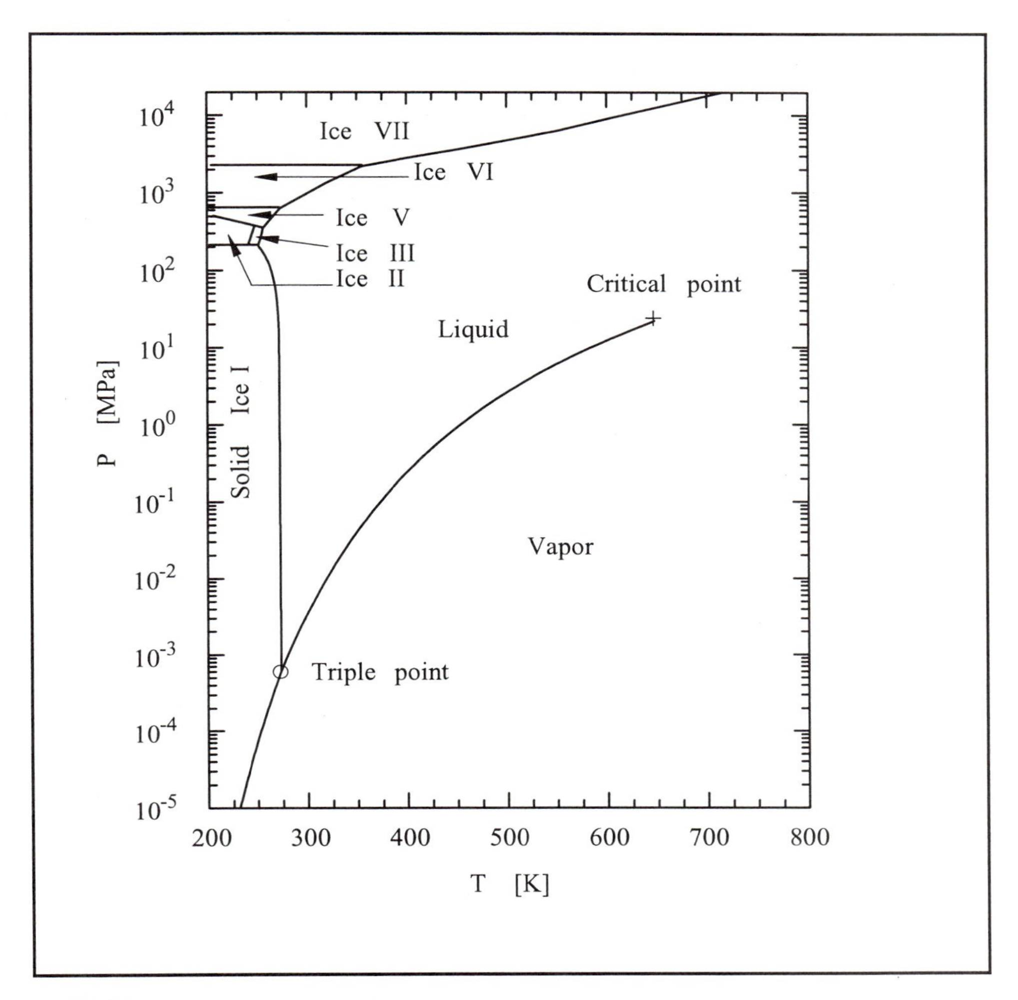

FIGURE A.4 Water phase diagram.

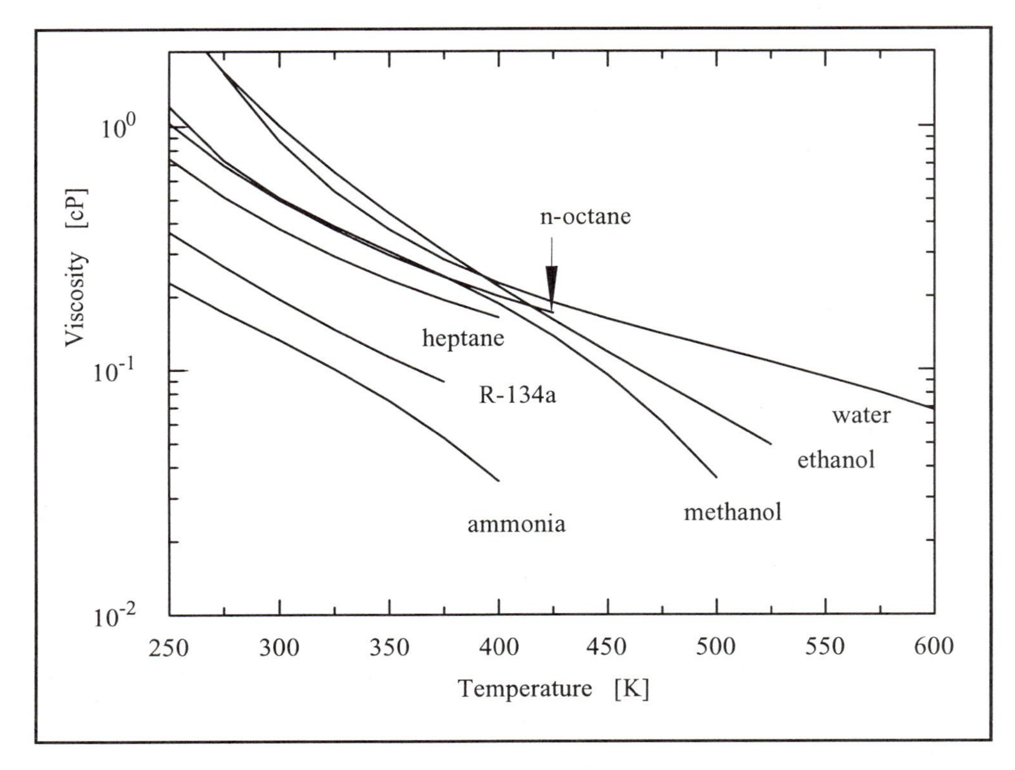

FIGURE A.5 Viscosity of liquids as a function of temperature. 1 cP = 10^{-3} kg/s-m

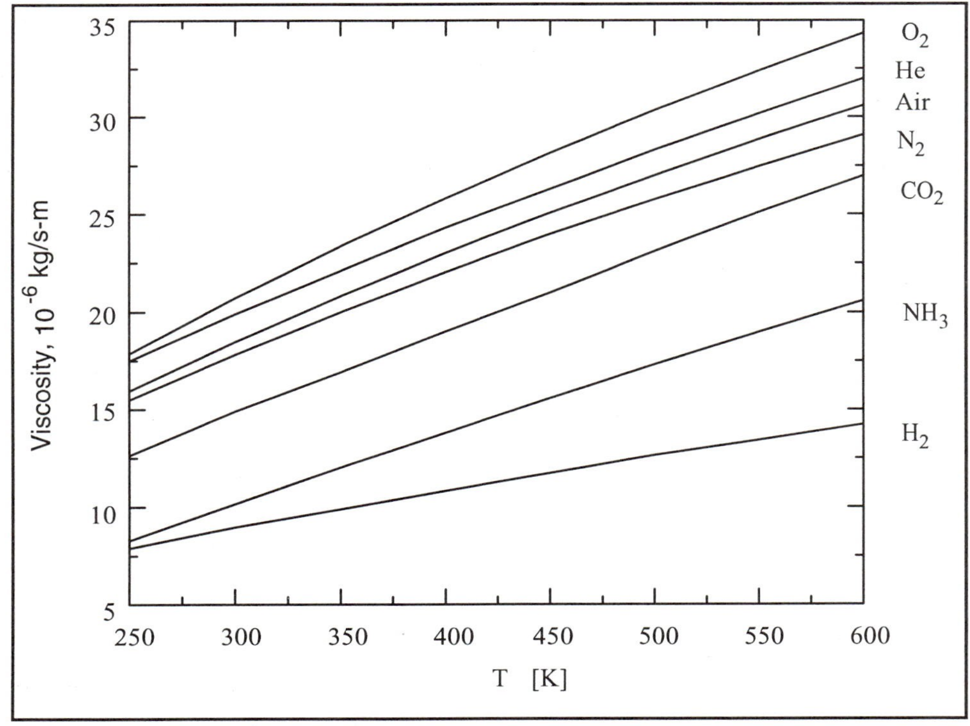

FIGURE A.6 Viscosity of gases as a function of temperature.

SECTION B

PROPERTIES OF PURE SUBSTANCES

TABLE B.1 SI *Thermodynamic Properties of Water*
TABLE B.1.1 SI *Saturated Water*

Temp.	Press.	Specific Volume, m³/kg			Internal Energy, kJ/kg		
C	kPa	Sat. Liquid	Evap.	Sat. Vapor	Sat. Liquid	Evap.	Sat. Vapor
T	P	v_f	v_{fg}	v_g	u_f	u_{fg}	u_g
0.01	0.6113	0.001000	206.131	206.132	0	2375.33	2375.33
5	0.8721	0.001000	147.117	147.118	20.97	2361.27	2382.24
10	1.2276	0.001000	106.376	106.377	41.99	2347.16	2389.15
15	1.705	0.001001	77.924	77.925	62.98	2333.06	2396.04
20	2.339	0.001002	57.7887	57.7897	83.94	2318.98	2402.91
25	3.169	0.001003	43.3583	43.3593	104.86	2304.90	2409.76
30	4.246	0.001004	32.8922	32.8932	125.77	2290.81	2416.58
35	5.628	0.001006	25.2148	25.2158	146.65	2276.71	2423.36
40	7.384	0.001008	19.5219	19.5229	167.53	2262.57	2430.11
45	9.593	0.001010	15.2571	15.2581	188.41	2248.40	2436.81
50	12.350	0.001012	12.0308	12.0318	209.30	2234.17	2443.47
55	15.758	0.001015	9.56734	9.56835	230.19	2219.89	2450.08
60	19.941	0.001017	7.66969	7.67071	251.09	2205.54	2456.63
65	25.03	0.001020	6.19554	6.19656	272.00	2191.12	2463.12
70	31.19	0.001023	5.04114	5.04217	292.93	2176.62	2469.55
75	38.58	0.001026	4.13021	4.13123	313.87	2162.03	2475.91
80	47.39	0.001029	3.40612	3.40715	334.84	2147.36	2482.19
85	57.83	0.001032	2.82654	2.82757	355.82	2132.58	2488.40
90	70.14	0.001036	2.35953	2.36056	376.82	2117.70	2494.52
95	84.55	0.001040	1.98082	1.98186	397.86	2102.70	2500.56
100	101.3	0.001044	1.67185	1.67290	418.91	2087.58	2506.50
105	120.8	0.001047	1.41831	1.41936	440.00	2072.34	2512.34
110	143.3	0.001052	1.20909	1.21014	461.12	2056.96	2518.09
115	169.1	0.001056	1.03552	1.03658	482.28	2041.44	2523.72
120	198.5	0.001060	0.89080	0.89186	503.48	2025.76	2529.24
125	232.1	0.001065	0.76953	0.77059	524.72	2009.91	2534.63
130	270.1	0.001070	0.66744	0.66850	546.00	1993.90	2539.90
135	313.0	0.001075	0.58110	0.58217	567.34	1977.69	2545.03
140	361.3	0.001080	0.50777	0.50885	588.72	1961.30	2550.02
145	415.4	0.001085	0.44524	0.44632	610.16	1944.69	2554.86
150	475.9	0.001090	0.39169	0.39278	631.66	1927.87	2559.54
155	543.1	0.001096	0.34566	0.34676	653.23	1910.82	2564.04
160	617.8	0.001102	0.30596	0.30706	674.85	1893.52	2568.37
165	700.5	0.001108	0.27158	0.27269	696.55	1875.97	2572.51
170	791.7	0.001114	0.24171	0.24283	718.31	1858.14	2576.46
175	892.0	0.001121	0.21568	0.21680	740.16	1840.03	2580.19
180	1002.2	0.001127	0.19292	0.19405	762.08	1821.62	2583.70
185	1122.7	0.001134	0.17295	0.17409	784.08	1802.90	2586.98
190	1254.4	0.001141	0.15539	0.15654	806.17	1783.84	2590.01

(Continued)

TABLE B.1.1 SI (Continued) ***Saturated Water***

Temp.	Press.	Enthalpy, kJ/kg			Entropy, kJ/kg K		
C	kPa	Sat. Liquid	Evap.	Sat. Vapor	Sat. Liquid	Evap.	Sat. Vapor
T	P	h_f	h_{fg}	h_g	s_f	s_{fg}	s_g
0.01	0.6113	0.00	2501.35	2501.35	0	9.1562	9.1562
5	0.8721	20.98	2489.57	2510.54	0.0761	8.9496	9.0257
10	1.2276	41.99	2477.75	2519.74	0.1510	8.7498	8.9007
15	1.705	62.98	2465.93	2528.91	0.2245	8.5569	8.7813
20	2.339	83.94	2454.12	2538.06	0.2966	8.3706	8.6671
25	3.169	104.87	2442.30	2547.17	0.3673	8.1905	8.5579
30	4.246	125.77	2430.48	2556.25	0.4369	8.0164	8.4533
35	5.628	146.66	2418.62	2565.28	0.5052	7.8478	8.3530
40	7.384	167.54	2406.72	2574.26	0.5724	7.6845	8.2569
45	9.593	188.42	2394.77	2583.19	0.6386	7.5261	8.1647
50	12.350	209.31	2382.75	2592.06	0.7037	7.3725	8.0762
55	15.758	230.20	2370.66	2600.86	0.7679	7.2234	7.9912
60	19.941	251.11	2358.48	2609.59	0.8311	7.0784	7.9095
65	25.03	272.03	2346.21	2618.24	0.8934	6.9375	7.8309
70	31.19	292.96	2333.85	2626.80	0.9548	6.8004	7.7552
75	38.58	313.91	2321.37	2635.28	1.0154	6.6670	7.6824
80	47.39	334.88	2308.77	2643.66	1.0752	6.5369	7.6121
85	57.83	355.88	2296.05	2651.93	1.1342	6.4102	7.5444
90	70.14	376.90	2283.19	2660.09	1.1924	6.2866	7.4790
95	84.55	397.94	2270.19	2668.13	1.2500	6.1659	7.4158
100	101.3	419.02	2257.03	2676.05	1.3068	6.0480	7.3548
105	120.8	440.13	2243.70	2683.83	1.3629	5.9328	7.2958
110	143.3	461.27	2230.20	2691.47	1.4184	5.8202	7.2386
115	169.1	482.46	2216.50	2698.96	1.4733	5.7100	7.1832
120	198.5	503.69	2202.61	2706.30	1.5275	5.6020	7.1295
125	232.1	524.96	2188.50	2713.46	1.5812	5.4962	7.0774
130	270.1	546.29	2174.16	2720.46	1.6343	5.3925	7.0269
135	313.0	567.67	2159.59	2727.26	1.6869	5.2907	6.9777
140	361.3	589.11	2144.75	2733.87	1.7390	5.1908	6.9298
145	415.4	610.61	2129.65	2740.26	1.7906	5.0926	6.8832
150	475.9	632.18	2114.26	2746.44	1.8417	4.9960	6.8378
155	543.1	653.82	2098.56	2752.39	1.8924	4.9010	6.7934
160	617.8	675.53	2082.55	2758.09	1.9426	4.8075	6.7501
165	700.5	697.32	2066.20	2763.53	1.9924	4.7153	6.7078
170	791.7	719.20	2049.50	2768.70	2.0418	4.6244	6.6663
175	892.0	741.16	2032.42	2773.58	2.0909	4.5347	6.6256
180	1002.2	763.21	2014.96	2778.16	2.1395	4.4461	6.5857
185	1122.7	785.36	1997.07	2782.43	2.1878	4.3586	6.5464
190	1254.4	807.61	1978.76	2786.37	2.2358	4.2720	6.5078

(Continued)

TABLE B.1.1 SI (Continued) *Saturated Water*

Temp.	Press.	Specific Volume, m^3/kg			Internal Energy, kJ/kg		
C	kPa	Sat. Liquid	Evap.	Sat. Vapor	Sat. Liquid	Evap.	Sat. Vapor
T	P	v_f	v_{fg}	v_g	u_f	u_{fg}	u_g
195	1397.8	0.001149	0.13990	0.14105	828.36	1764.43	2592.79
200	1553.8	0.001156	0.12620	0.12736	850.64	1744.66	2595.29
205	1723.0	0.001164	0.11405	0.11521	873.02	1724.49	2597.52
210	1906.3	0.001173	0.10324	0.10441	895.51	1703.93	2599.44
215	2104.2	0.001181	0.09361	0.09479	918.12	1682.94	2601.06
220	2317.8	0.001190	0.08500	0.08619	940.85	1661.49	2602.35
225	2547.7	0.001199	0.07729	0.07849	963.72	1639.58	2603.30
230	2794.9	0.001209	0.07037	0.07158	986.72	1617.17	2603.89
235	3060.1	0.001219	0.06415	0.06536	1009.88	1594.24	2604.11
240	3344.2	0.001229	0.05853	0.05976	1033.19	1570.75	2603.95
245	3648.2	0.001240	0.05346	0.05470	1056.69	1546.68	2603.37
250	3973.0	0.001251	0.04887	0.05013	1080.37	1522.00	2602.37
255	4319.5	0.001263	0.04471	0.04598	1104.26	1496.66	2600.93
260	4688.6	0.001276	0.04093	0.04220	1128.37	1470.64	2599.01
265	5081.3	0.001289	0.03748	0.03877	1152.72	1443.87	2596.60
270	5498.7	0.001302	0.03434	0.03564	1177.33	1416.33	2593.66
275	5941.8	0.001317	0.03147	0.03279	1202.23	1387.94	2590.17
280	6411.7	0.001332	0.02884	0.03017	1227.43	1358.66	2586.09
285	6909.4	0.001348	0.02642	0.02777	1252.98	1328.41	2581.38
290	7436.0	0.001366	0.02420	0.02557	1278.89	1297.11	2575.99
295	7992.8	0.001384	0.02216	0.02354	1305.21	1264.67	2569.87
300	8581.0	0.001404	0.02027	0.02167	1331.97	1230.99	2562.96
305	9201.8	0.001425	0.01852	0.01995	1359.22	1195.94	2555.16
310	9856.6	0.001447	0.01690	0.01835	1387.03	1159.37	2546.40
315	10547	0.001472	0.01539	0.01687	1415.44	1121.11	2536.55
320	11274	0.001499	0.01399	0.01549	1444.55	1080.93	2525.48
325	12040	0.001528	0.01267	0.01420	1474.44	1038.57	2513.01
330	12845	0.001561	0.01144	0.01300	1505.24	993.66	2498.91
335	13694	0.001597	0.01027	0.01186	1537.11	945.77	2482.88
340	14586	0.001638	0.00916	0.01080	1570.26	894.26	2464.53
345	15525	0.001685	0.00810	0.00978	1605.01	838.29	2443.30
350	16514	0.001740	0.00707	0.00881	1641.81	776.58	2418.39
355	17554	0.001807	0.00607	0.00787	1681.41	707.11	2388.52
360	18651	0.001892	0.00505	0.00694	1725.19	626.29	2351.47
365	19807	0.002011	0.00398	0.00599	1776.13	526.54	2302.67
370	21028	0.002213	0.00271	0.00493	1843.84	384.69	2228.53
374.1	22089	0.003155	0	0.00315	2029.58	0	2029.58

(Continued)

TABLE B.1.1 SI (Continued) *Saturated Water*

Temp.	Press.	Enthalpy, kJ/kg			Entropy, kJ/kg K		
C	kPa	Sat. Liquid	Evap.	Sat. Vapor	Sat. Liquid	Evap.	Sat. Vapor
T	P	h_f	h_{fg}	h_g	s_f	s_{fg}	s_g
195	1397.8	829.96	1959.99	2789.96	2.2835	4.1863	6.4697
200	1553.8	852.43	1940.75	2793.18	2.3308	4.1014	6.4322
205	1723.0	875.03	1921.00	2796.03	2.3779	4.0172	6.3951
210	1906.3	897.75	1900.73	2798.48	2.4247	3.9337	6.3584
215	2104.2	920.61	1879.91	2800.51	2.4713	3.8507	6.3221
220	2317.8	943.61	1858.51	2802.12	2.5177	3.7683	6.2860
225	2547.7	966.77	1836.50	2803.27	2.5639	3.6863	6.2502
230	2794.9	990.10	1813.85	2803.95	2.6099	3.6047	6.2146
235	3060.1	1013.61	1790.53	2804.13	2.6557	3.5233	6.1791
240	3344.2	1037.31	1766.50	2803.81	2.7015	3.4422	6.1436
245	3648.2	1061.21	1741.73	2802.95	2.7471	3.3612	6.1083
250	3973.0	1085.34	1716.18	2801.52	2.7927	3.2802	6.0729
255	4319.5	1109.72	1689.80	2799.51	2.8382	3.1992	6.0374
260	4688.6	1134.35	1662.54	2796.89	2.8837	3.1181	6.0018
265	5081.3	1159.27	1634.34	2793.61	2.9293	3.0368	5.9661
270	5498.7	1184.49	1605.16	2789.65	2.9750	2.9551	5.9301
275	5941.8	1210.05	1574.92	2784.97	3.0208	2.8730	5.8937
280	6411.7	1235.97	1543.55	2779.53	3.0667	2.7903	5.8570
285	6909.4	1262.29	1510.97	2773.27	3.1129	2.7069	5.8198
290	7436.0	1289.04	1477.08	2766.13	3.1593	2.6227	5.7821
295	7992.8	1316.27	1441.78	2758.05	3.2061	2.5375	5.7436
300	8581.0	1344.01	1404.93	2748.94	3.2533	2.4511	5.7044
305	9201.8	1372.33	1366.38	2738.72	3.3009	2.3633	5.6642
310	9856.6	1401.29	1325.97	2727.27	3.3492	2.2737	5.6229
315	10547	1430.97	1283.48	2714.44	3.3981	2.1821	5.5803
320	11274	1461.45	1238.64	2700.08	3.4479	2.0882	5.5361
325	12040	1492.84	1191.13	2683.97	3.4987	1.9913	5.4900
330	12845	1525.29	1140.56	2665.85	3.5506	1.8909	5.4416
335	13694	1558.98	1086.37	2645.35	3.6040	1.7863	5.3903
340	14586	1594.15	1027.86	2622.01	3.6593	1.6763	5.3356
345	15525	1631.17	964.02	2595.19	3.7169	1.5594	5.2763
350	16514	1670.54	893.38	2563.92	3.7776	1.4336	5.2111
355	17554	1713.13	813.59	2526.72	3.8427	1.2951	5.1378
360	18651	1760.48	720.52	2481.00	3.9146	1.1379	5.0525
365	19807	1815.96	605.44	2421.40	3.9983	0.9487	4.9470
370	21028	1890.37	441.75	2332.12	4.1104	0.6868	4.7972
374.1	22089	2099.26	0	2099.26	4.4297	0	4.4297

TABLE B.1.2 SI *Saturated Water Pressure Entry*

Press..	Temp.	Specific Volume, m³/kg			Internal Energy, kJ/kg		
kPa P	C T	Sat. Liquid v_f	Evap. v_{fg}	Sat. Vapor v_g	Sat. Liquid u_f	Evap. u_{fg}	Sat. Vapor u_g
0.6113	0.01	0.001000	206.131	206.132	0	2375.3	2375.3
1	6.98	0.001000	129.20702	129.20802	29.29	2355.69	2384.98
1.5	13.03	0.001001	87.97913	87.98013	54.70	2338.63	2393.32
2	17.50	0.001001	67.00285	67.00385	73.47	2326.02	2399.48
2.5	21.08	0.001002	54.25285	54.25385	88.47	2315.93	2404.40
3	24.08	0.001003	45.66402	45.66502	101.03	2307.48	2408.51
4	28.96	0.001004	34.79915	34.80015	121.44	2293.73	2415.17
5	32.88	0.001005	28.19150	28.19251	137.79	2282.70	2420.49
7.5	40.29	0.001008	19.23674	19.23775	168.76	2261.74	2430.50
10	45.81	0.001010	14.67254	14.67355	191.79	2246.10	2437.89
15	53.97	0.001014	10.02117	10.02218	225.90	2222.83	2448.73
20	60.06	0.001017	7.64835	7.64937	251.35	2205.36	2456.71
25	64.97	0.001020	6.20322	6.20424	271.88	2191.21	2463.08
30	69.10	0.001022	5.22816	5.22918	289.18	2179.22	2468.40
40	75.87	0.001026	3.99243	3.99345	317.51	2159.49	2477.00
50	81.33	0.001030	3.23931	3.24034	340.42	2143.43	2483.85
75	91.77	0.001037	2.21607	2.21711	384.29	2112.39	2496.67
100	99.62	0.001043	1.69296	1.69400	417.33	2088.72	2506.06
125	105.99	0.001048	1.37385	1.37490	444.16	2069.32	2513.48
150	111.37	0.001053	1.15828	1.15933	466.92	2052.72	2519.64
175	116.06	0.001057	1.00257	1.00363	486.78	2038.12	2524.90
200	120.23	0.001061	0.88467	0.88573	504.47	2025.02	2529.49
225	124.00	0.001064	0.79219	0.79325	520.45	2013.10	2533.56
250	127.43	0.001067	0.71765	0.71871	535.08	2002.14	2537.21
275	130.60	0.001070	0.65624	0.65731	548.57	1991.95	2540.53
300	133.55	0.001073	0.60475	0.60582	561.13	1982.43	2543.55
325	136.30	0.001076	0.56093	0.56201	572.88	1973.46	2546.34
350	138.88	0.001079	0.52317	0.52425	583.93	1964.98	2548.92
375	141.32	0.001081	0.49029	0.49137	594.38	1956.93	2551.31
400	143.63	0.001084	0.46138	0.46246	604.29	1949.26	2553.55
450	147.93	0.001088	0.41289	0.41398	622.75	1934.87	2557.62
500	151.86	0.001093	0.37380	0.37489	639.66	1921.57	2561.23
550	155.48	0.001097	0.34159	0.34268	655.30	1909.17	2564.47
600	158.85	0.001101	0.31457	0.31567	669.88	1897.52	2567.40
650	162.01	0.001104	0.29158	0.29268	683.55	1886.51	2570.06
700	164.97	0.001108	0.27176	0.27286	696.43	1876.07	2572.49
750	167.77	0.001111	0.25449	0.25560	708.62	1866.11	2574.73
800	170.43	0.001115	0.23931	0.24043	720.20	1856.58	2576.79

(Continued)

TABLE B.1.2 SI (Continued) ***Saturated Water Pressure Entry***

Press. kPa P	Temp. C T	Enthalpy, kJ/kg Sat. Liquid h_f	Evap. h_{fg}	Sat. Vapor h_g	Entropy, kJ/kg K Sat. Liquid s_f	Evap. s_{fg}	Sat. Vapor s_g
0.6113	0.01	0.00	2501.3	2501.3	0	9.1562	9.1562
1.0	6.98	29.29	2484.89	2514.18	0.1059	8.8697	8.9756
1.5	13.03	54.70	2470.59	2525.30	0.1956	8.6322	8.8278
2.0	17.50	73.47	2460.02	2533.49	0.2607	8.4629	8.7236
2.5	21.08	88.47	2451.56	2540.03	0.3120	8.3311	8.6431
3.0	24.08	101.03	2444.47	2545.50	0.3545	8.2231	8.5775
4.0	28.96	121.44	2432.93	2554.37	0.4226	8.0520	8.4746
5.0	32.88	137.79	2423.66	2561.45	0.4763	7.9187	8.3950
7.5	40.29	168.77	2406.02	2574.79	0.5763	7.6751	8.2514
10	45.81	191.81	2392.82	2584.63	0.6492	7.5010	8.1501
15	53.97	225.91	2373.14	2599.06	0.7548	7.2536	8.0084
20	60.06	251.38	2358.33	2609.70	0.8319	7.0766	7.9085
25	64.97	271.90	2346.29	2618.19	0.8930	6.9383	7.8313
30	69.10	289.21	2336.07	2625.28	0.9439	6.8247	7.7686
40	75.87	317.55	2319.19	2636.74	1.0258	6.6441	7.6700
50	81.33	340.47	2305.40	2645.87	1.0910	6.5029	7.5939
75	91.77	384.36	2278.59	2662.96	1.2129	6.2434	7.4563
100	99.62	417.44	2258.02	2675.46	1.3025	6.0568	7.3593
125	105.99	444.30	2241.05	2685.35	1.3739	5.9104	7.2843
150	111.37	467.08	2226.46	2693.54	1.4335	5.7897	7.2232
175	116.06	486.97	2213.57	2700.53	1.4848	5.6868	7.1717
200	120.23	504.68	2201.96	2706.63	1.5300	5.5970	7.1271
225	124.00	520.69	2191.35	2712.04	1.5705	5.5173	7.0878
250	127.43	535.34	2181.55	2716.89	1.6072	5.4455	7.0526
275	130.60	548.87	2172.42	2721.29	1.6407	5.3801	7.0208
300	133.55	561.45	2163.85	2725.30	1.6717	5.3201	6.9918
325	136.30	573.23	2155.76	2728.99	1.7005	5.2646	6.9651
350	138.88	584.31	2148.10	2732.40	1.7274	5.2130	6.9404
375	141.32	594.79	2140.79	2735.58	1.7527	5.1647	6.9174
400	143.63	604.73	2133.81	2738.53	1.7766	5.1193	6.8958
450	147.93	623.24	2120.67	2743.91	1.8206	5.0359	6.8565
500	151.86	640.21	2108.47	2748.67	1.8606	4.9606	6.8212
550	155.48	655.91	2097.04	2752.94	1.8972	4.8920	6.7892
600	158.85	670.54	2086.26	2756.80	1.9311	4.8289	6.7600
650	162.01	684.26	2076.04	2760.30	1.9627	4.7704	6.7330
700	164.97	697.20	2066.30	2763.50	1.9922	4.7158	6.7080
750	167.77	709.45	2056.98	2766.43	2.0199	4.6647	6.6846
800	170.43	721.10	2048.04	2769.13	2.0461	4.6166	6.6627

(Continued)

TABLE B.1.2 SI (Continued) ***Saturated Water Pressure Entry***

Press..	Temp.	Specific Volume, m^3/kg			Internal Energy, kJ/kg		
kPa P	C T	Sat. Liquid v_f	Evap. v_{fg}	Sat. Vapor v_g	Sat. Liquid u_f	Evap. u_{fg}	Sat. Vapor u_g
850	172.96	0.001118	0.22586	0.22698	731.25	1847.45	2578.69
900	175.38	0.001121	0.21385	0.21497	741.81	1838.65	2580.46
950	177.69	0.001124	0.20306	0.20419	751.94	1830.17	2582.11
1000	179.91	0.001127	0.19332	0.19444	761.67	1821.97	2583.64
1100	184.09	0.001133	0.17639	0.17753	780.08	1806.32	2586.40
1200	187.99	0.001139	0.16220	0.16333	797.27	1791.55	2588.82
1300	191.64	0.001144	0.15011	0.15125	813.42	1777.53	2590.95
1400	195.07	0.001149	0.13969	0.14084	828.68	1764.15	2592.83
1500	198.32	0.001154	0.13062	0.13177	843.14	1751.3	2594.5
1750	205.76	0.001166	0.11232	0.11349	876.44	1721.39	2597.83
2000	212.42	0.001177	0.09845	0.09963	906.42	1693.84	2600.26
2250	218.45	0.001187	0.08756	0.08875	933.81	1668.18	2601.98
2500	223.99	0.001197	0.07878	0.07998	959.09	1644.04	2603.13
2750	229.12	0.001207	0.07154	0.07275	982.65	1621.16	2603.81
3000	233.90	0.001216	0.06546	0.06668	1004.76	1599.34	2604.10
3250	238.38	0.001226	0.06029	0.06152	1025.62	1578.43	2604.04
3500	242.60	0.001235	0.05583	0.05707	1045.41	1558.29	2603.70
4000	250.40	0.001252	0.04853	0.04978	1082.28	1519.99	2602.27
5000	263.99	0.001286	0.03815	0.03944	1147.78	1449.34	2597.12
6000	275.64	0.001319	0.03112	0.03244	1205.41	1384.27	2589.69
7000	285.88	0.001351	0.02602	0.02737	1257.51	1322.97	2580.48
8000	295.06	0.001384	0.02213	0.02352	1305.54	1264.25	2569.79
9000	303.40	0.001418	0.01907	0.02048	1350.47	1207.28	2557.75
10000	311.06	0.001452	0.01657	0.01803	1393.00	1151.40	2544.41
11000	318.15	0.001489	0.01450	0.01599	1433.68	1096.06	2529.74
12000	324.75	0.001527	0.01274	0.01426	1472.92	1040.76	2513.67
13000	330.93	0.001567	0.01121	0.01278	1511.09	984.99	2496.08
14000	336.75	0.001611	0.00987	0.01149	1548.53	928.23	2476.76
15000	342.24	0.001658	0.00868	0.01034	1585.58	869.85	2455.43
16000	347.43	0.001711	0.00760	0.00931	1622.63	809.07	2431.70
17000	352.37	0.001770	0.00659	0.00836	1660.16	744.80	2404.96
18000	357.06	0.001840	0.00565	0.00749	1698.86	675.42	2374.28
19000	361.54	0.001924	0.00473	0.00666	1739.87	598.18	2338.05
20000	365.81	0.002035	0.00380	0.00583	1785.47	507.58	2293.05
21000	369.89	0.002206	0.00275	0.00495	1841.97	388.74	2230.71
22000	373.80	0.002808	0.00072	0.00353	1973.16	108.24	2081.39
22089	374.14	0.003155	0	0.00315	2029.58	0	2029.58

(Continued)

TABLE B.1.2 SI (Continued) *Saturated Water Pressure Entry*

Press.	Temp.	Enthalpy, kJ/kg			Entropy, kJ/kg K		
kPa	C	Sat. Liquid	Evap.	Sat. Vapor	Sat. Liquid	Evap.	Sat. Vapor
P	T	h_f	h_{fg}	h_g	s_f	s_{fg}	s_g
850	172.96	732.20	2039.43	2771.63	2.0709	4.5711	6.6421
900	175.38	742.82	2031.12	2773.94	2.0946	4.5280	6.6225
950	177.69	753.00	2023.08	2776.08	2.1171	4.4869	6.6040
1000	179.91	762.79	2015.29	2778.08	2.1386	4.4478	6.5864
1100	184.09	781.32	2000.36	2781.68	2.1791	4.3744	6.5535
1200	187.99	798.64	1986.19	2784.82	2.2165	4.3067	6.5233
1300	191.64	814.91	1972.67	2787.58	2.2514	4.2438	6.4953
1400	195.07	830.29	1959.72	2790.00	2.2842	4.1850	6.4692
1500	198.32	844.87	1947.28	2792.15	2.3150	4.1298	6.4448
1750	205.76	878.48	1917.95	2796.43	2.3851	4.0044	6.3895
2000	212.42	908.77	1890.74	2799.51	2.4473	3.8935	6.3408
2250	218.45	936.48	1865.19	2801.67	2.5034	3.7938	6.2971
2500	223.99	962.09	1840.98	2803.07	2.5546	3.7028	6.2574
2750	229.12	985.97	1817.89	2803.86	2.6018	3.6190	6.2208
3000	233.90	1008.41	1795.73	2804.14	2.6456	3.5412	6.1869
3250	238.38	1029.60	1774.37	2803.97	2.6866	3.4685	6.1551
3500	242.60	1049.73	1753.70	2803.43	2.7252	3.4000	6.1252
4000	250.40	1087.29	1714.09	2801.38	2.7963	3.2737	6.0700
5000	263.99	1154.21	1640.12	2794.33	2.9201	3.0532	5.9733
6000	275.64	1213.32	1571.00	2784.33	3.0266	2.8625	5.8891
7000	285.88	1266.97	1505.10	2772.07	3.1210	2.6922	5.8132
8000	295.06	1316.61	1441.33	2757.94	3.2067	2.5365	5.7431
9000	303.40	1363.23	1378.88	2742.11	3.2857	2.3915	5.6771
10000	311.06	1407.53	1317.14	2724.67	3.3595	2.2545	5.6140
11000	318.15	1450.05	1255.55	2705.60	3.4294	2.1233	5.5527
12000	324.75	1491.24	1193.59	2684.83	3.4961	1.9962	5.4923
13000	330.93	1531.46	1130.76	2662.22	3.5604	1.8718	5.4323
14000	336.75	1571.08	1066.47	2637.55	3.6231	1.7485	5.3716
15000	342.24	1610.45	1000.04	2610.49	3.6847	1.6250	5.3097
16000	347.43	1650.00	930.59	2580.59	3.7460	1.4995	5.2454
17000	352.37	1690.25	856.90	2547.15	3.8078	1.3698	5.1776
18000	357.06	1731.97	777.13	2509.09	3.8713	1.2330	5.1044
19000	361.54	1776.43	688.11	2464.54	3.9387	1.0841	5.0227
20000	365.81	1826.18	583.56	2409.74	4.0137	0.9132	4.9269
21000	369.89	1888.30	446.42	2334.72	4.1073	0.6942	4.8015
22000	373.80	2034.92	124.04	2158.97	4.3307	0.1917	4.5224
22089	374.14	2099.26	0	2099.26	4.4297	0	4.4297

TABLE B.1.3 SI *Superheated Vapor Water*

Temp. C	v m³/kg	u kJ/kg	h kJ/kg	s kJ/kg K	v m³/kg	u kJ/kg	h kJ/kg	s kJ/kg K
	P = 10 kPa (45.81)				P = 50 kPa (81.33)			
Sat.	14.67355	2437.89	2584.63	8.1501	3.24034	2483.85	2645.87	7.5939
50	14.86920	2443.87	2592.56	8.1749	--	--	--	--
100	17.19561	2515.50	2687.46	8.4479	3.41833	2511.61	2682.52	7.6947
150	19.51251	2587.86	2782.99	8.6881	3.88937	2585.61	2780.08	7.9400
200	21.82507	2661.27	2879.52	8.9037	4.35595	2659.85	2877.64	8.1579
250	24.13559	2735.95	2977.31	9.1002	4.82045	2734.97	2975.99	8.3555
300	26.44508	2812.06	3076.51	9.2812	5.28391	2811.33	3075.52	8.5372
400	31.06252	2968.89	3279.51	9.6076	6.20929	2968.43	3278.89	8.8641
500	35.67896	3132.26	3489.05	9.8977	7.13364	3131.94	3488.62	9.1545
600	40.29488	3302.45	3705.40	10.1608	8.05748	3302.22	3705.10	9.4177
700	44.91052	3479.63	3928.73	10.4028	8.98104	3479.45	3928.51	9.6599
800	49.52599	3663.84	4159.10	10.6281	9.90444	3663.70	4158.92	9.8852
900	54.14137	3855.03	4396.44	10.8395	10.82773	3854.91	4396.30	10.0967
1000	58.75669	4053.01	4640.58	11.0392	11.75097	4052.91	4640.46	10.2964
1100	63.37198	4257.47	4891.19	11.2287	12.67418	4257.37	4891.08	10.4858
1200	67.98724	4467.91	5147.78	11.4090	13.59737	4467.82	5147.69	10.6662
1300	72.60250	4683.68	5409.70	11.5810	14.52054	4683.58	5409.61	10.8382
	100 kPa (99.62)				200 kPa (120.23)			
Sat.	1.69400	2506.06	2675.46	7.3593	0.88573	2529.49	2706.63	7.1271
150	1.93636	2582.75	2776.38	7.6133	0.95964	2576.87	2768.80	7.2795
200	2.17226	2658.05	2875.27	7.8342	1.08034	2654.39	2870.46	7.5066
250	2.40604	2733.73	2974.33	8.0332	1.19880	2731.22	2970.98	7.7085
300	2.63876	2810.41	3074.28	8.2157	1.31616	2808.55	3071.79	7.8926
400	3.10263	2967.85	3278.11	8.5434	1.54930	2966.69	3276.55	8.2217
500	3.56547	3131.54	3488.09	8.8341	1.78139	3130.75	3487.03	8.5132
600	4.02781	3301.94	3704.72	9.0975	2.01297	3301.36	3703.96	8.7769
700	4.48986	3479.24	3928.23	9.3398	2.24426	3478.81	3927.66	9.0194
800	4.95174	3663.53	4158.71	9.5652	2.47539	3663.19	4158.27	9.2450
900	5.41353	3854.77	4396.12	9.7767	2.70643	3854.49	4395.77	9.4565
1000	5.87526	4052.78	4640.31	9.9764	2.93740	4052.53	4640.01	9.6563
1100	6.33696	4257.25	4890.95	10.1658	3.16834	4257.01	4890.68	9.8458
1200	6.79863	4467.70	5147.56	10.3462	3.39927	4467.46	5147.32	10.0262
1300	7.26030	4683.47	5409.49	10.5182	3.63018	4683.23	5409.26	10.1982
	300 kPa (133.55)				400 kPa (143.63)			
Sat.	0.60582	2543.55	2725.30	6.9918	0.46246	2553.55	2738.53	6.8958
150	0.63388	2570.79	2760.95	7.0778	0.47084	2564.48	2752.82	6.9299
200	0.71629	2650.65	2865.54	7.3115	0.53422	2646.83	2860.51	7.1706
250	0.79636	2728.69	2967.59	7.5165	0.59512	2726.11	2964.16	7.3788
300	0.87529	2806.69	3069.28	7.7022	0.65484	2804.81	3066.75	7.5661
400	1.03151	2965.53	3274.98	8.0329	0.77262	2964.36	3273.41	7.8984

(Continued)

TABLE B.1.3 SI (Continued) *Superheated Vapor Water*

Temp. C	*v* m³/kg	*u* kJ/kg	*h* kJ/kg	*s* kJ/kg K	*v* m³/kg	*u* kJ/kg	*h* kJ/kg	*s* kJ/kg K
	300 kPa (133.55)				400 kPa (143.63)			
500	1.18669	3129.95	3485.96	8.3250	0.88934	3129.15	3484.89	8.1912
600	1.34136	3300.79	3703.20	8.5892	1.00555	3300.22	3702.44	8.4557
700	1.49573	3478.38	3927.10	8.8319	1.12147	3477.95	3926.53	8.6987
800	1.64994	3662.85	4157.83	9.0575	1.23722	3662.51	4157.40	8.9244
900	1.80406	3854.20	4395.42	9.2691	1.35288	3853.91	4395.06	9.1361
1000	1.95812	4052.27	4639.71	9.4689	1.46847	4052.02	4639.41	9.3360
1100	2.11214	4256.77	4890.41	9.6585	1.58404	4256.53	4890.15	9.5255
1200	2.26614	4467.23	5147.07	9.8389	1.69958	4466.99	5146.83	9.7059
1300	2.42013	4682.99	5409.03	10.0109	1.81511	4682.75	5408.80	9.8780
	500 kPa (151.86)				600 kPa (158.85)			
Sat.	0.37489	2561.23	2748.67	6.8212	0.31567	2567.40	2756.80	6.7600
200	0.42492	2642.91	2855.37	7.0592	0.35202	2638.91	2850.12	6.9665
250	0.47436	2723.50	2960.68	7.2708	0.39383	2720.86	2957.16	7.1816
300	0.52256	2802.91	3064.20	7.4598	0.43437	2801.00	3061.63	7.3723
350	0.57012	2882.59	3167.65	7.6328	0.47424	2881.12	3165.66	7.5463
400	0.61728	2963.19	3271.83	7.7937	0.51372	2962.02	3270.25	7.7078
500	0.71093	3128.35	3483.82	8.0872	0.59199	3127.55	3482.75	8.0020
600	0.80406	3299.64	3701.67	8.3521	0.66974	3299.07	3700.91	8.2673
700	0.89691	3477.52	3925.97	8.5952	0.74720	3477.08	3925.41	8.5107
800	0.98959	3662.17	4156.96	8.8211	0.82450	3661.83	4156.52	8.7367
900	1.08217	3853.63	4394.71	9.0329	0.90169	3853.34	4394.36	8.9485
1000	1.17469	4051.76	4639.11	9.2328	0.97883	4051.51	4638.81	9.1484
1100	1.26718	4256.29	4889.88	9.4224	1.05594	4256.05	4889.61	9.3381
1200	1.35964	4466.76	5146.58	9.6028	1.13302	4466.52	5146.34	9.5185
1300	1.45210	4682.52	5408.57	9.7749	1.21009	4682.28	5408.34	9.6906
	800 kPa (170.43)				1000 kPa (179.91)			
Sat.	0.24043	2576.79	2769.13	6.6627	0.19444	2583.64	2778.08	6.5864
200	0.26080	2630.61	2839.25	6.8158	0.20596	2621.90	2827.86	6.6939
250	0.29314	2715.46	2949.97	7.0384	0.23268	2709.91	2942.59	6.9246
300	0.32411	2797.14	3056.43	7.2327	0.25794	2793.21	3051.15	7.1228
350	0.35439	2878.16	3161.68	7.4088	0.28247	2875.18	3157.65	7.3010
400	0.38426	2959.66	3267.07	7.5715	0.30659	2957.29	3263.88	7.4650
500	0.44331	3125.95	3480.60	7.8672	0.35411	3124.34	3478.44	7.7621
600	0.50184	3297.91	3699.38	8.1332	0.40109	3296.76	3697.85	8.0289
700	0.56007	3476.22	3924.27	8.3770	0.44779	3475.35	3923.14	8.2731
800	0.61813	3661.14	4155.65	8.6033	0.49432	3660.46	4154.78	8.4996
900	0.67610	3852.77	4393.65	8.8153	0.54075	3852.19	4392.94	8.7118
1000	0.73401	4051.00	4638.20	9.0153	0.58712	4050.49	4637.60	8.9119
1100	0.79188	4255.57	4889.08	9.2049	0.63345	4255.09	4888.55	9.1016
1200	0.84974	4466.05	5145.85	9.3854	0.67977	4465.58	5145.36	9.2821
1300	0.90758	4681.81	5407.87	9.5575	0.72608	4681.33	5407.41	9.4542

(Continued)

TABLE B.1.3 SI (Continued) *Superheated Vapor Water*

Temp. C	v m³/kg	u kJ/kg	h kJ/kg	s kJ/kg K	v m³/kg	u kJ/kg	h kJ/kg	s kJ/kg K
	1200 kPa (187.99)				1400 kPa (195.07)			
Sat.	0.16333	2588.82	2784.82	6.5233	0.14084	2592.83	2790.00	6.4692
200	0.16930	2612.74	2815.90	6.5898	0.14302	2603.09	2803.32	6.4975
250	0.19235	2704.20	2935.01	6.8293	0.16350	2698.32	2927.22	6.7467
300	0.21382	2789.22	3045.80	7.0316	0.18228	2785.16	3040.35	6.9533
350	0.23452	2872.16	3153.59	7.2120	0.20026	2869.12	3149.49	7.1359
400	0.25480	2954.90	3260.66	7.3773	0.21780	2952.50	3257.42	7.3025
500	0.29463	3122.72	3476.28	7.6758	0.25215	3121.10	3474.11	7.6026
600	0.33393	3295.60	3696.32	7.9434	0.28596	3294.44	3694.78	7.8710
700	0.37294	3474.48	3922.01	8.1881	0.31947	3473.61	3920.87	8.1160
800	0.41177	3659.77	4153.90	8.4149	0.35281	3659.09	4153.03	8.3431
900	0.45051	3851.62	4392.23	8.6272	0.38606	3851.05	4391.53	8.5555
1000	0.48919	4049.98	4637.00	8.8274	0.41924	4049.47	4636.41	8.7558
1100	0.52783	4254.61	4888.02	9.0171	0.45239	4254.14	4887.49	8.9456
1200	0.56646	4465.12	5144.87	9.1977	0.48552	4464.65	5144.38	9.1262
1300	0.60507	4680.86	5406.95	9.3698	0.51864	4680.39	5406.49	9.2983
	1600 kPa (201.40)				1800 kPa (207.15)			
Sat.	0.12380	2595.95	2794.02	6.4217	0.11042	2598.38	2797.13	6.3793
250	0.14184	2692.26	2919.20	6.6732	0.12497	2686.02	2910.96	6.6066
300	0.15862	2781.03	3034.83	6.8844	0.14021	2776.83	3029.21	6.8226
350	0.17456	2866.05	3145.35	7.0693	0.15457	2862.95	3141.18	7.0099
400	0.19005	2950.09	3254.17	7.2373	0.16847	2947.66	3250.90	7.1793
500	0.22029	3119.47	3471.93	7.5389	0.19550	3117.84	3469.75	7.4824
600	0.24998	3293.27	3693.23	7.8080	0.22199	3292.10	3691.69	7.7523
700	0.27937	3472.74	3919.73	8.0535	0.24818	3471.87	3918.59	7.9983
800	0.30859	3658.40	4152.15	8.2808	0.27420	3657.71	4151.27	8.2258
900	0.33772	3850.47	4390.82	8.4934	0.30012	3849.90	4390.11	8.4386
1000	0.36678	4048.96	4635.81	8.6938	0.32598	4048.45	4635.21	8.6390
1100	0.39581	4253.66	4886.95	8.8837	0.35180	4253.18	4886.42	8.8290
1200	0.42482	4464.18	5143.89	9.0642	0.37761	4463.71	5143.40	9.0096
1300	0.45382	4679.92	5406.02	9.2364	0.40340	4679.44	5405.56	9.1817

(Continued)

TABLE B.1.3 SI (Continued) *Superheated Vapor Water*

Temp. C	v m³/kg	u kJ/kg	h kJ/kg	s kJ/kg K	v m³/kg	u kJ/kg	h kJ/kg	s kJ/kg K
	2000 kPa (212.42)				2500 kPa (223.99)			
Sat.	0.09963	2600.26	2799.51	6.3408	0.07998	2603.13	2803.07	6.2574
250	0.11144	2679.58	2902.46	6.5452	0.08700	2662.55	2880.06	6.4084
300	0.12547	2772.56	3023.50	6.7663	0.09890	2761.56	3008.81	6.6437
350	0.13857	2859.81	3136.96	6.9562	0.10976	2851.84	3126.24	6.8402
400	0.15120	2945.21	3247.60	7.1270	0.12010	2939.03	3239.28	7.0147
450	0.16353	3030.41	3357.48	7.2844	0.13014	3025.43	3350.77	7.1745
500	0.17568	3116.20	3467.55	7.4316	0.13998	3112.08	3462.04	7.3233
600	0.19960	3290.93	3690.14	7.7023	0.15930	3287.99	3686.25	7.5960
700	0.22323	3470.99	3917.45	7.9487	0.17832	3468.80	3914.59	7.8435
800	0.24668	3657.03	4150.40	8.1766	0.19716	3655.30	4148.20	8.0720
900	0.27004	3849.33	4389.40	8.3895	0.21590	3847.89	4387.64	8.2853
1000	0.29333	4047.94	4634.61	8.5900	0.23458	4046.67	4633.12	8.4860
1100	0.31659	4252.71	4885.89	8.7800	0.25322	4251.52	4884.57	8.6761
1200	0.33984	4463.25	5142.92	8.9606	0.27185	4462.08	5141.70	8.8569
1300	0.36306	4678.97	5405.10	9.1328	0.29046	4677.80	5403.95	9.0291
	3000 kPa (233.90)				3500 kPa (242.60)			
Sat.	0.06668	2604.10	2804.14	6.1869	0.05707	2603.70	2803.43	6.1252
250	0.07058	2644.00	2855.75	6.2871	0.05873	2623.65	2829.19	6.1748
300	0.08114	2750.05	2993.48	6.5389	0.06842	2737.99	2977.46	6.4460
350	0.09053	2843.66	3115.25	6.7427	0.07678	2835.27	3103.99	6.6578
400	0.09936	2932.75	3230.82	6.9211	0.08453	2926.37	3222.24	6.8404
450	0.10787	3020.38	3344.00	7.0833	0.09196	3015.28	3337.15	7.0051
500	0.11619	3107.92	3456.48	7.2337	0.09918	3103.73	3450.87	7.1571
600	0.13243	3285.03	3682.34	7.5084	0.11324	3282.06	3678.40	7.4338
700	0.14838	3466.59	3911.72	7.7571	0.12699	3464.37	3908.84	7.6837
800	0.16414	3653.58	4146.00	7.9862	0.14056	3651.84	4143.80	7.9135
900	0.17980	3846.46	4385.87	8.1999	0.15402	3845.02	4384.11	8.1275
1000	0.19541	4045.40	4631.63	8.4009	0.16743	4044.14	4630.14	8.3288
1100	0.21098	4250.33	4883.26	8.5911	0.18080	4249.14	4881.94	8.5191
1200	0.22652	4460.92	5140.49	8.7719	0.19415	4459.76	5139.28	8.7000
1300	0.24206	4676.63	5402.81	8.9442	0.20749	4675.45	5401.66	8.8723

(Continued)

TABLE B.1.3 SI (Continued) *Superheated Vapor Water*

Temp. C	v m³/kg	u kJ/kg	h kJ/kg	s kJ/kg K	v m³/kg	u kJ/kg	h kJ/kg	s kJ/kg K
	4000 kPa (250.40)				4500 kPa (257.48)			
Sat.	0.04978	2602.27	2801.38	6.0700	0.04406	2600.03	2798.29	6.0198
300	0.05884	2725.33	2960.68	6.3614	0.05135	2712.00	2943.07	6.2827
350	0.06645	2826.65	3092.43	6.5820	0.05840	2817.78	3080.57	6.5130
400	0.07341	2919.88	3213.51	6.7689	0.06475	2913.29	3204.65	6.7046
450	0.08003	3010.13	3330.23	6.9362	0.07074	3004.91	3323.23	6.8745
500	0.08643	3099.49	3445.21	7.0900	0.07651	3095.23	3439.51	7.0300
600	0.09885	3279.06	3674.44	7.3688	0.08765	3276.04	3670.47	7.3109
700	0.11095	3462.15	3905.94	7.6198	0.09847	3459.91	3903.04	7.5631
800	0.12287	3650.11	4141.59	7.8502	0.10911	3648.37	4139.38	7.7942
900	0.13469	3843.59	4382.34	8.0647	0.11965	3842.15	4380.58	8.0091
1000	0.14645	4042.87	4628.65	8.2661	0.13013	4041.61	4627.17	8.2108
1100	0.15817	4247.96	4880.63	8.4566	0.14056	4246.78	4879.32	8.4014
1200	0.16987	4458.60	5138.07	8.6376	0.15098	4457.45	5136.87	8.5824
1300	0.18156	4674.29	5400.52	8.8099	0.16139	4673.12	5399.38	8.7548
	5000 kPa (263.99)				6000 kPa (275.64)			
Sat.	0.03944	2597.12	2794.33	5.9733	0.03244	2589.69	2784.33	5.8891
300	0.04532	2697.94	2924.53	6.2083	0.03616	2667.22	2884.19	6.0673
350	0.05194	2808.67	3068.39	6.4492	0.04223	2789.61	3042.97	6.3334
400	0.05781	2906.58	3195.64	6.6458	0.04739	2892.81	3177.17	6.5407
450	0.06330	2999.64	3316.15	6.8185	0.05214	2988.90	3301.76	6.7192
500	0.06857	3090.92	3433.76	6.9758	0.05665	3082.20	3422.12	6.8802
550	0.07368	3181.82	3550.23	7.1217	0.06101	3174.57	3540.62	7.0287
600	0.07869	3273.01	3666.47	7.2588	0.06525	3266.89	3658.40	7.1676
700	0.08849	3457.67	3900.13	7.5122	0.07352	3453.15	3894.28	7.4234
800	0.09811	3646.62	4137.17	7.7440	0.08160	3643.12	4132.74	7.6566
900	0.10762	3840.71	4378.82	7.9593	0.08958	3837.84	4375.29	7.8727
1000	0.11707	4040.35	4625.69	8.1612	0.09749	4037.83	4622.74	8.0751
1100	0.12648	4245.61	4878.02	8.3519	0.10536	4243.26	4875.42	8.2661
1200	0.13587	4456.30	5135.67	8.5330	0.11321	4454.00	5133.28	8.4473
1300	0.14526	4671.96	5398.24	8.7055	0.12106	4669.64	5395.97	8.6199

(Continued)

TABLE B.1.3 SI (Continued) ***Superheated Vapor Water***

Temp. C	v m³/kg	u kJ/kg	h kJ/kg	s kJ/kg K	v m³/kg	u kJ/kg	h kJ/kg	s kJ/kg K
	7000 kPa (285.88)				8000 kPa (295.06)			
Sat.	0.02737	2580.48	2772.07	5.8132	0.02352	2569.79	2757.94	5.7431
300	0.02947	2632.13	2838.40	5.9304	0.02426	2590.93	2784.98	5.7905
350	0.03524	2769.34	3016.02	6.2282	0.02995	2747.67	2987.30	6.1300
400	0.03993	2878.55	3158.07	6.4477	0.03432	2863.75	3138.28	6.3633
450	0.04416	2977.91	3287.04	6.6326	0.03817	2966.66	3271.99	6.5550
500	0.04814	3073.33	3410.29	6.7974	0.04175	3064.30	3398.27	6.7239
550	0.05195	3167.21	3530.87	6.9486	0.04516	3159.76	3521.01	6.8778
600	0.05565	3260.69	3650.26	7.0894	0.04845	3254.43	3642.03	7.0205
700	0.06283	3448.60	3888.39	7.3476	0.05481	3444.00	3882.47	7.2812
800	0.06981	3639.61	4128.30	7.5822	0.06097	3636.08	4123.84	7.5173
900	0.07669	3834.96	4371.77	7.7991	0.06702	3832.08	4368.26	7.7350
1000	0.08350	4035.31	4619.80	8.0020	0.07301	4032.81	4616.87	7.9384
1100	0.09027	4240.92	4872.83	8.1933	0.07896	4238.60	4870.25	8.1299
1200	0.09703	4451.72	5130.90	8.3747	0.08489	4449.45	5128.54	8.3115
1300	0.10377	4667.33	5393.71	8.5472	0.09080	4665.02	5391.46	8.4842
	9000 kPa (303.40)				10000 kPa (311.06)			
Sat.	0.02048	2557.75	2742.11	5.6771	0.01803	2544.41	2724.67	5.6140
350	0.02580	2724.38	2956.55	6.0361	0.02242	2699.16	2923.39	5.9442
400	0.02993	2848.38	3117.76	6.2853	0.02641	2832.38	3096.46	6.2119
450	0.03350	2955.13	3256.59	6.4843	0.02975	2943.32	3240.83	6.4189
500	0.03677	3055.12	3386.05	6.6575	0.03279	3045.77	3373.63	6.5965
550	0.03987	3152.20	3511.02	6.8141	0.03564	3144.54	3500.92	6.7561
600	0.04285	3248.09	3633.73	6.9588	0.03837	3241.68	3625.34	6.9028
650	0.04574	3343.65	3755.32	7.0943	0.04101	3338.22	3748.27	7.0397
700	0.04857	3439.38	3876.51	7.2221	0.04358	3434.72	3870.52	7.1687
800	0.05409	3632.53	4119.38	7.4597	0.04859	3628.97	4114.91	7.4077
900	0.05950	3829.20	4364.74	7.6782	0.05349	3826.32	4361.24	7.6272
1000	0.06485	4030.30	4613.95	7.8821	0.05832	4027.81	4611.04	7.8315
1100	0.07016	4236.28	4867.69	8.0739	0.06312	4233.97	4865.14	8.0236
1200	0.07544	4447.18	5126.18	8.2556	0.06789	4444.93	5123.84	8.2054
1300	0.08072	4662.73	5389.22	8.4283	0.07265	4660.44	5386.99	8.3783

(Continued)

TABLE B.1.3 SI (Continued) ***Superheated Vapor Water***

Temp. C	*v* m^3/kg	*u* kJ/kg	*h* kJ/kg	*s* kJ/kg K	*v* m^3/kg	*u* kJ/kg	*h* kJ/kg	*s* kJ/kg K
	12500 kPa (327.89)				15000 kPa (342.24)			
Sat.	0.01350	2505.08	2673.77	5.4623	0.01034	2455.43	2610.49	5.3097
350	0.01613	2624.57	2826.15	5.7117	0.01147	2520.36	2692.41	5.4420
400	0.02000	2789.25	3039.30	6.0416	0.01565	2740.70	2975.44	5.8810
450	0.02299	2912.44	3199.78	6.2718	0.01845	2879.47	3156.15	6.1403
500	0.02560	3021.68	3341.72	6.4617	0.02080	2996.52	3308.53	6.3442
550	0.02801	3124.94	3475.13	6.6289	0.02293	3104.71	3448.61	6.5198
600	0.03029	3225.37	3604.05	6.7810	0.02491	3208.64	3582.30	6.6775
650	0.03248	3324.43	3730.44	6.9218	0.02680	3310.37	3712.32	6.8223
700	0.03460	3422.93	3855.41	7.0536	0.02861	3410.94	3840.12	6.9572
800	0.03869	3620.02	4103.69	7.2965	0.03210	3610.99	4092.43	7.2040
900	0.04267	3819.11	4352.48	7.5181	0.03546	3811.89	4343.75	7.4279
1000	0.04658	4021.59	4603.81	7.7237	0.03875	4015.41	4596.63	7.6347
1100	0.05045	4228.23	4858.82	7.9165	0.04200	4222.55	4852.56	7.8282
1200	0.05430	4439.33	5118.02	8.0987	0.04523	4433.78	5112.27	8.0108
1300	0.05813	4654.76	5381.44	8.2717	0.04845	4649.12	5375.94	8.1839
	17500 kPa (354.75)				20000 kPa (365.81)			
Sat.	0.00792	2390.19	2528.79	5.1418	0.00583	2293.05	2409.74	4.9269
400	0.01245	2684.98	2902.82	5.7212	0.00994	2619.22	2818.07	5.5539
450	0.01517	2844.15	3109.69	6.0182	0.01270	2806.16	3060.06	5.9016
500	0.01736	2970.25	3274.02	6.2382	0.01477	2942.82	3238.18	6.1400
550	0.01929	3083.84	3421.37	6.4229	0.01656	3062.34	3393.45	6.3347
600	0.02106	3191.51	3560.13	6.5866	0.01818	3174.00	3537.57	6.5048
650	0.02274	3296.04	3693.94	6.7356	0.01969	3281.46	3675.32	6.6582
700	0.02434	3398.78	3824.67	6.8736	0.02113	3386.46	3809.09	6.7993
750	0.02588	3500.56	3953.48	7.0026	0.02251	3490.01	3940.27	6.9308
800	0.02738	3601.89	4081.13	7.1245	0.02385	3592.73	4069.80	7.0544
900	0.03031	3804.67	4335.05	7.3507	0.02645	3797.44	4326.37	7.2830
1000	0.03316	4009.25	4589.52	7.5588	0.02897	4003.12	4582.45	7.4925
1100	0.03597	4216.90	4846.37	7.7530	0.03145	4211.30	4840.24	7.6874
1200	0.03876	4428.28	5106.59	7.9359	0.03391	4422.81	5100.96	7.8706
1300	0.04154	4643.52	5370.50	8.1093	0.03636	4637.95	5365.10	8.0441
	25000 kPa				30000 kPa			
375	0.001973	1798.60	1847.93	4.0319	0.001789	1737.75	1791.43	3.9303
400	0.006004	2430.05	2580.16	5.1418	0.002790	2067.34	2151.04	4.4728
425	0.007882	2609.21	2806.25	5.4722	0.005304	2455.06	2614.17	5.1503
450	0.009162	2720.65	2949.70	5.6743	0.006735	2619.30	2821.35	5.4423
500	0.011124	2884.29	3162.39	5.9592	0.008679	2820.67	3081.03	5.7904
550	0.012724	3017.51	3335.62	6.1764	0.010168	2970.31	3275.36	6.0342
600	0.014138	3137.92	3491.36	6.3602	0.011446	3100.53	3443.91	6.2330
650	0.015433	3251.64	3637.46	6.5229	0.012596	3221.04	3598.93	6.4057
700	0.016647	3361.39	3777.56	6.6707	0.013661	3335.84	3745.67	6.5606

(Continued)

TABLE B.1.3 SI (Continued) *Superheated Vapor Water*

Temp. C	v m³/kg	u kJ/kg	h kJ/kg	s kJ/kg K	v m³/kg	u kJ/kg	h kJ/kg	s kJ/kg K
	25000 kPa				30000 kPa			
800	0.018913	3574.26	4047.08	6.9345	0.015623	3555.60	4024.31	6.8332
900	0.021045	3782.97	4309.09	7.1679	0.017448	3768.48	4291.93	7.0717
1000	0.023102	3990.92	4568.47	7.3801	0.019196	3978.79	4554.68	7.2867
1100	0.025119	4200.18	4828.15	7.5765	0.020903	4189.18	4816.28	7.4845
1200	0.027115	4412.00	5089.86	7.7604	0.022589	4401.29	5078.97	7.6691
1300	0.029101	4626.91	5354.44	7.9342	0.024266	4615.96	5343.95	7.8432
	35000 kPa				40000 kPa			
375	0.001700	1702.86	1762.37	3.8721	0.001641	1677.09	1742.71	3.8289
400	0.002100	1914.02	1987.52	4.2124	0.001908	1854.52	1930.83	4.1134
425	0.003428	2253.42	2373.41	4.7747	0.002532	2096.83	2198.11	4.5028
450	0.004962	2498.71	2672.36	5.1962	0.003693	2365.07	2512.79	4.9459
500	0.006927	2751.88	2994.34	5.6281	0.005623	2678.36	2903.26	5.4699
550	0.008345	2920.94	3213.01	5.9025	0.006984	2869.69	3149.05	5.7784
600	0.009527	3062.03	3395.49	6.1178	0.008094	3022.61	3346.38	6.0113
650	0.010575	3189.79	3559.91	6.3010	0.009064	3158.04	3520.58	6.2054
700	0.011533	3309.89	3713.54	6.4631	0.009942	3283.63	3681.29	6.3750
800	0.013278	3536.81	4001.54	6.7450	0.011523	3517.89	3978.80	6.6662
900	0.014883	3753.96	4274.87	6.9886	0.012963	3739.42	4257.93	6.9150
1000	0.016410	3966.70	4541.05	7.2063	0.014324	3954.64	4527.59	7.1356
1100	0.017895	4178.25	4804.59	7.4056	0.015643	4167.38	4793.08	7.3364
1200	0.019360	4390.67	5068.26	7.5910	0.016940	4380.11	5057.72	7.5224
1300	0.020815	4605.09	5333.62	7.7652	0.018229	4594.28	5323.45	7.6969
	50000 kPa				60000 kPa			
375	0.001559	1638.55	1716.52	3.7638	0.001503	1609.34	1699.51	3.7140
400	0.001731	1788.04	1874.58	4.0030	0.001633	1745.34	1843.35	3.9317
425	0.002007	1959.63	2059.98	4.2733	0.001817	1892.66	2001.65	4.1625
450	0.002486	2159.60	2283.91	4.5883	0.002085	2053.86	2178.96	4.4119
500	0.003892	2525.45	2720.07	5.1725	0.002956	2390.53	2567.88	4.9320
550	0.005118	2763.61	3019.51	5.5485	0.003957	2658.76	2896.16	5.3440
600	0.006112	2941.98	3247.59	5.8177	0.004835	2861.14	3151.21	5.6451
650	0.006966	3093.56	3441.84	6.0342	0.005595	3028.83	3364.55	5.8829
700	0.007727	3230.54	3616.91	6.2189	0.006272	3177.25	3553.56	6.0824
800	0.009076	3479.82	3933.62	6.5290	0.007459	3441.60	3889.12	6.4110
900	0.010283	3710.26	4224.41	6.7882	0.008508	3680.97	4191.47	6.6805
1000	0.011411	3930.53	4501.09	7.0146	0.009480	3906.36	4475.16	6.9126
1100	0.012497	4145.72	4770.55	7.2183	0.010409	4124.07	4748.61	7.1194
1200	0.013561	4359.12	5037.15	7.4058	0.011317	4338.18	5017.19	7.3082
1300	0.014616	4572.77	5303.56	7.5807	0.012215	4551.35	5284.28	7.4837

TABLE B.1.4 SI *Compressed Liquid Water*

Temp. C	v m³/kg	u kJ/kg	h kJ/kg	s kJ/kg K	v m³/kg	u kJ/kg	h kJ/kg	s kJ/kg K
	5000 kPa (263.99)				10000 kPa (311.06)			
Sat.	0.00129	1147.78	1154.21	2.9201	0.001452	1393.00	1407.53	3.3595
0	0.000998	0.03	5.02	0.0001	0.000995	0.10	10.05	0.0003
20	0.00100	83.64	88.64	0.2955	0.00997	83.35	93.32	0.2945
40	0.00101	166.93	171.95	0.5705	0.00100	166.33	176.36	0.5685
60	0.00101	250.21	255.28	0.8284	0.00101	249.34	259.47	0.8258
80	0.00103	333.69	338.83	1.0719	0.00102	332.56	342.81	1.0687
100	0.00104	417.50	422.71	1.3030	0.00104	416.09	426.48	1.2992
120	0.00106	501.79	507.07	1.5232	0.00105	500.07	510.61	1.5188
140	0.00108	586.74	592.13	1.7342	0.00107	584.67	595.40	1.7291
160	0.00110	672.61	678.10	1.9374	0.00110	670.11	681.07	1.9316
180	0.00112	759.62	765.24	2.1341	0.00112	756.63	767.83	2.1274
200	0.00115	848.08	853.85	2.3254	0.00115	844.49	855.97	2.3178
220	0.00119	938.43	944.36	2.5128	0.00118	934.07	945.88	2.5038
240	0.00123	1031.34	1037.47	2.6978	0.00122	1025.94	1038.13	2.6872
260	0.00127	1127.92	1134.30	2.8829	0.00126	1121.03	1133.68	2.8698
280					0.00132	1220.90	1234.11	3.0547
300					0.00140	1328.34	1342.31	3.2468
	15000 kPa (342.24)				20000 kPa (365.81)			
Sat.	0.001658	1585.58	1610.45	3.6847	0.002035	1785.47	1826.18	4.0137
0	0.000993	0.15	15.04	0.0004	0.00099	0.20	20.00	0.0004
20	0.000995	83.05	97.97	0.2934	0.000993	82.75	102.61	0.2922
40	0.00100	165.73	180.75	0.5665	0.00100	165.15	185.14	0.5646
60	0.00101	248.49	263.65	0.8231	0.00101	247.66	267.82	0.8205
80	0.00102	331.46	346.79	1.0655	0.00102	330.38	350.78	1.0623
100	0.00104	414.72	430.26	1.2954	0.00103	413.37	434.04	1.2917
120	0.00105	498.39	514.17	1.5144	0.00105	496.75	517.74	1.5101
140	0.00107	582.64	598.70	1.7241	0.00107	580.67	602.03	1.7192
160	0.00109	667.69	684.07	1.9259	0.00109	665.34	687.11	1.9203
180	0.00112	753.74	770.48	2.1209	0.00111	750.94	773.18	2.1146
200	0.00114	841.04	858.18	2.3103	0.00114	837.70	860.47	2.3031
220	0.00117	929.89	947.52	2.4952	0.00117	925.89	949.27	2.4869
240	0.00121	1020.82	1038.99	2.6770	0.00120	1015.94	1040.04	2.6673
260	0.00126	1114.59	1133.41	2.8575	0.00125	1108.53	1133.45	2.8459
280	0.00131	1212.47	1232.09	3.0392	0.00130	1204.69	1230.62	3.0248
300	0.00138	1316.58	1337.23	3.2259	0.00136	1306.10	1333.29	3.2071
320	0.00147	1431.05	1453.13	3.4246	0.00144	1415.66	1444.53	3.3978
340	0.00163	1567.42	1591.88	3.6545	0.00157	1539.64	1571.01	3.6074
360					0.00182	1702.78	1739.23	3.8770

(Continued)

TABLE B.1.4 SI (Continued) ***Compressed Liquid Water***

Temp. C	v m³/kg	u kJ/kg	h kJ/kg	s kJ/kg K	v m³/kg	u kJ/kg	h kJ/kg	s kJ/kg K
	30000 kPa				50000 kPa			
0	0.000986	0.25	29.82	0.0001	0.000977	0.20	49.03	-0.0014
20	0.000989	82.16	111.82	0.2898	0.000980	80.98	130.00	0.2847
40	0.000995	164.01	193.87	0.5606	0.000987	161.84	211.20	0.5526
60	0.001004	246.03	276.16	0.8153	0.000996	242.96	292.77	0.8051
80	0.001016	328.28	358.75	1.0561	0.001007	324.32	374.68	1.0439
100	0.001029	410.76	441.63	1.2844	0.001020	405.86	456.87	1.2703
120	0.001044	493.58	524.91	1.5017	0.001035	487.63	539.37	1.4857
140	0.001062	576.86	608.73	1.7097	0.001052	569.76	622.33	1.6915
160	0.001082	660.81	693.27	1.9095	0.001070	652.39	705.91	1.8890
180	0.001105	745.57	778.71	2.1024	0.001091	735.68	790.24	2.0793
200	0.001130	831.34	865.24	2.2892	0.001115	819.73	875.46	2.2634
220	0.001159	918.32	953.09	2.4710	0.001141	904.67	961.71	2.4419
240	0.001192	1006.84	1042.60	2.6489	0.001170	990.69	1049.20	2.6158
260	0.001230	1097.38	1134.29	2.8242	0.001203	1078.06	1138.23	2.7860
280	0.001275	1190.69	1228.96	2.9985	0.001242	1167.19	1229.26	2.9536
300	0.001330	1287.89	1327.80	3.1740	0.001286	1258.66	1322.95	3.1200
320	0.001400	1390.64	1432.63	3.3538	0.001339	1353.23	1420.17	3.2867
340	0.001492	1501.71	1546.47	3.5425	0.001403	1451.91	1522.07	3.4556
360	0.001627	1626.57	1675.36	3.7492	0.001484	1555.97	1630.16	3.6290
380	0.001869	1781.35	1837.43	4.0010	0.001588	1667.13	1746.54	3.8100

TABLE B.1.5 SI ***Saturated Solid-Saturated Vapor Water***

Temp.	Press.	Specific Volume, m³/kg			Internal Energy, kJ/kg		
C	kPa	Sat. Solid	Evap.	Sat. Vapor	Sat. Solid	Evap.	Sat. Vapor
T	P	v_i	v_{ig}	v_g	u_i	u_{ig}	u_g
0.01	0.6113	0.0010908	206.152	206.153	−333.40	2708.7	2375.3
0	0.6108	0.0010908	206.314	206.315	−333.42	2708.7	2375.3
−2	0.5177	0.0010905	241.662	241.663	−337.61	2710.2	2372.5
−4	0.4376	0.0010901	283.798	283.799	−341.78	2711.5	2369.8
−6	0.3689	0.0010898	334.138	334.139	−345.91	2712.9	2367.0
−8	0.3102	0.0010894	394.413	394.414	−350.02	2714.2	2364.2
−10	0.2601	0.0010891	466.756	466.757	−354.09	2715.5	2361.4
−12	0.2176	0.0010888	553.802	553.803	−358.14	2716.8	2358.7
−14	0.1815	0.0010884	658.824	658.824	−362.16	2718.0	2355.9
−16	0.1510	0.0010881	785.906	785.907	−366.14	2719.2	2353.1
−18	0.1252	0.0010878	940.182	940.183	−370.10	2720.4	2350.3
−20	0.10355	0.0010874	1128.112	1128.113	−374.03	2721.6	2347.5
−22	0.08535	0.0010871	1357.863	1357.864	−377.93	2722.7	2344.7
−24	0.07012	0.0010868	1639.752	1639.753	−381.80	2723.7	2342.0
−26	0.05741	0.0010864	1986.775	1986.776	−385.64	2724.8	2339.2
−28	0.04684	0.0010861	2415.200	2415.201	−389.45	2725.8	2336.4
−30	0.03810	0.0010858	2945.227	2945.228	−393.23	2726.8	2333.6
−32	0.03090	0.0010854	3601.822	3601.823	−396.98	2727.8	2330.8
−34	0.02499	0.0010851	4416.252	4416.253	−400.71	2728.7	2328.0
−36	0.02016	0.0010848	5430.115	5430.116	−404.40	2729.6	2325.2
−38	0.01618	0.0010844	6707.021	6707.022	−408.06	2730.5	2322.4
−40	0.01286	0.0010841	8366.395	8366.396	−411.70	2731.3	2319.6

(Continued)

TABLE B.1.5 SI (Continued) ***Saturated Solid-Saturated Vapor Water***

Temp.	Press.	Enthalpy, kJ/kg			Entropy, kJ/kg K		
C	kPa	Sat. Solid	Evap.	Sat. Vapor	Sat. Solid	Evap.	Sat. Vapor
T	P	h_i	h_{ig}	h_g	s_i	s_{ig}	s_g
0.01	0.6113	–333.40	2834.7	2501.3	–1.2210	10.3772	9.1562
0	0.6108	–333.42	2834.8	2501.3	–1.2211	10.3776	9.1565
–2	0.5177	–337.61	2835.3	2497.6	–1.2369	10.4562	9.2193
–4	0.4376	–341.78	2835.7	2494.0	–1.2526	10.5358	9.2832
–6	0.3689	–345.91	2836.2	2490.3	–1.2683	10.6165	9.3482
–8	0.3102	–350.02	2836.6	2486.6	–1.2839	10.6982	9.4143
–10	0.2601	–354.09	2837.0	2482.9	–1.2995	10.7809	9.4815
–12	0.2176	–358.14	2837.3	2479.2	–1.3150	10.8648	9.5498
–14	0.1815	–362.16	2837.6	2475.5	–1.3306	10.9498	9.6192
–16	0.1510	–366.14	2837.9	2471.8	–1.3461	11.0359	9.6898
–18	0.1252	–370.10	2838.2	2468.1	–1.3617	11.1233	9.7616
–20	0.10355	–374.03	2838.4	2464.3	–1.3772	11.2120	9.8348
–22	0.08535	–377.93	2838.6	2460.6	–1.3928	11.3020	9.9093
–24	0.07012	–381.80	2838.7	2456.9	–1.4083	11.3935	9.9852
–26	0.05741	–385.64	2838.9	2453.2	–1.4239	11.4864	10.0625
–28	0.04684	–389.45	2839.0	2449.5	–1.4394	11.5808	10.1413
–30	0.03810	–393.23	2839.0	2445.8	–1.4550	11.6765	10.2215
–32	0.03090	–396.98	2839.1	2442.1	–1.4705	11.7733	10.3028
–34	0.02499	–400.71	2839.1	2438.4	–1.4860	11.8713	10.3853
–36	0.02016	–404.40	2839.1	2434.7	–1.5014	11.9704	10.4690
–38	0.01618	–408.06	2839.0	2431.0	–1.5168	12.0714	10.5546
–40	0.01286	–411.70	2838.9	2427.2	–1.5321	12.1768	10.6447

TABLE B.2 SI *Thermodynamic Properties of Ammonia*
TABLE B.2.1 SI *Saturated Ammonia*

Temp.	Press.	Specific Volume, m^3/kg			Internal Energy, kJ/kg		
C	kPa	Sat. Liquid	Evap.	Sat. Vapor	Sat. Liquid	Evap.	Sat. Vapor
T	P	v_f	v_{fg}	v_g	u_f	u_{fg}	u_g
-50	40.9	0.001424	2.62557	2.62700	-43.82	1309.1	1265.2
-45	54.5	0.001437	2.00489	2.00632	-22.01	1293.5	1271.4
-40	71.7	0.001450	1.55111	1.55256	-0.10	1277.6	1277.4
-35	93.2	0.001463	1.21466	1.21613	21.93	1261.3	1283.3
-30	119.5	0.001476	0.96192	0.96339	44.08	1244.8	1288.9
-25	151.6	0.001490	0.76970	0.77119	66.36	1227.9	1294.3
-20	190.2	0.001504	0.62184	0.62334	88.76	1210.7	1299.5
-15	236.3	0.001519	0.50686	0.50838	111.30	1193.2	1304.5
-10	290.9	0.001534	0.41655	0.41808	133.96	1175.2	1309.2
-5	354.9	0.001550	0.34493	0.34648	156.76	1157.0	1313.7
0	429.6	0.001566	0.28763	0.28920	179.69	1138.3	1318.0
5	515.9	0.001583	0.24140	0.24299	202.77	1119.2	1322.0
10	615.2	0.001600	0.20381	0.20541	225.99	1099.7	1325.7
15	728.6	0.001619	0.17300	0.17462	249.36	1079.7	1329.1
20	857.5	0.001638	0.14758	0.14922	272.89	1059.3	1332.2
25	1003.2	0.001658	0.12647	0.12813	296.59	1038.4	1335.0
30	1167.0	0.001680	0.10881	0.11049	320.46	1016.9	1337.4
35	1350.4	0.001702	0.09397	0.09567	344.50	994.9	1339.4
40	1554.9	0.001725	0.08141	0.08313	368.74	972.2	1341.0
45	1782.0	0.001750	0.07073	0.07248	393.19	948.9	1342.1
50	2033.1	0.001777	0.06159	0.06337	417.87	924.8	1342.7
55	2310.1	0.001804	0.05375	0.05555	442.79	899.9	1342.7
60	2614.4	0.001834	0.04697	0.04880	467.99	874.2	1342.1
65	2947.8	0.001866	0.04109	0.04296	493.51	847.4	1340.9
70	3312.0	0.001900	0.03597	0.03787	519.39	819.5	1338.9
75	3709.0	0.001937	0.03148	0.03341	545.70	790.4	1336.1
80	4140.5	0.001978	0.02753	0.02951	572.50	759.9	1332.4
85	4608.6	0.002022	0.02404	0.02606	599.90	727.8	1327.7
90	5115.3	0.002071	0.02093	0.02300	627.99	693.7	1321.7
95	5662.9	0.002126	0.01815	0.02028	656.95	657.4	1314.4
100	6253.7	0.002188	0.01565	0.01784	686.96	618.4	1305.3
105	6890.4	0.002261	0.01337	0.01564	718.30	575.9	1294.2
110	7575.7	0.002347	0.01128	0.01363	751.37	529.1	1280.5
115	8313.3	0.002452	0.00933	0.01178	786.82	476.2	1263.1
120	9107.2	0.002589	0.00744	0.01003	825.77	414.5	1240.3
125	9963.5	0.002783	0.00554	0.00833	870.69	337.7	1208.4
130	10891.6	0.003122	0.00337	0.00649	929.29	226.9	1156.2
132.3	11333.2	0.004255	0	0.00426	1037.62	0	1037.6

(Continued)

TABLE B.2.1 SI (Continued) *Saturated Ammonia*

Temp.	Press.	Enthalpy, kJ/kg			Entropy, kJ/kg K		
C T	kPa P	Sat. Liquid h_f	Evap. h_{fg}	Sat. Vapor h_g	Sat. Liquid s_f	Evap. s_{fg}	Sat. Vapor s_g
-50	40.9	-43.76	1416.3	1372.6	-0.1916	6.3470	6.1554
-45	54.5	-21.94	1402.8	1380.8	-0.0950	6.1484	6.0534
-40	71.7	0	1388.8	1388.8	0	5.9567	5.9567
-35	93.2	22.06	1374.5	1396.5	0.0935	5.7715	5.8650
-30	119.5	44.26	1359.8	1404.0	0.1856	5.5922	5.7778
-25	151.6	66.58	1344.6	1411.2	0.2763	5.4185	5.6947
-20	190.2	89.05	1329.0	1418.0	0.3657	5.2498	5.6155
-15	236.3	111.66	1312.9	1424.6	0.4538	5.0859	5.5397
-10	290.9	134.41	1296.4	1430.8	0.5408	4.9265	5.4673
-5	354.9	157.31	1279.4	1436.7	0.6266	4.7711	5.3977
0	429.6	180.36	1261.8	1442.2	0.7114	4.6195	5.3309
5	515.9	203.58	1243.7	1447.3	0.7951	4.4715	5.2666
10	615.2	226.97	1225.1	1452.0	0.8779	4.3266	5.2045
15	728.6	250.54	1205.8	1456.3	0.9598	4.1846	5.1444
20	857.5	274.30	1185.9	1460.2	1.0408	4.0452	5.0860
25	1003.2	298.25	1165.2	1463.5	1.1210	3.9083	5.0293
30	1167.0	322.42	1143.9	1466.3	1.2005	3.7734	4.9738
35	1350.4	346.80	1121.8	1468.6	1.2792	3.6403	4.9196
40	1554.9	371.43	1098.8	1470.2	1.3574	3.5088	4.8662
45	1782.0	396.31	1074.9	1471.2	1.4350	3.3786	4.8136
50	2033.1	421.48	1050.0	1471.5	1.5121	3.2493	4.7614
55	2310.1	446.96	1024.1	1471.0	1.5888	3.1208	4.7095
60	2614.4	472.79	997.0	1469.7	1.6652	2.9925	4.6577
65	2947.8	499.01	968.5	1467.5	1.7415	2.8642	4.6057
70	3312.0	525.69	938.7	1464.4	1.8178	2.7354	4.5533
75	3709.0	552.88	907.2	1460.1	1.8943	2.6058	4.5001
80	4140.5	580.69	873.9	1454.6	1.9712	2.4746	4.4458
85	4608.6	609.21	838.6	1447.8	2.0488	2.3413	4.3901
90	5115.3	638.59	800.8	1439.4	2.1273	2.2051	4.3325
95	5662.9	668.99	760.2	1429.2	2.2073	2.0650	4.2723
100	6253.7	700.64	716.2	1416.9	2.2893	1.9195	4.2088
105	6890.4	733.87	668.1	1402.0	2.3740	1.7667	4.1407
110	7575.7	769.15	614.6	1383.7	2.4625	1.6040	4.0665
115	8313.3	807.21	553.8	1361.0	2.5566	1.4267	3.9833
120	9107.2	849.36	482.3	1331.7	2.6593	1.2268	3.8861
125	9963.5	898.42	393.0	1291.4	2.7775	0.9870	3.7645
130	10892	963.29	263.7	1227.0	2.9326	0.6540	3.5866
132.3	11333	1085.85	0	1085.9	3.2316	0	3.2316

TABLE B.2.2 SI *Superheated Ammonia*

Temp. C	v m³/kg	h kJ/kg	s kJ/kg K	v m³/kg	h kJ/kg	s kJ/kg K	v m³/kg	h kJ/kg	s kJ/kg K
	50 kPa (-46.53)			75 kPa (-39.16)			100 kPa (-33.60)		
Sat.	2.17521	1378.3	6.0839	1.48922	1390.1	5.9411	1.13806	1398.7	5.8401
-30	2.34484	1413.4	6.2333	1.55321	1410.1	6.0247	1.15727	1406.7	5.8734
-20	2.44631	1434.6	6.3187	1.62221	1431.7	6.1120	1.21007	1428.8	5.9626
-10	2.54711	1455.7	6.4006	1.69050	1453.3	6.1954	1.26213	1450.8	6.0477
0	2.64736	1476.9	6.4795	1.75823	1474.8	6.2756	1.31362	1472.6	6.1291
10	2.74716	1498.1	6.5556	1.82551	1496.2	6.3527	1.36465	1494.4	6.2073
20	2.84661	1519.3	6.6293	1.89243	1517.7	6.4272	1.41532	1516.1	6.2826
30	2.94578	1540.6	6.7008	1.95906	1539.2	6.4993	1.46569	1537.7	6.3553
40	3.04472	1562.0	6.7703	2.02547	1560.7	6.5693	1.51582	1559.5	6.4258
50	3.14348	1583.5	6.8379	2.09168	1582.4	6.6373	1.56577	1581.2	6.4943
60	3.24209	1605.1	6.9038	2.15775	1604.1	6.7036	1.61557	1603.1	6.5609
70	3.34058	1626.9	6.9682	2.22369	1626.0	6.7683	1.66525	1625.1	6.6258
80	3.43897	1648.8	7.0312	2.28954	1648.0	6.8315	1.71482	1647.1	6.6892
100	3.63551	1693.2	7.1533	2.42099	1692.4	6.9539	1.81373	1691.7	6.8120
120	3.83183	1738.2	7.2708	2.55221	1737.5	7.0716	1.91240	1736.9	6.9300
140	4.02797	1783.9	7.3842	2.68326	1783.4	7.1853	2.01091	1782.8	7.0439
160	4.22398	1830.4	7.4941	2.81418	1829.9	7.2953	2.10927	1829.4	7.1540
180	4.41988	1877.7	7.6008	2.94499	1877.2	7.4021	2.20754	1876.8	7.2609
200	4.61570	1925.7	7.7045	3.07571	1925.3	7.5059	2.30571	1924.9	7.3648
	125 kPa (-29.07)			150 kPa (-25.22)			200 kPa (-18.86)		
Sat.	0.92365	1405.4	5.7620	0.77870	1410.9	5.6983	0.59460	1419.6	5.5979
-20	0.96271	1425.9	5.8446	0.79774	1422.9	5.7465	—	—	—
-10	1.00506	1448.3	5.9314	0.83364	1445.7	5.8349	0.61926	1440.6	5.6791
0	1.04682	1470.5	6.0141	0.86892	1468.3	5.9189	0.64648	1463.8	5.7659
10	1.08811	1492.5	6.0933	0.90373	1490.6	5.9992	0.67319	1486.8	5.8484
20	1.12903	1514.4	6.1694	0.93815	1512.8	6.0761	0.69951	1509.4	5.9270
30	1.16964	1536.3	6.2428	0.97227	1534.8	6.1502	0.72553	1531.9	6.0025
40	1.21003	1558.2	6.3138	1.00615	1556.9	6.2217	0.75129	1554.3	6.0751
50	1.25022	1580.1	6.3827	1.03984	1578.9	6.2910	0.77685	1576.6	6.1453
60	1.29026	1602.1	6.4496	1.07338	1601.0	6.3583	0.80226	1598.9	6.2133
70	1.33017	1624.1	6.5149	1.10678	1623.2	6.4238	0.82754	1621.3	6.2794
80	1.36998	1646.3	6.5785	1.14009	1645.4	6.4877	0.85271	1643.7	6.3437
100	1.44937	1691.0	6.7017	1.20646	1690.2	6.6112	0.90282	1688.8	6.4679
120	1.52852	1736.3	6.8199	1.27259	1735.6	6.7297	0.95268	1734.4	6.5869
140	1.60749	1782.2	6.9339	1.33855	1781.7	6.8439	1.00237	1780.6	6.7015
160	1.68633	1828.9	7.0443	1.40437	1828.4	6.9544	1.05192	1827.4	6.8123
180	1.76507	1876.3	7.1513	1.47009	1875.9	7.0615	1.10136	1875.0	6.9196
200	1.84371	1924.5	7.2553	1.53572	1924.1	7.1656	1.15072	1923.3	7.0239
220	1.92229	1973.4	7.3566	1.60127	1973.1	7.2670	1.20000	1972.4	7.1255

(Continued)

TABLE B.2.2 SI (Continued) ***Superheated Ammonia***

Temp. C	v m³/kg	h kJ/kg	s kJ/kg K	v m³/kg	h kJ/kg	s kJ/kg K	v m³/kg	h kJ/kg	s kJ/kg K
	250 kPa (-13.66)			300 kPa (-9.24)			350 kPa (-5.36)		
Sat.	0.48213	1426.3	5.5201	0.40607	1431.7	5.4565	0.35108	1436.3	5.4026
0	0.51293	1459.3	5.6441	0.42382	1454.7	5.5420	0.36011	1449.9	5.4532
10	0.53481	1482.9	5.7288	0.44251	1478.9	5.6290	0.37654	1474.9	5.5427
20	0.55629	1506.0	5.8093	0.46077	1502.6	5.7113	0.39251	1499.1	5.6270
30	0.57745	1529.0	5.8861	0.47870	1525.9	5.7896	0.40814	1522.9	5.7068
40	0.59835	1551.7	5.9599	0.49636	1549.0	5.8645	0.42350	1546.3	5.7828
50	0.61904	1574.3	6.0309	0.51382	1571.9	5.9365	0.43865	1569.5	5.8557
60	0.63958	1596.8	6.0997	0.53111	1594.7	6.0060	0.45362	1592.6	5.9259
70	0.65998	1619.4	6.1663	0.54827	1617.5	6.0732	0.46846	1615.5	5.9938
80	0.68028	1641.9	6.2312	0.56532	1640.2	6.1385	0.48319	1638.4	6.0596
100	0.72063	1687.3	6.3561	0.59916	1685.8	6.2642	0.51240	1684.3	6.1860
120	0.76073	1733.1	6.4756	0.63276	1731.8	6.3842	0.54135	1730.5	6.3066
140	0.80065	1779.4	6.5906	0.66618	1778.3	6.4996	0.57012	1777.2	6.4223
160	0.84044	1826.4	6.7016	0.69946	1825.4	6.6109	0.59876	1824.4	6.5340
180	0.88012	1874.1	6.8093	0.73263	1873.2	6.7188	0.62728	1872.3	6.6421
200	0.91972	1922.5	6.9138	0.76572	1921.7	6.8235	0.65571	1920.9	6.7470
220	0.95923	1971.6	7.0155	0.79872	1970.9	6.9254	0.68407	1970.2	6.8491
240	0.99868	2021.5	7.1147	0.83167	2020.9	7.0247	0.71237	2020.3	6.9486
260	1.03808	2072.2	7.2115	0.86455	2071.6	7.1217	0.74060	2071.0	7.0456
	400 kPa (-1.89)			500 kPa (4.13)			600 kPa (9.28)		
Sat.	0.30942	1440.2	5.3559	0.25035	1446.5	5.2776	0.21038	1451.4	5.2133
10	0.32701	1470.7	5.4663	0.25757	1462.3	5.3340	0.21115	1453.4	5.2205
20	0.34129	1495.6	5.5525	0.26949	1488.3	5.4244	0.22154	1480.8	5.3156
30	0.35520	1519.8	5.6338	0.28103	1513.5	5.5090	0.23152	1507.1	5.4037
40	0.36884	1543.6	5.7111	0.29227	1538.1	5.5889	0.24118	1532.5	5.4862
50	0.38226	1567.1	5.7850	0.30328	1562.3	5.6647	0.25059	1557.3	5.5641
60	0.39550	1590.4	5.8560	0.31410	1586.1	5.7373	0.25981	1581.6	5.6383
70	0.40860	1613.6	5.9244	0.32478	1609.6	5.8070	0.26888	1605.7	5.7094
80	0.42160	1636.7	5.9907	0.33535	1633.1	5.8744	0.27783	1629.5	5.7778
100	0.44732	1682.8	6.1179	0.35621	1679.8	6.0031	0.29545	1676.8	5.9081
120	0.47279	1729.2	6.2390	0.37681	1726.6	6.1253	0.31281	1724.0	6.0314
140	0.49808	1776.0	6.3552	0.39722	1773.8	6.2422	0.32997	1771.5	6.1491
160	0.52323	1823.4	6.4671	0.41748	1821.4	6.3548	0.34699	1819.4	6.2623
180	0.54827	1871.4	6.5755	0.43764	1869.6	6.4636	0.36389	1867.8	6.3717
200	0.57321	1920.1	6.6806	0.45771	1918.5	6.5691	0.38071	1916.9	6.4776
220	0.59809	1969.5	6.7828	0.47770	1968.1	6.6717	0.39745	1966.6	6.5806
240	0.62289	2019.6	6.8825	0.49763	2018.3	6.7717	0.41412	2017.1	6.6808
260	0.64764	2070.5	6.9797	0.51749	2069.3	6.8692	0.43073	2068.2	6.7786
280	0.67234	2122.1	7.0747	0.53731	2121.1	6.9644	0.44729	2120.1	6.8741

(Continued)

TABLE B.2 .2 SI (Continued) ***Superheated Ammonia***

Temp. C	v m³/kg	h kJ/kg	s kJ/kg K	v m³/kg	h kJ/kg	s kJ/kg K	v m³/kg	h kJ/kg	s kJ/kg K
	700 kPa (13.80)			800 kPa (17.85)			900 kPa (21.52)		
Sat.	0.18148	1455.3	5.1586	0.15958	1458.6	5:1110	0.14239	1461.2	5.0686
20	0.18721	1473.0	5.2196	0.16138	1464.9	5.1328	—	—	—
30	0.19610	1500.4	5.3115	0.16947	1493.5	5.2287	0.14872	1486.5	5.1530
40	0.20464	1526.7	5.3968	0.17720	1520.8	5.3171	0.15582	1514.7	5.2447
50	0.21293	1552.2	5.4770	0.18465	1547.0	5.3996	0.16263	1541.7	5.3296
60	0.22101	1577.1	5.5529	0.19189	1572.5	5.4774	0.16922	1567.9	5.4093
70	0.22894	1601.6	5.6254	0.19896	1597.5	5.5513	0.17563	1593.3	5.4847
80	0.23674	1625.8	5.6949	0.20590	1622.1	5.6219	0.18191	1618.4	5.5565
100	0.25205	1673.7	5.8268	0.21949	1670.6	5.7555	0.19416	1667.5	5.6919
120	0.26709	1721.4	5.9512	0.23280	1718.7	5.8811	0.20612	1716.1	5.8187
140	0.28193	1769.2	6.0698	0.24590	1766.9	6.0006	0.21787	1764.5	5.9389
160	0.29663	1817.3	6.1837	0.25886	1815.3	6.1150	0.22948	1813.2	6.0541
180	0.31121	1866.0	6.2935	0.27170	1864.2	6.2254	0.24097	1862.4	6.1649
200	0.32570	1915.3	6.3999	0.28445	1913.6	6.3322	0.25236	1912.0	6.2721
220	0.34012	1965.2	6.5032	0.29712	1963.7	6.4358	0.26368	1962.3	6.3762
240	0.35447	2015.8	6.6037	0.30973	2014.5	6.5367	0.27493	2013.2	6.4774
260	0.36876	2067.1	6.7018	0.32228	2065.9	6.6350	0.28612	2064.8	6.5760
280	0.38299	2119.1	6.7975	0.33477	2118.0	6.7310	0.29726	2117.0	6.6722
300	0.39718	2171.8	6.8911	0.34722	2170.9	6.8248	0.30835	2170.0	6.7662
	1000 kPa (24.90)			1200 kPa (30.94)			1400 kPa (36.26)		
Sat.	0.12852	1463.4	5.0304	0.10751	1466.8	4.9635	0.09231	1469.0	4.9060
30	0.13206	1479.1	5.0826	—	—	—	—	—	—
40	0.13868	1508.5	5.1778	0.11287	1495.4	5.0564	0.09432	1481.6	4.9463
50	0.14499	1536.3	5.2654	0.11846	1525.1	5.1497	0.09942	1513.4	5.0462
60	0.15106	1563.1	5.3471	0.12378	1553.3	5.2357	0.10423	1543.1	5.1370
70	0.15695	1589.1	5.4240	0.12890	1580.5	5.3159	0.10882	1571.5	5.2209
80	0.16270	1614.6	5.4971	0.13387	1606.8	5.3916	0.11324	1598.8	5.2994
100	0.17389	1664.3	5.6342	0.14347	1658.0	5.5325	0.12172	1651.4	5.4443
120	0.18477	1713.4	5.7622	0.15275	1708.0	5.6631	0.12986	1702.5	5.5775
140	0.19545	1762.2	5.8834	0.16181	1757.5	5.7860	0.13777	1752.8	5.7023
160	0.20597	1811.2	5.9992	0.17071	1807.1	5.9031	0.14552	1802.9	5.8208
180	0.21638	1860.5	6.1105	0.17950	1856.9	6.0156	0.15315	1853.2	5.9343
200	0.22669	1910.4	6.2182	0.18819	1907.1	6.1241	0.16068	1903.8	6.0437
220	0.23693	1960.8	6.3226	0.19680	1957.9	6.2292	0.16813	1955.0	6.1495
240	0.24710	2011.9	6.4241	0.20534	2009.3	6.3313	0.17551	2006.7	6.2523
260	0.25720	2063.6	6.5229	0.21382	2061.3	6.4308	0.18283	2059.0	6.3523
280	0.26726	2116.0	6.6194	0.22225	2114.0	6.5278	0.19010	2111.9	6.4498
300	0.27726	2169.1	6.7137	0.23063	2167.3	6.6225	0.19732	2165.5	6.5450
320	0.28723	2222.9	6.8059	0.23897	2221.3	6.7151	0.20450	2219.8	6.6380

(Continued)

TABLE B.2.2 SI (Continued) *Superheated Ammonia*

Temp. C	v m³/kg	h kJ/kg	s kJ/kg K	v m³/kg	h kJ/kg	s kJ/kg K	v m³/kg	h kJ/kg	s kJ/kg K
	1600 kPa (41.03)			1800 kPa (45.38)			2000 kPa (49.37)		
Sat.	0.08079	1470.5	4.8553	0.07174	1471.3	4.8096	0.06444	1471.5	4.7680
50	0.08506	1501.0	4.9510	0.07381	1487.9	4.8614	0.06471	1473.9	4.7754
60	0.08951	1532.5	5.0472	0.07801	1521.4	4.9637	0.06875	1509.8	4.8848
70	0.09372	1562.3	5.1351	0.08193	1552.7	5.0561	0.07246	1542.7	4.9821
80	0.09774	1590.7	5.2167	0.08565	1582.2	5.1410	0.07595	1573.5	5.0707
100	0.10539	1644.8	5.3659	0.09267	1638.0	5.2948	0.08248	1631.1	5.2294
120	0.11268	1696.9	5.5018	0.09931	1691.2	5.4337	0.08861	1685.5	5.3714
140	0.11974	1748.0	5.6286	0.10570	1743.1	5.5624	0.09447	1738.2	5.5022
160	0.12662	1798.7	5.7485	0.11192	1794.5	5.6838	0.10016	1790.2	5.6251
180	0.13339	1849.5	5.8631	0.11801	1845.7	5.7995	0.10571	1842.0	5.7420
200	0.14005	1900.5	5.9734	0.12400	1897.2	5.9107	0.11116	1893.9	5.8540
220	0.14663	1952.0	6.0800	0.12990	1949.1	6.0180	0.11652	1946.1	5.9621
240	0.15314	2004.1	6.1834	0.13574	2001.4	6.1221	0.12182	1998.8	6.0668
260	0.15959	2056.7	6.2839	0.14152	2054.3	6.2232	0.12705	2052.0	6.1685
280	0.16599	2109.9	6.3819	0.14724	2107.8	6.3217	0.13224	2105.8	6.2675
300	0.17234	2163.7	6.4775	0.15291	2161.9	6.4178	0.13737	2160.1	6.3641
320	0.17865	2218.2	6.5710	0.15854	2216.7	6.5116	0.14246	2215.1	6.4583
340	0.18492	2273.3	6.6624	0.16414	2272.0	6.6034	0.14751	2270.7	6.5505
360	0.19115	2329.1	6.7519	0.16969	2328.0	6.6932	0.15253	2326.8	6.6406
	5000 kPa (88.90)			10000 kPa (125.20)			20000 kPa		
Sat.	0.02365	1441.4	4.3454	0.00826	1289.4	3.7587	—	—	—
100	0.02636	1501.5	4.5091	—	—	—	—	—	—
120	0.03024	1586.3	4.7306	—	—	—	—	—	—
140	0.03350	1657.3	4.9068	0.01195	1461.3	4.1839	0.00251	918.9	2.7630
160	0.03643	1721.7	5.0591	0.01461	1578.3	4.4610	0.00323	1097.2	3.1838
180	0.03916	1782.7	5.1968	0.01666	1667.2	4.6617	0.00490	1329.7	3.7087
200	0.04174	1841.8	5.3245	0.01842	1744.5	4.8287	0.00653	1497.7	4.0721
220	0.04422	1900.0	5.4450	0.02001	1816.0	4.9767	0.00782	1618.7	4.3228
240	0.04662	1957.9	5.5600	0.02150	1884.2	5.1123	0.00891	1718.6	4.5214
260	0.04895	2015.6	5.6704	0.02290	1950.6	5.2392	0.00988	1807.6	4.6916
280	0.05123	2073.6	5.7771	0.02424	2015.9	5.3596	0.01077	1890.5	4.8442
300	0.05346	2131.8	5.8805	0.02552	2080.7	5.4746	0.01159	1969.6	4.9847
320	0.05565	2190.3	5.9809	0.02676	2145.2	5.5852	0.01237	2046.3	5.1164
340	0.05779	2249.2	6.0786	0.02796	2209.6	5.6921	0.01312	2121.6	5.2412
360	0.05990	2308.6	6.1738	0.02913	2274.1	5.7955	0.01382	2195.8	5.3603
380	0.06198	2368.4	6.2668	0.03026	2338.7	5.8960	0.01450	2269.4	5.4748
400	0.06403	2428.6	6.3576	0.03137	2403.5	5.9937	0.01516	2342.6	5.5851
420	0.06606	2489.3	6.4464	0.03245	2468.5	6.0888	0.01579	2415.4	5.6917
440	0.06806	2550.4	6.5334	0.03351	2533.7	6.1815	0.01641	2488.1	5.7950

TABLE B.3 SI *Thermodynamic Properties of Argon*
TABLE B.3.1 SI *Saturated Argon*

Temp.	Press.	Specific Volume, m³/kg			Internal Energy, kJ/kg		
K	kPa	Sat. Liquid	Evap.	Sat. Vapor	Sat. Liquid	Evap.	Sat. Vapor
T	P	v_f	v_{fg}	v_g	u_f	u_{fg}	u_g
83.8	69.0	0.000706	0.24583	0.24653	-121.10	146.69	25.59
85	79.0	0.000709	0.21693	0.21764	-119.83	145.62	25.79
87.3	101.3	0.000716	0.17240	0.17312	-117.37	143.52	26.15
90	133.6	0.000725	0.13350	0.13423	-114.45	141.01	26.56
95	213.2	0.000742	0.08648	0.08722	-108.98	136.22	27.24
100	324.0	0.000760	0.05838	0.05914	-103.40	131.21	27.81
105	472.6	0.000781	0.04073	0.04151	-97.70	125.91	28.21
110	665.7	0.000804	0.02915	0.02996	-91.86	120.26	28.40
115	910.5	0.000830	0.02127	0.02210	-85.84	114.15	28.31
120	1213.9	0.000860	0.01573	0.01659	-79.60	107.47	27.87
125	1583.5	0.000894	0.01171	0.01260	-73.10	100.07	26.97
130	2027.0	0.000936	0.00870	0.00964	-66.23	91.70	25.47
135	2552.9	0.000989	0.00638	0.00737	-58.82	81.97	23.16
140	3171.0	0.001061	0.00451	0.00558	-50.51	70.10	19.58
145	3892.9	0.001178	0.00290	0.00408	-40.45	54.16	13.71
150	4736.0	0.001475	0.00109	0.00256	-24.67	24.80	0.13
150.7	4860.0	0.001884	0	0.00188	-12.72	0	-12.72

TABLE B.3.2 SI *Superheated Argon*

Temp. K	v m³/kg	h kJ/kg	s kJ/kg K	v m³/kg	h kJ/kg	s kJ/kg K	v m³/kg	h kJ/kg	s kJ/kg K
	100 kPa (87.17)			200 kPa (94.28)			300 kPa (99.04)		
Sat.	0.17522	43.66	3.2242	0.09254	45.66	3.1084	0.06354	46.77	3.0411
100	0.20329	50.51	3.2973	0.09912	48.95	3.1423	0.06431	47.36	3.0470
125	0.25709	64.14	3.4191	0.12697	63.23	3.2700	0.08358	62.30	3.1807
150	0.31003	77.41	3.5158	0.15392	76.76	3.3687	0.10188	76.11	3.2815
175	0.36263	90.56	3.5970	0.18052	90.07	3.4508	0.11981	89.58	3.3645
200	0.41506	103.67	3.6670	0.20693	103.28	3.5214	0.13756	102.89	3.4356
225	0.46739	116.75	3.7286	0.23324	116.43	3.5833	0.15519	116.12	3.4979
250	0.51964	129.81	3.7837	0.25948	129.55	3.6386	0.17276	129.29	3.5535
275	0.57184	142.86	3.8334	0.28566	142.64	3.6885	0.19027	142.42	3.6035
300	0.62401	155.90	3.8788	0.31182	155.72	3.7340	0.20775	155.53	3.6492
325	0.67615	168.94	3.9205	0.33795	168.78	3.7759	0.22521	168.62	3.6911
350	0.72828	181.97	3.9592	0.36405	181.83	3.8146	0.24265	181.70	3.7298
375	0.78039	195.00	3.9951	0.39015	194.88	3.8506	0.26007	194.76	3.7659

(Continued)

TABLE B.3.1 SI (Continued) ***Saturated Argon***

Temp.	Press.	Enthalpy, kJ/kg			Entropy, kJ/kg K		
K T	kPa P	Sat. Liquid h_f	Evap. h_{fg}	Sat. Vapor h_g	Sat. Liquid s_f	Evap. s_{fg}	Sat. Vapor s_g
83.8	69.0	-121.05	163.64	42.59	1.3340	1.9527	3.2867
85	79.0	-119.77	162.75	42.98	1.3491	1.9147	3.2639
87.3	101.3	-117.30	160.99	43.69	1.3777	1.8443	3.2220
90	133.6	-114.35	158.85	44.50	1.4107	1.7650	3.1756
95	213.2	-108.82	154.66	45.84	1.4698	1.6280	3.0978
100	324.0	-103.16	150.13	46.97	1.5271	1.5013	3.0283
105	472.6	-97.33	145.16	47.83	1.5828	1.3825	2.9653
110	665.7	-91.32	139.67	48.35	1.6373	1.2697	2.9070
115	910.5	-85.08	133.52	48.44	1.6910	1.1610	2.8520
120	1213.9	-78.56	126.57	48.01	1.7443	1.0547	2.7990
125	1583.5	-71.69	118.61	46.92	1.7978	0.9489	2.7466
130	2027.0	-64.33	109.34	45.01	1.8522	0.8411	2.6933
135	2552.9	-56.29	98.26	41.97	1.9091	0.7279	2.6370
140	3171.0	-47.15	84.41	37.26	1.9710	0.6029	2.5739
145	3892.9	-35.87	65.44	29.57	2.0444	0.4513	2.4958
150	4736.0	-17.68	29.95	12.27	2.1601	0.1997	2.3597
150.7	4860.0	-3.56	0	-3.56	2.2526	0	2.2526

TABLE B.3.2 SI (Continued) ***Superheated Argon***

Temp. K	v m³/kg	h kJ/kg	s kJ/kg K	v m³/kg	h kJ/kg	s kJ/kg K	v m³/kg	h kJ/kg	s kJ/kg K
	100 kPa (87.17)			200 kPa (94.28)			300 kPa (99.04)		
400	0.83249	208.02	4.0287	0.41623	207.92	3.8842	0.27748	207.82	3.7996
425	0.88458	221.04	4.0603	0.44230	220.96	3.9158	0.29488	220.87	3.8312
450	0.93666	234.06	4.0901	0.46837	233.99	3.9456	0.31227	233.91	3.8611
475	0.98873	247.08	4.1182	0.49443	247.02	3.9738	0.32966	246.95	3.8893
500	1.04080	260.10	4.1449	0.52048	260.04	4.0005	0.34704	259.99	3.9160
525	1.09287	273.11	4.1703	0.54653	273.07	4.0260	0.36442	273.02	3.9414
550	1.14493	286.13	4.1946	0.57258	286.09	4.0502	0.38179	286.05	3.9657
575	1.19699	299.14	4.2177	0.59862	299.11	4.0733	0.39916	299.08	3.9889
600	1.24905	312.16	4.2399	0.62466	312.13	4.0955	0.41653	312.10	4.0110
625	1.30110	325.17	4.2611	0.65070	325.15	4.1168	0.43390	325.13	4.0323
650	1.35315	338.19	4.2815	0.67673	338.17	4.1372	0.45126	338.15	4.0527
675	1.40520	351.20	4.3012	0.70277	351.19	4.1568	0.46862	351.17	4.0724
700	1.45725	364.21	4.3201	0.72880	364.20	4.1758	0.48598	364.19	4.0913

(Continued)

TABLE B.3.2 SI (Continued) *Superheated Argon*

Temp. K	v m³/kg	h kJ/kg	s kJ/kg K	v m³/kg	h kJ/kg	s kJ/kg K	v m³/kg	h kJ/kg	s kJ/kg K
	400 kPa (102.73)			500 kPa (105.79)			600 kPa (108.43)		
Sat.	0.04856	47.48	2.9932	0.03936	47.94	2.9558	0.03309	48.23	2.9248
125	0.06186	61.34	3.1157	0.04882	60.36	3.0639	0.04011	59.35	3.0205
150	0.07585	75.46	3.2187	0.06023	74.79	3.1693	0.04981	74.11	3.1283
175	0.08945	89.08	3.3027	0.07124	88.58	3.2543	0.05909	88.08	3.2144
200	0.10287	102.50	3.3744	0.08205	102.11	3.3266	0.06818	101.71	3.2873
225	0.11617	115.80	3.4371	0.09275	115.48	3.3896	0.07714	115.16	3.3507
250	0.12940	129.03	3.4928	0.10338	128.76	3.4456	0.08604	128.50	3.4069
275	0.14258	142.20	3.5431	0.11396	141.99	3.4960	0.09488	141.77	3.4574
300	0.15572	155.35	3.5888	0.12450	155.16	3.5419	0.10369	154.98	3.5034
325	0.16884	168.46	3.6308	0.13502	168.31	3.5839	0.11248	168.15	3.5456
350	0.18194	181.56	3.6696	0.14552	181.43	3.6228	0.12124	181.29	3.5846
375	0.19503	194.65	3.7057	0.15600	194.53	3.6590	0.12999	194.41	3.6208
400	0.20810	207.72	3.7395	0.16648	207.62	3.6928	0.13873	207.52	3.6546
450	0.23422	233.84	3.8010	0.18740	233.76	3.7544	0.15618	233.69	3.7162
500	0.26032	259.93	3.8560	0.20829	259.88	3.8094	0.17360	259.82	3.7713
550	0.28640	286.01	3.9057	0.22917	285.97	3.8591	0.19101	285.93	3.8211
600	0.31247	312.08	3.9511	0.25003	312.05	3.9045	0.20840	312.02	3.8665
650	0.33852	338.13	3.9928	0.27088	338.12	3.9463	0.22579	338.10	3.9082
700	0.36457	364.19	4.0314	0.29173	364.18	3.9849	0.24317	364.17	3.9469
	700 kPa (110.77)			800 kPa (112.88)			1000 kPa (116.59)		
Sat.	0.02855	48.39	2.8983	0.02508	48.46	2.8750	0.02015	48.36	2.8350
125	0.03387	58.30	2.9826	0.02918	57.22	2.9488	0.02258	54.94	2.8895
150	0.04236	73.43	3.0931	0.03678	72.74	3.0622	0.02895	71.32	3.0093
175	0.05042	87.57	3.1804	0.04391	87.06	3.1506	0.03480	86.03	3.1001
200	0.05826	101.32	3.2538	0.05083	100.92	3.2246	0.04042	100.12	3.1754
225	0.06599	114.84	3.3176	0.05763	114.52	3.2887	0.04593	113.88	3.2403
250	0.07365	128.24	3.3740	0.06436	127.98	3.3454	0.05135	127.45	3.2974
275	0.08126	141.55	3.4247	0.07104	141.33	3.3963	0.05673	140.89	3.3487
300	0.08883	154.79	3.4708	0.07768	154.61	3.4426	0.06207	154.24	3.3951
325	0.09637	167.99	3.5131	0.08429	167.83	3.4849	0.06738	167.52	3.4377
350	0.10390	181.16	3.5521	0.09089	181.02	3.5240	0.07268	180.75	3.4769
375	0.11141	194.30	3.5884	0.09747	194.18	3.5603	0.07796	193.95	3.5133
400	0.11891	207.42	3.6223	0.10404	207.32	3.5942	0.08323	207.12	3.5473
450	0.13388	233.62	3.6840	0.11715	233.54	3.6560	0.09374	233.39	3.6092
500	0.14882	259.77	3.7391	0.13024	259.72	3.7112	0.10423	259.61	3.6644
550	0.16375	285.89	3.7889	0.14331	285.86	3.7610	0.11470	285.78	3.7143
600	0.17867	312.00	3.8343	0.15637	311.97	3.8064	0.12515	311.92	3.7598
650	0.19358	338.09	3.8761	0.16942	338.07	3.8482	0.13560	338.04	3.8016
700	0.20848	364.16	3.9147	0.18246	364.15	3.8869	0.14604	364.14	3.8403

(Continued)

TABLE B.3.2 SI (Continued) *Superheated Argon*

Temp. K	v m³/kg	h kJ/kg	s kJ/kg K	v m³/kg	h kJ/kg	s kJ/kg K	v m³/kg	h kJ/kg	s kJ/kg K
	1250 kPa (120.53)			1500 kPa (123.95)			2000 kPa (129.72)		
Sat.	0.01610	47.92	2.7934	0.01334	47.21	2.7576	0.00978	45.14	2.6964
150	0.02267	69.49	2.9545	0.01847	67.58	2.9077	0.01318	63.49	2.8285
175	0.02750	84.72	3.0485	0.02263	83.38	3.0053	0.01654	80.63	2.9344
200	0.03209	99.11	3.1254	0.02654	98.09	3.0839	0.01960	96.02	3.0167
225	0.03656	113.08	3.1912	0.03032	112.27	3.1507	0.02251	110.64	3.0856
250	0.04095	126.79	3.2490	0.03401	126.13	3.2091	0.02534	124.81	3.1453
275	0.04528	140.34	3.3007	0.03765	139.79	3.2612	0.02812	138.69	3.1983
300	0.04958	153.77	3.3475	0.04126	153.31	3.3083	0.03086	152.39	3.2459
325	0.05386	167.13	3.3902	0.04484	166.74	3.3512	0.03357	165.95	3.2894
350	0.05811	180.42	3.4296	0.04840	180.08	3.3908	0.03627	179.42	3.3293
375	0.06235	193.66	3.4662	0.05195	193.38	3.4275	0.03894	192.81	3.3662
400	0.06658	206.87	3.5003	0.05548	206.63	3.4617	0.04161	206.14	3.4006
450	0.07501	233.21	3.5623	0.06252	233.03	3.5239	0.04692	232.67	3.4631
500	0.08342	259.47	3.6176	0.06954	259.34	3.5793	0.05220	259.07	3.5188
550	0.09180	285.68	3.6676	0.07654	285.59	3.6294	0.05747	285.40	3.5690
600	0.10018	311.85	3.7132	0.08353	311.79	3.6750	0.06272	311.66	3.6147
650	0.10854	338.00	3.7550	0.09051	337.96	3.7169	0.06796	337.88	3.6566
700	0.11690	364.12	3.7937	0.09748	364.10	3.7556	0.07320	364.07	3.6954
	4000 kPa (145.68)			6000 kPa			10000 kPa		
Sat.	0.00388	28.13	2.4829	—	—	—	—	—	—
150	0.00477	39.67	2.5611	0.00118	-30.76	2.0622	0.00104	-37.59	1.9875
175	0.00733	68.27	2.7392	0.00417	52.88	2.5878	0.00177	14.88	2.3089
200	0.00917	87.32	2.8411	0.00569	77.87	2.7219	0.00296	57.48	2.5383
225	0.01081	103.96	2.9196	0.00692	97.06	2.8124	0.00386	83.00	2.6589
250	0.01235	119.46	2.9850	0.00804	114.08	2.8842	0.00462	103.41	2.7450
275	0.01383	134.31	3.0416	0.00908	129.95	2.9448	0.00532	121.47	2.8139
300	0.01527	148.74	3.0918	0.01009	145.14	2.9976	0.00597	138.20	2.8722
325	0.01668	162.87	3.1371	0.01106	159.85	3.0448	0.00659	154.09	2.9230
350	0.01807	176.80	3.1784	0.01202	174.25	3.0874	0.00720	169.40	2.9684
375	0.01945	190.57	3.2164	0.01296	188.40	3.1265	0.00779	184.30	3.0096
400	0.02081	204.22	3.2516	0.01389	202.37	3.1625	0.00837	198.89	3.0472
450	0.02352	231.26	3.3153	0.01572	229.91	3.2274	0.00951	227.40	3.1144
500	0.02620	258.05	3.3718	0.01753	257.08	3.2847	0.01062	255.30	3.1732
550	0.02886	284.67	3.4225	0.01933	284.00	3.3360	0.01171	282.79	3.2256
600	0.03151	311.18	3.4686	0.02111	310.74	3.3826	0.01280	309.98	3.2729
650	0.03415	337.60	3.5109	0.02288	337.35	3.4252	0.01388	336.96	3.3161
700	0.03678	363.95	3.5500	0.02465	363.86	3.4644	0.01495	363.78	3.3559

TABLE B.4 SI *Thermodynamic Properties of Iso-Butane*
TABLE B.4.1 SI *Saturated Iso-Butane*

Temp.	Press.	Specific Volume, m³/kg			Internal Energy, kJ/kg		
K	kPa	Sat. Liquid	Evap.	Sat. Vapor	Sat. Liquid	Evap.	Sat. Vapor
T	P	v_f	v_{fg}	v_g	u_f	u_{fg}	u_g
113.6	1.96E-5	0.001349	8.29E5	8.29E5	-712.66	468.30	-244.36
120	9.54E-5	0.001360	1.80E5	1.80E5	-701.62	462.08	-239.54
140	4.83E-3	0.001397	4.15E3	4.15E3	-666.65	443.26	-223.38
160	8.32E-2	0.001436	2.75E2	2.75E2	-630.53	425.02	-205.52
180	0.707	0.001477	36.36171	36.36319	-593.13	407.03	-186.09
200	3.71	0.001521	7.68778	7.68930	-554.32	389.05	-165.27
220	13.8	0.001568	2.26099	2.26256	-514.00	370.82	-143.19
240	39.8	0.001621	0.84250	0.84412	-472.06	352.07	-119.98
260	95.3	0.001680	0.37274	0.37442	-428.33	332.54	-95.79
261.5	101.3	0.001684	0.35205	0.35373	-424.88	330.99	-93.89
270	139.6	0.001712	0.26044	0.26215	-405.74	322.38	-83.36
280	198.3	0.001746	0.18684	0.18859	-382.64	311.90	-70.73
290	274.0	0.001783	0.13707	0.13885	-358.99	301.05	-57.94
300	369.6	0.001824	0.10244	0.10427	-334.77	289.77	-45.00
310	488.2	0.001868	0.07774	0.07961	-309.94	277.97	-31.97
320	632.8	0.001916	0.05970	0.06162	-284.45	265.55	-18.90
340	1013.6	0.002032	0.03601	0.03805	-231.26	238.29	7.03
360	1540.9	0.002188	0.02178	0.02397	-174.50	206.17	31.67
380	2249.7	0.002421	0.01248	0.01490	-112.62	165.01	52.39
400	3189.7	0.002925	0.00537	0.00829	-39.20	98.82	59.62
407.9	3640.0	0.004457	0	0.00446	24.42	0	24.42

TABLE B.4.2 SI *Superheated Iso-Butane*

Temp. K	v m³/kg	h kJ/kg	s kJ/kg K	v m³/kg	h kJ/kg	s kJ/kg K	v m³/kg	h kJ/kg	s kJ/kg K
	50 kPa (244.88)			100 kPa (261.21)			200 kPa (280.26)		
Sat.	0.68255	-80.03	-0.1237	0.35810	-58.49	-0.1344	0.18703	-33.00	-0.1341
280	0.78838	-26.30	0.0811	0.38762	-28.57	-0.0239	—	—	—
300	0.84758	6.59	0.1945	0.41830	4.74	0.0910	0.20340	0.84	-0.0175
320	0.90632	41.25	0.3064	0.44848	39.71	0.2038	0.21939	36.49	0.0975
340	0.96473	77.74	0.4169	0.47832	76.42	0.3151	0.23500	73.71	0.2103
360	1.02291	116.06	0.5264	0.50790	114.93	0.4251	0.25033	112.60	0.3215
380	1.08091	156.23	0.6350	0.53731	155.24	0.5341	0.26546	153.21	0.4312
400	1.13877	198.23	0.7427	0.56658	197.35	0.6420	0.28044	195.56	0.5398
420	1.19654	242.03	0.8495	0.59575	241.24	0.7491	0.29532	239.64	0.6473

(Continued)

TABLE B.4.1 SI (Continued) *Saturated Iso-Butane*

Temp. K T	Press. kPa P	Enthalpy, kJ/kg Sat. Liquid h_f	Evap. h_{fg}	Sat. Vapor h_g	Entropy, kJ/kg K Sat. Liquid s_f	Evap. s_{fg}	Sat. Vapor s_g
113.6	1.96E-5	-712.66	484.55	-228.12	-3.1289	4.2672	1.1384
120	9.54E-5	-701.62	479.24	-222.37	-3.0342	3.9937	0.9595
140	4.83E-3	-666.65	463.29	-203.36	-2.7648	3.3092	0.5444
160	8.32E-2	-630.53	447.90	-182.63	-2.5238	2.7994	0.2756
180	0.707	-593.13	432.75	-160.38	-2.3036	2.4042	0.1006
200	3.71	-554.32	417.54	-136.77	-2.0992	2.0877	-0.0115
220	13.8	-513.98	401.94	-112.04	-1.9071	1.8270	-0.0801
240	39.8	-471.99	385.59	-86.40	-1.7247	1.6066	-0.1181
260	95.3	-428.17	368.07	-60.09	-1.5497	1.4157	-0.1340
261.5	101.3	-424.71	366.66	-58.04	-1.5365	1.4019	-0.1346
270	139.6	-405.50	358.74	-46.76	-1.4645	1.3287	-0.1358
280	198.3	-382.29	348.94	-33.35	-1.3804	1.2462	-0.1342
290	274.0	-358.50	338.60	-19.90	-1.2974	1.1676	-0.1298
300	369.6	-334.10	327.63	-6.47	-1.2153	1.0921	-0.1232
310	488.2	-309.03	315.92	6.89	-1.1338	1.0191	-0.1147
320	632.8	-283.24	303.34	20.10	-1.0528	0.9479	-0.1049
340	1013.6	-229.20	274.80	45.60	-0.8913	0.8082	-0.0831
360	1540.9	-171.13	239.73	68.61	-0.7286	0.6659	-0.0627
380	2249.7	-107.17	193.09	85.92	-0.5602	0.5081	-0.0520
400	3189.7	-29.87	115.94	86.07	-0.3685	0.2899	-0.0787
407.9	3640.0	40.64	0	40.64	-0.1982	0	-0.1982

TABLE B.4.2 SI (Continued) *Superheated Iso-Butane*

Temp. K	v m³/kg	h kJ/kg	s kJ/kg K	v m³/kg	h kJ/kg	s kJ/kg K	v m³/kg	h kJ/kg	s kJ/kg K
	50 kPa (244.88)			100 kPa (261.21)			200 kPa (280.26)		
440	1.25423	287.59	0.9555	0.62483	286.88	0.8552	0.31011	285.44	0.7538
460	1.31186	334.87	1.0606	0.65385	334.22	0.9604	0.32484	332.91	0.8593
480	1.36944	383.83	1.1647	0.68283	383.23	1.0647	0.33951	382.03	0.9639
500	1.42697	434.40	1.2679	0.71176	433.85	1.1680	0.35414	432.74	1.0674
520	1.48447	486.55	1.3702	0.74065	486.04	1.2704	0.36874	485.01	1.1698
540	1.54195	540.23	1.4715	0.76952	539.75	1.3717	0.38330	538.79	1.2713
560	1.59940	595.38	1.5718	0.79836	594.93	1.4720	0.39784	594.04	1.3718
580	1.65683	651.96	1.6710	0.82718	651.55	1.5714	0.41236	650.70	1.4712
600	1.71424	709.94	1.7693	0.85599	709.54	1.6697	0.42686	708.75	1.5696

(Continued)

TABLE B.4.2 SI (Continued) ***Superheated Iso-Butane***

Temp. K	v m³/kg	h kJ/kg	s kJ/kg K	v m³/kg	h kJ/kg	s kJ/kg K	v m³/kg	h kJ/kg	s kJ/kg K
	300 kPa (292.95)			400 kPa (302.77)			500 kPa (310.89)		
Sat.	0.12734	-15.93	-0.1281	0.09661	-2.76	-0.1210	0.07777	8.08	-0.1139
300	0.13150	-3.35	-0.0857	—	—	—	—	—	—
320	0.14287	33.10	0.0319	0.10446	29.50	-0.0174	0.08128	25.66	-0.0581
340	0.15378	70.89	0.1465	0.11308	67.93	0.0991	0.08858	64.83	0.0606
360	0.16439	110.20	0.2588	0.12137	107.71	0.2128	0.09550	105.12	0.1757
380	0.17479	151.13	0.3694	0.12941	148.99	0.3243	0.10216	146.78	0.2883
400	0.18503	193.73	0.4787	0.13729	191.86	0.4343	0.10863	189.94	0.3990
420	0.19515	238.02	0.5867	0.14505	236.36	0.5428	0.11497	234.66	0.5081
440	0.20519	283.97	0.6936	0.15271	282.49	0.6501	0.12122	280.97	0.6158
460	0.21515	331.58	0.7994	0.16030	330.24	0.7562	0.12739	328.87	0.7222
480	0.22507	380.81	0.9041	0.16784	379.58	0.8612	0.13349	378.34	0.8275
500	0.23493	431.62	1.0078	0.17532	430.50	0.9651	0.13955	429.36	0.9316
520	0.24476	483.98	1.1105	0.18277	482.93	1.0679	0.14558	481.88	1.0346
540	0.25456	537.83	1.2121	0.19019	536.86	1.1697	0.15157	535.88	1.1365
560	0.26434	593.14	1.3126	0.19758	592.23	1.2704	0.15753	591.32	1.2373
580	0.27409	649.86	1.4122	0.20495	649.01	1.3700	0.16347	648.16	1.3370
600	0.28382	707.95	1.5106	0.21231	707.15	1.4685	0.16940	706.35	1.4356
	600 kPa (317.89)			700 kPa (324.07)			800 kPa (329.64)		
Sat.	0.06498	17.33	-0.1070	0.05570	25.41	-0.1006	0.04864	32.59	-0.0945
320	0.06570	21.53	-0.0939	—	—	—	—	—	—
340	0.07218	61.56	0.0275	0.06038	58.11	-0.0021	0.05146	54.42	-0.0293
360	0.07821	102.43	0.1443	0.06581	99.63	0.1166	0.05648	96.70	0.0915
380	0.08395	144.51	0.2580	0.07092	142.16	0.2315	0.06113	139.73	0.2078
400	0.08950	187.98	0.3695	0.07582	185.96	0.3438	0.06554	183.89	0.3211
420	0.09491	232.94	0.4791	0.08056	231.18	0.4541	0.06980	229.38	0.4320
440	0.10021	279.44	0.5873	0.08520	277.88	0.5627	0.07393	276.29	0.5411
460	0.10543	327.49	0.6941	0.08975	326.09	0.6699	0.07798	324.67	0.6486
480	0.11059	377.09	0.7996	0.09423	375.82	0.7757	0.08196	374.53	0.7547
500	0.11571	428.21	0.9039	0.09867	427.05	0.8802	0.08589	425.87	0.8595
520	0.12078	480.82	1.0071	0.10306	479.75	0.9836	0.08978	478.68	0.9630
540	0.12582	534.90	1.1091	0.10743	533.91	1.0858	0.09363	532.92	1.0654
560	0.13083	590.40	1.2100	0.11176	589.48	1.1868	0.09745	588.56	1.1666
580	0.13582	647.30	1.3099	0.11607	646.44	1.2868	0.10126	645.57	1.2666
600	0.14079	705.54	1.4086	0.12036	704.74	1.3856	0.10504	703.93	1.3655

(Continued)

TABLE B.4.2 SI (Continued) ***Superheated Iso-Butane***

Temp. K	v m³/kg	h kJ/kg	s kJ/kg K	v m³/kg	h kJ/kg	s kJ/kg K	v m³/kg	h kJ/kg	s kJ/kg K
	1000 kPa (339.39)			1250 kPa (349.73)			1500 kPa (358.65)		
Sat.	0.03859	44.85	-0.0838	0.03036	57.24	-0.0726	0.02473	67.18	-0.0639
360	0.04328	90.39	0.0465	0.03250	81.41	-0.0045	0.02501	70.62	-0.0543
380	0.04734	134.61	0.1660	0.03619	127.60	0.1204	0.02862	119.75	0.0786
400	0.05111	179.58	0.2814	0.03950	173.83	0.2390	0.03168	167.60	0.2013
420	0.05469	225.67	0.3938	0.04257	220.79	0.3535	0.03444	215.61	0.3184
440	0.05814	273.03	0.5040	0.04548	268.79	0.4651	0.03702	264.36	0.4318
460	0.06149	321.77	0.6123	0.04829	318.03	0.5746	0.03947	314.15	0.5424
480	0.06477	371.92	0.7190	0.05101	368.57	0.6821	0.04183	365.12	0.6509
500	0.06800	423.50	0.8242	0.05368	420.47	0.7880	0.04412	417.37	0.7575
520	0.07118	476.50	0.9282	0.05629	473.73	0.8925	0.04637	470.91	0.8625
540	0.07432	530.91	1.0308	0.05887	528.36	0.9955	0.04857	525.78	0.9660
560	0.07743	586.69	1.1323	0.06141	584.34	1.0973	0.05074	581.96	1.0682
580	0.08052	643.84	1.2325	0.06393	641.65	1.1979	0.05288	639.43	1.1690
600	0.08359	702.30	1.3316	0.06643	700.25	1.2972	0.05500	698.19	1.2686
	1750 kPa (366.52)			2000 kPa (373.59)			2500 kPa (385.90)		
Sat.	0.02061	75.13	-0.0575	0.01744	81.34	-0.0535	0.01280	88.87	-0.0534
380	0.02305	110.72	0.0379	0.01865	99.92	-0.0042	—	—	—
400	0.02601	160.79	0.1663	0.02168	153.23	0.1326	0.01532	134.78	0.0635
420	0.02859	210.09	0.2866	0.02416	204.18	0.2569	0.01785	190.87	0.2004
440	0.03094	259.71	0.4020	0.02637	254.83	0.3747	0.01991	244.24	0.3246
460	0.03315	310.13	0.5140	0.02841	305.95	0.4883	0.02173	297.11	0.4421
480	0.03526	361.58	0.6235	0.03033	357.93	0.5989	0.02342	350.31	0.5553
500	0.03730	414.19	0.7309	0.03218	410.95	0.7071	0.02500	404.25	0.6654
520	0.03928	468.04	0.8365	0.03396	465.13	0.8133	0.02652	459.14	0.7730
540	0.04121	523.16	0.9405	0.03570	520.50	0.9178	0.02798	515.09	0.8786
560	0.04312	579.55	1.0430	0.03740	577.11	1.0208	0.02941	572.17	0.9824
580	0.04499	637.20	1.1442	0.03907	634.96	1.1222	0.03080	630.41	1.0846
600	0.04684	696.11	1.2440	0.04072	694.03	1.2224	0.03216	689.82	1.1853
	3000 kPa (396.39)			10000 kPa			20000 kPa		
Sat.	0.00942	88.95	-0.0672	—	—	—	—	—	—
420	0.01345	174.74	0.1435	0.00247	2.15	-0.3342	0.00218	-5.68	-0.4076
440	0.01552	232.33	0.2775	0.00273	68.26	-0.1804	0.00230	53.26	-0.2705
460	0.01725	287.52	0.4002	0.00309	138.24	-0.0249	0.00243	113.98	-0.1356
480	0.01879	342.24	0.5166	0.00358	211.11	0.1301	0.00257	176.33	-0.0029
500	0.02022	397.26	0.6289	0.00417	284.84	0.2806	0.00273	240.12	0.1273
520	0.02156	452.96	0.7382	0.00481	357.09	0.4223	0.00291	305.15	0.2548
540	0.02285	509.55	0.8449	0.00543	426.90	0.5541	0.00310	371.22	0.3795
560	0.02409	567.15	0.9497	0.00601	494.74	0.6774	0.00330	438.17	0.5012
580	0.02529	625.82	1.0526	0.00655	561.42	0.7944	0.00351	505.88	0.6200
600	0.02647	685.59	1.1539	0.00705	627.60	0.9066	0.00373	574.22	0.7358

TABLE B.5 SI ***Thermodynamic Properties of Ethane***
TABLE B.5.1 SI ***Saturated Ethane***

Temp.	Press.	Specific Volume, m³/kg			Internal Energy, kJ/kg		
K	kPa	Sat. Liquid	Evap.	Sat. Vapor	Sat. Liquid	Evap.	Sat. Vapor
T	P	v_f	v_{fg}	v_g	u_f	u_{fg}	u_g
90.4	1.13E-3	0.001535	2.21E4	2.21E4	-495.07	571.27	76.20
100	1.11E-2	0.001559	2.49E3	2.49E3	-473.07	557.99	84.91
110	7.47E-2	0.001586	4.073E2	4.073E2	-449.59	543.75	94.16
120	0.355	0.001615	93.57746	93.57908	-426.07	529.73	103.66
130	1.29	0.001644	27.82363	27.82528	-402.69	516.08	113.39
140	3.83	0.001675	10.08838	10.09005	-379.39	502.66	123.27
150	9.67	0.001708	4.27002	4.27173	-356.09	489.25	133.16
160	21.5	0.001743	2.04200	2.04374	-332.68	475.64	142.96
170	42.9	0.001780	1.07625	1.07803	-309.09	461.65	152.57
180	78.7	0.001819	0.61328	0.61510	-285.21	447.13	161.92
184.6	101.3	0.001838	0.48515	0.48698	-274.22	440.31	166.09
190	134.7	0.001862	0.37217	0.37404	-260.97	431.95	170.98
200	217.4	0.001909	0.23764	0.23954	-236.30	416.01	179.71
210	334.0	0.001959	0.15808	0.16004	-211.12	399.17	188.05
220	492.3	0.002016	0.10864	0.11066	-185.34	381.28	195.93
230	700.6	0.002079	0.07659	0.07867	-158.87	362.13	203.26
240	967.3	0.002150	0.05502	0.05717	-131.58	341.47	209.89
250	1301.5	0.002233	0.04000	0.04223	-103.31	318.92	215.62
260	1712.5	0.002331	0.02921	0.03154	-73.80	293.91	220.11
270	2210.7	0.002452	0.02121	0.02367	-42.67	265.52	222.85
280	2807.6	0.002610	0.01506	0.01767	-9.19	232.05	222.87
290	3516.7	0.002843	0.01007	0.01291	28.35	189.66	218.01
300	4356.5	0.003298	0.00544	0.00873	75.40	125.50	200.90
305.3	4871.8	0.004841	0	0.00484	140.64	0	140.64

TABLE B.5.2 SI ***Superheated Ethane***

Temp. K	v m³/kg	h kJ/kg	s kJ/kg K	v m³/kg	h kJ/kg	s kJ/kg K	v m³/kg	h kJ/kg	s kJ/kg K
	50 kPa (172.41)			100 kPa (184.31)			200 kPa (198.18)		
Sat.	0.93550	201.62	6.9854	0.49296	215.16	6.8746	0.25893	229.93	6.7685
200	1.09188	239.26	7.1877	0.53870	237.11	6.9888	0.26174	232.61	6.7819
220	1.20500	268.28	7.3259	0.59658	266.48	7.1287	0.29219	262.75	6.9255
240	1.31752	298.46	7.4572	0.65384	296.96	7.2613	0.32190	293.89	7.0609
260	1.42960	329.92	7.5831	0.71065	328.65	7.3881	0.35111	326.07	7.1897
280	1.54137	362.79	7.7048	0.76713	361.70	7.5105	0.37998	359.49	7.3135
300	1.65292	397.17	7.8234	0.82339	396.22	7.6296	0.40860	394.30	7.4336
320	1.76431	433.15	7.9395	0.87948	432.31	7.7460	0.43705	430.62	7.5507
340	1.87557	470.79	8.0535	0.93544	470.04	7.8604	0.46537	468.53	7.6656
360	1.98673	510.12	8.1659	0.99131	509.44	7.9729	0.49359	508.09	7.7787

(Continued)

TABLE B.5.1 SI (Continued) *Saturated Ethane*

Temp. K T	Press. kPa P	Enthalpy, kJ/kg Sat. Liquid h_f	Evap. h_{fg}	Sat. Vapor h_g	Entropy, kJ/kg K Sat. Liquid s_f	Evap. s_{fg}	Sat. Vapor s_g
90.4	1.13E-3	-495.07	596.25	101.18	2.5492	6.5992	9.1484
100	1.11E-2	-473.07	585.64	112.57	2.7804	5.8564	8.6368
110	7.47E-2	-449.59	574.17	124.58	3.0043	5.2197	8.2240
120	0.355	-426.07	562.92	136.85	3.2088	4.6910	7.8998
130	1.29	-402.68	552.02	149.33	3.3960	4.2463	7.6423
140	3.83	-379.38	541.31	161.93	3.5687	3.8665	7.4352
150	9.67	-356.07	530.54	174.47	3.7295	3.5370	7.2664
160	21.5	-332.65	519.45	186.80	3.8805	3.2466	7.1271
170	42.9	-309.01	507.80	198.79	4.0236	2.9871	7.0106
180	78.7	-285.06	495.40	210.34	4.1601	2.7522	6.9123
184.6	101.3	-274.04	489.47	215.43	4.2203	2.6522	6.8725
190	134.7	-260.72	482.08	221.37	4.2911	2.5373	6.8284
200	217.4	-235.88	467.67	231.78	4.4177	2.3383	6.7560
210	334.0	-210.46	451.97	241.50	4.5406	2.1522	6.6928
220	492.3	-184.35	434.77	250.42	4.6606	1.9762	6.6368
230	700.6	-157.42	415.79	258.38	4.7784	1.8078	6.5862
240	967.3	-129.50	394.69	265.19	4.8948	1.6446	6.5394
250	1301.5	-100.40	370.98	270.58	5.0106	1.4839	6.4945
260	1712.5	-69.81	343.94	274.13	5.1269	1.3228	6.4497
270	2210.7	-37.25	312.42	275.17	5.2453	1.1571	6.4024
280	2807.6	-1.86	274.34	272.48	5.3684	0.9798	6.3482
290	3516.7	38.34	225.08	263.42	5.5027	0.7761	6.2788
300	4356.5	89.77	149.18	238.94	5.6682	0.4973	6.1654
305.3	4871.8	164.23	0	164.23	5.9071	0	5.9071

TABLE B.5.2 SI (Continued) *Superheated Ethane*

Temp. K	v m^3/kg	h kJ/kg	s kJ/kg K	v m^3/kg	h kJ/kg	s kJ/kg K	v m^3/kg	h kJ/kg	s kJ/kg K
	50 kPa (172.41)			100 kPa (184.31)			200 kPa (198.18)		
380	2.09781	551.16	8.2768	1.04709	550.55	8.0841	0.52173	549.33	7.8901
400	2.20884	593.92	8.3865	1.10281	593.37	8.1938	0.54980	592.26	8.0002
420	2.31981	638.40	8.4950	1.15849	637.89	8.3024	0.57782	636.87	8.1090
440	2.43074	684.57	8.6023	1.21411	684.10	8.4099	0.60580	683.17	8.2167
460	2.54163	732.41	8.7087	1.26971	731.98	8.5163	0.63375	731.12	8.3233
480	2.65250	781.90	8.8140	1.32527	781.50	8.6217	0.66166	780.71	8.4288
500	2.76333	833.00	8.9183	1.38081	832.63	8.7260	0.68955	831.90	8.5333
520	2.87415	885.68	9.0216	1.43632	885.34	8.8294	0.71741	884.66	8.6367
540	2.98495	939.91	9.1239	1.49182	939.59	8.9318	0.74526	938.96	8.7392

(Continued)

TABLE B.5.2 SI (Continued) *Superheated Ethane*

Temp. K	v m³/kg	h kJ/kg	s kJ/kg K	v m³/kg	h kJ/kg	s kJ/kg K	v m³/kg	h kJ/kg	s kJ/kg K
	300 kPa (207.40)			400 kPa (214.52)			500 kPa (220.42)		
Sat.	0.17709	239.05	6.7085	0.13491	245.65	6.6667	0.10904	250.77	6.6346
220	0.19053	258.82	6.8010	0.13952	254.62	6.7080	—	—	—
240	0.21115	290.71	6.9397	0.15569	287.41	6.8507	0.12234	283.97	6.7789
260	0.23120	323.43	7.0707	0.17120	320.71	6.9839	0.13516	317.93	6.9148
280	0.25089	357.24	7.1959	0.18631	354.95	7.1108	0.14755	352.61	7.0433
300	0.27032	392.35	7.3170	0.20116	390.38	7.2330	0.15965	388.38	7.1667
320	0.28956	428.91	7.4350	0.21581	427.18	7.3517	0.17154	425.44	7.2862
340	0.30867	467.02	7.5504	0.23032	465.49	7.4678	0.18330	463.94	7.4029
360	0.32768	506.73	7.6639	0.24472	505.36	7.5817	0.19494	503.99	7.5174
380	0.34660	548.10	7.7757	0.25904	546.87	7.6939	0.20650	545.63	7.6299
400	0.36546	591.14	7.8861	0.27329	590.02	7.8046	0.21799	588.90	7.7409
420	0.38427	635.86	7.9952	0.28749	634.84	7.9139	0.22943	633.81	7.8504
440	0.40303	682.24	8.1030	0.30165	681.30	8.0219	0.24082	680.36	7.9587
460	0.42176	730.26	8.2098	0.31577	729.40	8.1288	0.25217	728.54	8.0657
480	0.44046	779.92	8.3154	0.32986	779.12	8.2346	0.26350	778.33	8.1717
500	0.45913	831.17	8.4200	0.34392	830.43	8.3393	0.27480	829.70	8.2765
520	0.47778	883.98	8.5236	0.35796	883.30	8.4430	0.28607	882.62	8.3803
540	0.49641	938.33	8.6261	0.37198	937.69	8.5456	0.29733	937.06	8.4830
560	0.51502	994.17	8.7277	0.38598	993.58	8.6473	0.30856	992.99	8.5847
580	0.53361	1051.48	8.8282	0.39997	1050.93	8.7479	0.31979	1050.38	8.6854
	600 kPa (225.49)			700 kPa (229.97)			800 kPa (234.01)		
Sat.	0.09147	254.92	6.6085	0.07874	258.36	6.5864	0.06906	261.26	6.5671
240	0.10003	280.36	6.7179	0.08403	276.57	6.6639	0.07195	272.55	6.6147
260	0.11110	315.06	6.8567	0.09388	312.11	6.8061	0.08093	309.06	6.7609
280	0.12168	350.23	6.9870	0.10319	347.80	6.9384	0.08930	345.32	6.8952
300	0.13196	386.35	7.1116	0.11217	384.28	7.0642	0.09732	382.19	7.0224
320	0.14203	423.67	7.2320	0.12094	421.89	7.1855	0.10511	420.09	7.1447
340	0.15195	462.39	7.3493	0.12955	460.82	7.3035	0.11275	459.24	7.2633
360	0.16175	502.60	7.4643	0.13804	501.21	7.4189	0.12026	499.81	7.3792
380	0.17147	544.38	7.5772	0.14645	543.13	7.5322	0.12768	541.88	7.4930
400	0.18112	587.77	7.6884	0.15478	586.64	7.6438	0.13503	585.51	7.6048
420	0.19071	632.79	7.7982	0.16306	631.76	7.7538	0.14232	630.72	7.7151
440	0.20026	679.42	7.9067	0.17130	678.48	7.8625	0.14957	677.54	7.8240
460	0.20978	727.68	8.0139	0.17949	726.82	7.9699	0.15678	725.95	7.9316
480	0.21926	777.53	8.1200	0.18766	776.74	8.0761	0.16396	775.94	8.0380
500	0.22871	828.96	8.2250	0.19580	828.22	8.1812	0.17111	827.49	8.1432
520	0.23814	881.93	8.3288	0.20391	881.25	8.2852	0.17824	880.57	8.2472
540	0.24756	936.43	8.4317	0.21201	935.79	8.3881	0.18535	935.16	8.3502
560	0.25695	992.40	8.5334	0.22009	991.81	8.4900	0.19244	991.22	8.4522
580	0.26633	1049.83	8.6342	0.22815	1049.28	8.5908	0.19952	1048.73	8.5531
600	0.27570	1108.68	8.7339	0.23620	1108.17	8.6906	0.20658	1107.66	8.6530

(Continued)

TABLE B.5.2 SI (Continued) *Superheated Ethane*

Temp. K	v m³/kg	h kJ/kg	s kJ/kg K	v m³/kg	h kJ/kg	s kJ/kg K	v m³/kg	h kJ/kg	s kJ/kg K
	1000 kPa (241.08)			1500 kPa (255.08)			2000 kPa (265.99)		
Sat.	0.05529	265.85	6.5345	0.03638	272.65	6.4719	0.02655	275.11	6.4219
260	0.06271	302.64	6.6814	0.03796	283.91	6.5156	—	—	—
280	0.06981	340.19	6.8206	0.04362	326.17	6.6723	0.03021	309.77	6.5489
300	0.07651	377.91	6.9507	0.04864	366.55	6.8117	0.03456	354.05	6.7017
320	0.08295	416.42	7.0749	0.05333	406.85	6.9417	0.03844	396.63	6.8392
340	0.08921	456.04	7.1950	0.05780	447.78	7.0657	0.04205	439.13	6.9680
360	0.09536	496.97	7.3120	0.06213	489.72	7.1856	0.04550	482.22	7.0911
380	0.10140	539.34	7.4265	0.06635	532.90	7.3023	0.04882	526.29	7.2102
400	0.10737	583.22	7.5390	0.07050	577.44	7.4165	0.05205	571.54	7.3263
420	0.11329	628.65	7.6498	0.07458	623.42	7.5287	0.05522	618.11	7.4399
440	0.11916	675.65	7.7592	0.07861	670.88	7.6391	0.05833	666.07	7.5514
460	0.12499	724.21	7.8671	0.08260	719.86	7.7479	0.06140	715.47	7.6612
480	0.13078	774.34	7.9737	0.08655	770.34	7.8553	0.06444	766.32	7.7694
500	0.13655	826.01	8.0792	0.09048	822.32	7.9614	0.06745	818.62	7.8761
520	0.14230	879.20	8.1835	0.09438	875.79	8.0662	0.07043	872.37	7.9815
540	0.14802	933.89	8.2867	0.09827	930.73	8.1699	0.07339	927.56	8.0857
560	0.15373	990.04	8.3888	0.10213	987.10	8.2724	0.07634	984.17	8.1886
580	0.15943	1047.63	8.4898	0.10598	1044.90	8.3738	0.07926	1042.17	8.2903
600	0.16511	1106.63	8.5898	0.10982	1104.08	8.4741	0.08218	1101.54	8.3910
620	0.17078	1167.01	8.6888	0.11364	1164.63	8.5734	0.08508	1162.26	8.4905
	4000 kPa (295.96)			6000 kPa			10000 kPa		
Sat.	0.01041	252.13	6.2211	—	—	—	—	—	—
300	0.01178	276.04	6.3014	—	—	—	—	—	—
320	0.01554	345.67	6.5267	0.00631	243.24	6.1403	0.00304	125.93	5.7275
340	0.01817	399.35	6.6895	0.00983	346.14	6.4536	0.00391	218.21	6.0069
360	0.02042	449.25	6.8321	0.01193	410.15	6.6367	0.00534	315.90	6.2863
380	0.02246	498.04	6.9640	0.01364	466.54	6.7893	0.00673	395.76	6.5024
400	0.02437	546.84	7.0892	0.01515	520.29	6.9271	0.00792	463.49	6.6762
420	0.02619	596.18	7.2095	0.01654	573.20	7.0562	0.00897	525.59	6.8277
440	0.02794	646.40	7.3263	0.01785	626.13	7.1793	0.00991	585.09	6.9661
460	0.02965	697.66	7.4403	0.01910	679.57	7.2981	0.01079	643.50	7.0960
480	0.03131	750.10	7.5519	0.02031	733.79	7.4134	0.01163	701.65	7.2197
500	0.03294	803.78	7.6614	0.02148	788.96	7.5260	0.01242	760.04	7.3389
520	0.03454	858.73	7.7692	0.02262	845.19	7.6363	0.01318	818.96	7.4544
540	0.03612	914.97	7.8753	0.02374	902.55	7.7445	0.01392	878.61	7.5670
560	0.03768	972.52	7.9799	0.02483	961.07	7.8510	0.01464	939.12	7.6770
580	0.03922	1031.37	8.0832	0.02591	1020.79	7.9557	0.01535	1000.58	7.7848
600	0.04075	1091.50	8.1851	0.02698	1081.70	8.0590	0.01604	1063.04	7.8907
620	0.04227	1152.91	8.2858	0.02804	1143.81	8.1608	0.01672	1126.53	7.9948

TABLE B.6 SI ***Thermodynamic Properties of Ethylene***

TABLE B.6.1 SI ***Saturated Ethylene***

Temp.	Press.	Specific Volume, m³/kg			Internal Energy, kJ/kg		
K	kPa	Sat. Liquid	Evap.	Sat. Vapor	Sat. Liquid	Evap.	Sat. Vapor
T	P	v_f	v_{fg}	v_g	u_f	u_{fg}	u_g
104.0	0.123	0.001527	251.19162	251.19315	235.71	537.35	773.06
110	0.334	0.001545	97.57094	97.57248	249.97	528.42	778.39
120	1.38	0.001576	25.74459	25.74616	275.34	511.88	787.22
130	4.46	0.001609	8.61762	8.61923	300.38	495.60	795.97
140	11.9	0.001644	3.45816	3.45981	324.69	479.89	804.58
150	27.5	0.001681	1.59601	1.59769	348.56	464.42	812.98
160	56.4	0.001721	0.82150	0.82322	372.29	448.81	821.10
169.4	101.3	0.001761	0.47731	0.47907	394.55	433.82	828.37
170	105.3	0.001763	0.46075	0.46251	396.10	432.76	828.86
180	182.1	0.001810	0.27657	0.27838	420.13	416.04	836.17
190	295.7	0.001861	0.17518	0.17704	444.50	398.44	842.94
200	456.0	0.001918	0.11574	0.11765	469.30	379.77	849.07
210	673.0	0.001981	0.07899	0.08097	494.65	359.78	854.43
220	957.5	0.002054	0.05521	0.05726	520.70	338.16	858.86
230	1320.7	0.002139	0.03918	0.04132	547.64	314.46	862.10
240	1774.2	0.002241	0.02798	0.03022	575.77	288.03	863.79
250	2330.7	0.002369	0.01985	0.02222	605.54	257.76	863.29
260	3004.6	0.002541	0.01370	0.01624	637.79	221.61	859.40
270	3813.2	0.002804	0.00872	0.01152	674.61	174.67	849.28
280	4782.4	0.003431	0.00374	0.00717	726.02	93.68	819.70
282.3	5040.1	0.004669	0	0.00467	771.84	0	771.84

TABLE B.6.2 SI ***Superheated Ethylene***

Temp. K	v m³/kg	h kJ/kg	s kJ/kg K	v m³/kg	h kJ/kg	s kJ/kg K	v m³/kg	h kJ/kg	s kJ/kg K
	50 kPa (158.23)			100 kPa (169.13)			200 kPa (181.85)		
Sat.	0.91941	865.66	7.1754	0.48494	876.69	7.0427	0.25511	888.48	6.9132
180	1.05301	892.95	7.3370	0.51930	890.73	7.1232	—	—	—
200	1.17429	918.27	7.4704	0.58143	916.55	7.2592	0.28480	912.97	7.0415
220	1.29479	944.16	7.5938	0.64271	942.77	7.3841	0.31657	939.90	7.1699
240	1.41480	970.87	7.7099	0.70348	969.70	7.5012	0.34776	967.32	7.2892
260	1.53450	998.57	7.8208	0.76392	997.57	7.6128	0.37859	995.55	7.4021
280	1.65397	1027.42	7.9276	0.82412	1026.55	7.7201	0.40918	1024.79	7.5104

(Continued)

TABLE B.6.1 SI (Continued) ***Saturated Ethylene***

Temp.	Press.	Enthalpy, kJ/kg			Entropy, kJ/kg K		
K T	kPa P	Sat. Liquid h_f	Evap. h_{fg}	Sat. Vapor h_g	Sat. Liquid s_f	Evap. s_{fg}	Sat. Vapor s_g
104.0	0.123	235.71	568.16	803.88	3.0053	5.4636	8.4689
110	0.334	249.97	561.01	810.98	3.1385	5.1001	8.2386
120	1.38	275.34	547.40	822.74	3.3592	4.5617	7.9209
130	4.46	300.38	533.99	834.37	3.5597	4.1076	7.6673
140	11.9	324.71	521.08	845.79	3.7399	3.7220	7.4619
150	27.5	348.61	508.26	856.88	3.9046	3.3884	7.2931
160	56.4	372.39	495.11	867.50	4.0578	3.0945	7.1522
169.4	101.3	394.72	482.19	876.91	4.1930	2.8473	7.0402
170	105.3	396.28	481.26	877.54	4.2021	2.8309	7.0330
180	182.1	420.46	466.40	886.85	4.3395	2.5911	6.9306
190	295.7	445.05	450.25	895.30	4.4713	2.3697	6.8410
200	456.0	470.17	432.54	902.71	4.5986	2.1627	6.7613
210	673.0	495.99	412.93	908.92	4.7225	1.9663	6.6888
220	957.5	522.67	391.01	913.68	4.8439	1.7773	6.6213
230	1320.7	550.47	366.21	916.68	4.9641	1.5922	6.5563
240	1774.2	579.74	337.67	917.41	5.0844	1.4069	6.4914
250	2330.7	611.06	304.02	915.08	5.2070	1.2161	6.4231
260	3004.6	645.42	262.76	908.19	5.3353	1.0106	6.3459
270	3813.2	685.30	207.91	893.21	5.4775	0.7700	6.2476
280	4782.4	742.43	111.56	853.99	5.6742	0.3984	6.0726

TABLE B.6.2 SI (Continued) ***Superheated Ethylene***

Temp. K	v m³/kg	h kJ/kg	s kJ/kg K	v m³/kg	h kJ/kg	s kJ/kg K	v m³/kg	h kJ/kg	s kJ/kg K
	50 kPa (158.23)			100 kPa (169.13)			200 kPa (181.85)		
300	1.77328	1057.53	8.0315	0.88416	1056.76	7.8243	0.43959	1055.21	7.6154
320	1.89246	1088.97	8.1329	0.94408	1088.29	7.9260	0.46988	1086.91	7.7176
340	2.01156	1121.81	8.2324	1.00390	1121.19	8.0257	0.50006	1119.96	7.8178
360	2.13057	1156.06	8.3303	1.06364	1155.50	8.1238	0.53017	1154.39	7.9161
380	2.24953	1191.74	8.4267	1.12332	1191.24	8.2203	0.56022	1190.22	8.0130
400	2.36843	1228.85	8.5219	1.18296	1228.39	8.3156	0.59022	1227.46	8.1085
420	2.48730	1267.37	8.6158	1.24255	1266.95	8.4097	0.62017	1266.10	8.2027
440	2.60613	1307.28	8.7086	1.30211	1306.89	8.5025	0.65009	1306.10	8.2958

(Continued)

TABLE B.6.2 SI (Continued) ***Superheated Ethylene***

Temp. K	v m³/kg	h kJ/kg	s kJ/kg K	v m³/kg	h kJ/kg	s kJ/kg K	v m³/kg	h kJ/kg	s kJ/kg K
	300 kPa (190.31)			400 kPa (196.86)			500 kPa (202.28)		
Sat.	0.17468	895.54	6.8384	0.13321	900.50	6.7855	0.10776	904.24	6.7443
200	0.18573	909.16	6.9082	0.13603	905.11	6.8087	—	—	—
220	0.20776	936.93	7.0405	0.15327	933.84	6.9456	0.12051	930.62	6.8693
240	0.22913	964.89	7.1621	0.16978	962.39	7.0698	0.13413	959.82	6.9963
260	0.25012	993.48	7.2766	0.18586	991.39	7.1859	0.14728	989.25	7.1141
280	0.27084	1023.01	7.3860	0.20166	1021.21	7.2963	0.16014	1019.38	7.2257
300	0.29139	1053.65	7.4916	0.21728	1052.07	7.4028	0.17281	1050.48	7.3330
320	0.31180	1085.52	7.5945	0.23276	1084.13	7.5062	0.18533	1082.72	7.4370
340	0.33211	1118.71	7.6950	0.24814	1117.46	7.6072	0.19775	1116.21	7.5385
360	0.35235	1153.26	7.7938	0.26343	1152.14	7.7063	0.21008	1151.01	7.6379
380	0.37252	1189.20	7.8909	0.27867	1188.18	7.8037	0.22235	1187.15	7.7356
400	0.39264	1226.53	7.9866	0.29385	1225.60	7.8997	0.23457	1224.66	7.8318
420	0.41271	1265.24	8.0811	0.30899	1264.39	7.9943	0.24675	1263.53	7.9266
440	0.43276	1305.32	8.1743	0.32409	1304.54	8.0877	0.25889	1303.75	8.0202
	600 kPa (206.94)			700 kPa (211.07)			800 kPa (214.78)		
Sat.	0.09047	907.16	6.7104	0.07793	909.50	6.6814	0.06840	911.40	6.6560
220	0.09860	927.26	6.8045	0.08288	923.72	6.7474	0.07103	920.00	6.6956
240	0.11032	957.18	6.9347	0.09329	954.46	6.8812	0.08049	951.66	6.8334
260	0.12154	987.07	7.0544	0.10314	984.85	7.0028	0.08933	982.59	6.9572
280	0.13245	1017.53	7.1672	0.11266	1015.65	7.1169	0.09781	1013.75	7.0726
300	0.14315	1048.87	7.2753	0.12196	1047.25	7.2259	0.10607	1045.61	7.1825
320	0.15371	1081.30	7.3799	0.13111	1079.88	7.3311	0.11417	1078.44	7.2884
340	0.16415	1114.94	7.4819	0.14015	1113.67	7.4336	0.12215	1112.40	7.3914
360	0.17451	1149.87	7.5817	0.14911	1148.73	7.5337	0.13005	1147.59	7.4919
380	0.18481	1186.13	7.6797	0.15799	1185.09	7.6320	0.13788	1184.06	7.5905
400	0.19506	1223.73	7.7761	0.16683	1222.79	7.7287	0.14566	1221.85	7.6874
420	0.20526	1262.68	7.8711	0.17562	1261.82	7.8239	0.15339	1260.96	7.7828
440	0.21542	1302.96	7.9648	0.18437	1302.18	7.9178	0.16109	1301.39	7.8768
	1000 kPa (221.30)			1200 kPa (226.93)			1400 kPa (231.91)		
Sat.	0.05483	914.18	6.6127	0.04560	915.97	6.5761	0.03889	917.01	6.5440
240	0.06249	945.76	6.7498	0.05038	939.41	6.6766	0.04163	932.50	6.6097
260	0.06995	977.91	6.8785	0.05698	973.02	6.8112	0.04767	967.88	6.7513
280	0.07700	1009.86	6.9969	0.06310	1005.85	6.9328	0.05314	1001.72	6.8767
300	0.08380	1042.28	7.1087	0.06894	1038.88	7.0467	0.05831	1035.40	6.9929
320	0.09043	1075.53	7.2160	0.07460	1072.58	7.1555	0.06329	1069.58	7.1032
340	0.09695	1109.82	7.3199	0.08014	1107.22	7.2605	0.06813	1104.59	7.2093
360	0.10337	1145.28	7.4212	0.08558	1142.96	7.3626	0.07287	1140.62	7.3122
380	0.10972	1181.98	7.5204	0.09095	1179.89	7.4624	0.07754	1177.79	7.4127
400	0.11602	1219.96	7.6178	0.09626	1218.07	7.5603	0.08215	1216.17	7.5111
420	0.12227	1259.24	7.7136	0.10153	1257.52	7.6565	0.08671	1255.78	7.6078
440	0.12849	1299.81	7.8080	0.10676	1298.23	7.7512	0.09123	1296.65	7.7028

(Continued)

TABLE B.6.2 SI (Continued) *Superheated Ethylene*

Temp. K	v m³/kg	h kJ/kg	s kJ/kg K	v m³/kg	h kJ/kg	s kJ/kg K	v m³/kg	h kJ/kg	s kJ/kg K
	1600 kPa (236.41)			1800 kPa (240.51)			2000 kPa (244.30)		
Sat.	0.03378	917.45	6.5149	0.02975	917.37	6.4880	0.02648	916.85	6.4627
260	0.04064	962.46	6.6966	0.03512	956.70	6.6454	0.03065	950.54	6.5965
280	0.04566	997.44	6.8263	0.03981	993.00	6.7800	0.03511	988.39	6.7368
300	0.05033	1031.84	6.9450	0.04411	1028.20	6.9014	0.03913	1024.47	6.8613
320	0.05480	1066.53	7.0569	0.04819	1063.43	7.0151	0.04289	1060.28	6.9768
340	0.05912	1101.92	7.1641	0.05211	1099.23	7.1236	0.04649	1096.50	7.0866
360	0.06334	1138.26	7.2680	0.05592	1135.88	7.2283	0.04999	1133.48	7.1923
380	0.06748	1175.68	7.3691	0.05966	1173.55	7.3301	0.05340	1171.41	7.2948
400	0.07156	1214.26	7.4680	0.06333	1212.34	7.4296	0.05674	1210.41	7.3948
420	0.07560	1254.05	7.5651	0.06695	1252.31	7.5271	0.06004	1250.56	7.4928
440	0.07959	1295.06	7.6605	0.07054	1293.48	7.6228	0.06329	1291.88	7.5889
	3000 kPa (259.94)			4000 kPa (272.08)			5000 kPa (281.99)		
Sat.	0.01627	908.25	6.3464	0.01064	888.31	6.2219	0.00578	826.88	5.9711
280	0.02073	961.82	6.5454	0.01288	924.18	6.3520	—	—	—
300	0.02404	1004.16	6.6916	0.01630	980.15	6.5455	0.01139	949.94	6.3991
320	0.02695	1043.64	6.8190	0.01889	1025.24	6.6911	0.01398	1004.56	6.5756
340	0.02963	1082.35	6.9364	0.02116	1067.24	6.8184	0.01605	1051.05	6.7166
360	0.03217	1121.16	7.0473	0.02325	1108.29	6.9358	0.01789	1094.85	6.8418
380	0.03461	1160.51	7.1537	0.02522	1149.30	7.0466	0.01958	1137.77	6.9579
400	0.03698	1200.67	7.2566	0.02711	1190.74	7.1529	0.02119	1180.63	7.0678
420	0.03930	1241.77	7.3569	0.02894	1232.87	7.2557	0.02273	1223.89	7.1733
440	0.04157	1283.89	7.4549	0.03072	1275.86	7.3556	0.02422	1267.79	7.2754
	10000 kPa			20000 kPa			50000 kPa		
300	0.003095	769.71	5.7101	0.002508	741.19	5.5244	0.002122	748.61	5.3218
320	0.004285	866.06	6.0207	0.002763	799.42	5.7122	0.002218	796.72	5.4771
340	0.005907	953.78	6.2870	0.003090	860.58	5.8976	0.002321	845.76	5.6257
360	0.007286	1020.25	6.4772	0.003493	923.30	6.0768	0.002433	895.76	5.7686
380	0.008454	1076.78	6.6301	0.003955	985.71	6.2455	0.002552	946.72	5.9064
400	0.009492	1128.82	6.7636	0.004447	1046.54	6.4016	0.002679	998.65	6.0395
420	0.010445	1178.76	6.8854	0.004941	1104.98	6.5442	0.002813	1051.49	6.1684
440	0.011338	1227.81	6.9995	0.005425	1161.38	6.6754	0.002953	1105.16	6.2932

TABLE B.7 SI *Thermodynamic Properties of Methane*
TABLE B.7.1 SI *Saturated Methane*

Temp.	Press.	Specific Volume, m^3/kg			Internal Energy, kJ/kg		
K	kPa	Sat. Liquid	Evap.	Sat. Vapor	Sat. Liquid	Evap.	Sat. Vapor
T	P	v_f	v_{fg}	v_g	u_f	u_{fg}	u_g
90.7	11.7	0.002215	3.97941	3.98163	-358.10	496.59	138.49
95	19.8	0.002243	2.44845	2.45069	-343.79	488.62	144.83
100	34.4	0.002278	1.47657	1.47885	-326.90	478.96	152.06
105	56.4	0.002315	0.93780	0.94012	-309.79	468.89	159.11
110	88.2	0.002353	0.62208	0.62443	-292.50	458.41	165.91
111.7	101.3	0.002367	0.54760	0.54997	-286.74	454.85	168.10
115	132.3	0.002395	0.42800	0.43040	-275.05	447.48	172.42
120	191.6	0.002439	0.30367	0.30610	-257.45	436.02	178.57
125	269.0	0.002486	0.22108	0.22357	-239.66	423.97	184.32
130	367.6	0.002537	0.16448	0.16701	-221.65	411.25	189.60
135	490.7	0.002592	0.12458	0.12717	-203.40	397.77	194.37
140	641.6	0.002653	0.09575	0.09841	-184.86	383.42	198.56
145	823.7	0.002719	0.07445	0.07717	-165.97	368.06	202.09
150	1040.5	0.002794	0.05839	0.06118	-146.65	351.53	204.88
155	1295.6	0.002877	0.04605	0.04892	-126.82	333.61	206.79
160	1592.8	0.002974	0.03638	0.03936	-106.35	314.01	207.66
165	1935.9	0.003086	0.02868	0.03177	-85.06	292.30	207.24
170	2329.3	0.003222	0.02241	0.02563	-62.67	267.81	205.14
175	2777.6	0.003393	0.01718	0.02058	-38.75	239.47	200.72
180	3286.4	0.003623	0.01266	0.01629	-12.43	205.16	192.73
185	3863.2	0.003977	0.00846	0.01243	18.47	159.49	177.96
190	4520.5	0.004968	0.00300	0.00797	69.10	67.01	136.11
190.6	4599.2	0.006148	0	0.00615	101.46	0	101.46

TABLE B.7.2 SI *Superheated Methane*

Temp. K	v m^3/kg	h kJ/kg	s kJ/kg K	v m^3/kg	h kJ/kg	s kJ/kg K	v m^3/kg	h kJ/kg	s kJ/kg K
	50 kPa (103.73)			100 kPa (111.50)			200 kPa (120.61)		
Sat.	1.05026	209.85	9.7295	0.55665	223.56	9.5084	0.29422	238.14	9.2918
125	1.27928	255.82	10.1331	0.63126	253.33	9.7606	0.30695	248.19	9.3736
150	1.54333	308.45	10.5170	0.76586	306.77	10.1504	0.37700	303.31	9.7759
175	1.80540	360.81	10.8399	0.89840	359.56	10.4759	0.44486	357.02	10.1071
200	2.06648	413.17	11.1196	1.02994	412.19	10.7570	0.51165	410.21	10.3912
225	2.32702	465.79	11.3674	1.16092	464.99	11.0058	0.57786	463.38	10.6417
250	2.58720	518.94	11.5914	1.29154	518.27	11.2303	0.64370	516.93	10.8674

(Continued)

TABLE B.7.1 SI (Continued) *Saturated Methane*

Temp. K T	Press. kPa P	Enthalpy, kJ/kg Sat. Liquid h_f	Evap. h_{fg}	Sat. Vapor h_g	Entropy, kJ/kg K Sat. Liquid s_f	Evap. s_{fg}	Sat. Vapor s_g
90.7	11.7	-358.07	543.12	185.05	4.2264	5.9891	10.2155
95	19.8	-343.75	537.18	193.43	4.3805	5.6545	10.0350
100	34.4	-326.83	529.77	202.94	4.5538	5.2977	9.8514
105	56.4	-309.66	521.82	212.16	4.7208	4.9697	9.6905
110	88.2	-292.29	513.29	221.00	4.8817	4.6663	9.5480
111.7	101.3	-286.50	510.33	223.83	4.9336	4.5706	9.5042
115	132.3	-274.74	504.12	229.38	5.0368	4.3836	9.4205
120	191.6	-256.98	494.20	237.23	5.1867	4.1184	9.3051
125	269.0	-238.99	483.44	244.45	5.3321	3.8675	9.1996
130	367.6	-220.72	471.72	251.00	5.4734	3.6286	9.1020
135	490.7	-202.13	458.90	256.77	5.6113	3.3993	9.0106
140	641.6	-183.16	444.85	261.69	5.7464	3.1775	8.9239
145	823.7	-163.73	429.38	265.66	5.8794	2.9613	8.8406
150	1040.5	-143.74	412.29	268.54	6.0108	2.7486	8.7594
155	1295.6	-123.09	393.27	270.18	6.1415	2.5372	8.6787
160	1592.8	-101.61	371.96	270.35	6.2724	2.3248	8.5971
165	1935.9	-79.08	347.82	268.74	6.4046	2.1080	8.5126
170	2329.3	-55.17	320.02	264.85	6.5399	1.8824	8.4224
175	2777.6	-29.33	287.20	257.87	6.6811	1.6411	8.3223
180	3286.4	-0.53	246.77	246.25	6.8333	1.3710	8.2043
185	3863.2	33.83	192.16	226.00	7.0095	1.0387	8.0483
190	4520.5	91.56	80.58	172.14	7.3015	0.4241	7.7256
190.6	4599.2	129.74	0	129.74	7.4999	0	7.4999

TABLE B.7.2 SI (Continued) *Superheated Methane*

Temp. K	v m³/kg	h kJ/kg	s kJ/kg K	v m³/kg	h kJ/kg	s kJ/kg K	v m³/kg	h kJ/kg	s kJ/kg K
	50 kPa (103.73)			100 kPa (111.50)			200 kPa (120.61)		
275	2.84715	572.92	11.7972	1.42193	572.36	11.4365	0.70931	571.22	11.0743
300	3.10694	628.06	11.9891	1.55215	627.58	11.6286	0.77475	626.60	11.2670
325	3.36661	684.65	12.1702	1.68225	684.23	11.8100	0.84008	683.38	11.4488
350	3.62619	742.94	12.3429	1.81226	742.57	11.9829	0.90530	741.83	11.6220
375	3.88569	803.13	12.5090	1.94220	802.80	12.1491	0.97046	802.16	11.7885
400	4.14515	865.38	12.6697	2.07209	865.10	12.3099	1.03557	864.53	11.9495
425	4.40455	929.80	12.8259	2.20193	929.55	12.4661	1.10062	929.05	12.1059

(Continued)

TABLE B.7.2 SI (Continued) *Superheated Methane*

Temp. K	v m³/kg	h kJ/kg	s kJ/kg K	v m³/kg	h kJ/kg	s kJ/kg K	v m³/kg	h kJ/kg	s kJ/kg K
	400 kPa (131.42)			600 kPa (138.72)			800 kPa (144.40)		
Sat.	0.15427	252.72	9.0754	0.10496	260.51	8.9458	0.07941	265.23	8.8505
150	0.18233	296.09	9.3843	0.11717	288.38	9.1390	0.08434	280.00	8.9509
175	0.21799	351.81	9.7280	0.14227	346.39	9.4970	0.10433	340.76	9.3260
200	0.25246	406.18	10.0185	0.16603	402.06	9.7944	0.12278	397.85	9.6310
225	0.28631	460.13	10.2726	0.18911	456.84	10.0525	0.14050	453.50	9.8932
250	0.31978	514.23	10.5007	0.21180	511.52	10.2830	0.15781	508.78	10.1262
275	0.35301	568.94	10.7092	0.23424	566.66	10.4931	0.17485	564.35	10.3381
300	0.38606	624.65	10.9031	0.25650	622.69	10.6882	0.19172	620.73	10.5343
325	0.41899	681.69	11.0857	0.27863	680.00	10.8716	0.20845	678.31	10.7186
350	0.45183	740.36	11.2595	0.30067	738.88	11.0461	0.22510	737.41	10.8938
375	0.48460	800.87	11.4265	0.32264	799.57	11.2136	0.24167	798.28	11.0617
400	0.51731	863.39	11.5879	0.34456	862.25	11.3754	0.25818	861.12	11.2239
425	0.54997	928.04	11.7446	0.36643	927.04	11.5324	0.27465	926.03	11.3813
450	0.58260	994.89	11.8974	0.38826	994.00	11.6855	0.29109	993.11	11.5346
475	0.61520	1063.97	12.0468	0.41006	1063.18	11.8351	0.30749	1062.40	11.6845
500	0.64778	1135.29	12.1931	0.43184	1134.59	11.9816	0.32387	1133.89	11.8311
525	0.68033	1208.81	12.3366	0.45360	1208.18	12.1252	0.34023	1207.56	11.9749
	1000 kPa (149.13)			1500 kPa (158.52)			2000 kPa (165.86)		
Sat.	0.06367	268.12	8.7735	0.04196	270.47	8.6215	0.03062	268.25	8.4975
175	0.08149	334.87	9.1871	0.05078	318.81	8.9121	0.03504	299.97	8.6839
200	0.09681	393.53	9.5006	0.06209	382.26	9.2514	0.04463	370.17	9.0596
225	0.11132	450.11	9.7672	0.07239	441.44	9.5303	0.05289	432.43	9.3532
250	0.12541	506.01	10.0028	0.08220	499.00	9.7730	0.06059	491.84	9.6036
275	0.13922	562.04	10.2164	0.09171	556.21	9.9911	0.06796	550.31	9.8266
300	0.15285	618.76	10.4138	0.10103	613.82	10.1916	0.07513	608.85	10.0303
325	0.16635	676.61	10.5990	0.11022	672.37	10.3790	0.08216	668.12	10.2200
350	0.17976	735.94	10.7748	0.11931	732.26	10.5565	0.08909	728.58	10.3992
375	0.19309	797.00	10.9433	0.12832	793.78	10.7263	0.09594	790.57	10.5703
400	0.20636	859.98	11.1059	0.13728	857.16	10.8899	0.10274	854.34	10.7349
425	0.21959	925.03	11.2636	0.14619	922.54	11.0484	0.10949	920.06	10.8942
450	0.23279	992.23	11.4172	0.15506	990.02	11.2027	0.11620	987.84	11.0491
475	0.24595	1061.61	11.5672	0.16391	1059.66	11.3532	0.12289	1057.72	11.2003
500	0.25909	1133.19	11.7141	0.17273	1131.46	11.5005	0.12955	1129.74	11.3480
525	0.27221	1206.95	11.8580	0.18152	1205.41	11.6448	0.13619	1203.88	11.4927
550	0.28531	1282.84	11.9992	0.19031	1281.48	11.7864	0.14281	1280.13	11.6346

(Continued)

TABLE B7.2 SI (Continued) *Superheated Methane*

Temp. K	v m³/kg	h kJ/kg	s kJ/kg K	v m³/kg	h kJ/kg	s kJ/kg K	v m³/kg	h kJ/kg	s kJ/kg K
	3000 kPa (177.26)			4000 kPa (186.10)			5000 kPa		
Sat.	0.01856	253.33	8.2718	0.01160	219.34	8.0035	—	—	—
200	0.02690	342.70	8.7492	0.01763	308.23	8.4675	0.01142	258.30	8.1459
225	0.03333	413.29	9.0823	0.02347	392.39	8.8653	0.01749	369.34	8.6728
250	0.03896	477.06	9.3512	0.02814	461.63	9.1574	0.02165	445.55	8.9945
275	0.04421	538.32	9.5848	0.03235	526.07	9.4031	0.02525	513.60	9.2540
300	0.04924	598.83	9.7954	0.03631	588.73	9.6212	0.02857	578.57	9.4802
325	0.05412	659.59	9.9899	0.04011	651.07	9.8208	0.03173	642.56	9.6851
350	0.05889	721.23	10.1726	0.04381	713.93	10.0071	0.03477	706.67	9.8751
375	0.06358	784.19	10.3464	0.04742	777.86	10.1835	0.03774	771.60	10.0543
400	0.06822	848.76	10.5130	0.05097	843.24	10.3523	0.04064	837.78	10.2251
425	0.07281	915.15	10.6740	0.05448	910.31	10.5149	0.04350	905.53	10.3894
450	0.07736	983.50	10.8303	0.05795	979.23	10.6725	0.04632	975.04	10.5483
475	0.08188	1053.89	10.9825	0.06139	1050.12	10.8258	0.04911	1046.42	10.7026
500	0.08638	1126.34	11.1311	0.06481	1123.01	10.9753	0.05187	1119.74	10.8531
525	0.09086	1200.88	11.2765	0.06820	1197.93	11.1215	0.05462	1195.04	11.0000
550	0.09532	1277.47	11.4190	0.07158	1274.86	11.2646	0.05735	1272.30	11.1437
575	0.09976	1356.07	11.5588	0.07495	1353.77	11.4049	0.06006	1351.51	11.2846
	6000 kPa			8000 kPa			10000 kPa		
200	0.006109	160.30	7.6125	0.004120	88.54	7.2069	0.003756	72.22	7.0862
225	0.013466	343.74	8.4907	0.008460	284.98	8.1344	0.005945	229.32	7.8245
250	0.017325	428.84	8.8502	0.011983	393.92	8.5954	0.008915	358.60	8.3716
275	0.020529	500.95	9.1253	0.014688	475.39	8.9064	0.011272	450.09	8.7210
300	0.023430	568.39	9.3601	0.017055	548.15	9.1598	0.013297	528.37	8.9936
325	0.026156	634.09	9.5705	0.019235	617.40	9.3815	0.015137	601.22	9.2270
350	0.028767	699.48	9.7643	0.021297	685.39	9.5831	0.016861	671.80	9.4362
375	0.031295	765.41	9.9463	0.023277	753.34	9.7706	0.018505	741.72	9.6292
400	0.033763	832.42	10.1192	0.025197	821.95	9.9477	0.020090	811.92	9.8104
425	0.036185	900.84	10.2851	0.027072	891.71	10.1169	0.021632	882.98	9.9827
450	0.038571	970.91	10.4453	0.028911	962.92	10.2796	0.023140	955.27	10.1480
475	0.040927	1042.79	10.6007	0.030721	1035.75	10.4372	0.024620	1029.04	10.3075
500	0.043259	1116.54	10.7520	0.032508	1110.34	10.5902	0.026077	1104.44	10.4622
525	0.045571	1192.21	10.8997	0.034276	1186.74	10.7393	0.027516	1181.55	10.6126
550	0.047866	1269.81	11.0441	0.036027	1264.99	10.8849	0.028939	1260.42	10.7594
575	0.050147	1349.31	11.1854	0.037765	1345.07	11.0272	0.030350	1341.06	10.9028

TABLE B.8 SI *Thermodynamic Properties of Nitrogen*

TABLE B.8.1 SI *Saturated Nitrogen*

Temp. K T	Press. kPa P	Specific Volume, m³/kg Sat. Liquid v_f	Evap. v_{fg}	Sat. Vapor v_g	Internal Energy, kJ/kg Sat. Liquid u_f	Evap. u_{fg}	Sat. Vapor u_g
63.1	12.5	0.001150	1.48074	1.48189	-150.92	196.86	45.94
65	17.4	0.001160	1.09231	1.09347	-147.19	194.37	47.17
70	38.6	0.001191	0.52513	0.52632	-137.13	187.54	50.40
75	76.1	0.001223	0.28052	0.28174	-127.04	180.47	53.43
77.3	101.3	0.001240	0.21515	0.21639	-122.27	177.04	54.76
80	137.0	0.001259	0.16249	0.16375	-116.86	173.06	56.20
85	229.1	0.001299	0.10018	0.10148	-106.55	165.20	58.65
90	360.8	0.001343	0.06477	0.06611	-96.06	156.76	60.70
95	541.1	0.001393	0.04337	0.04476	-85.35	147.60	62.25
100	779.2	0.001452	0.02975	0.03120	-74.33	137.50	63.17
105	1084.6	0.001522	0.02066	0.02218	-62.89	126.18	63.29
110	1467.6	0.001610	0.01434	0.01595	-50.81	113.11	62.31
115	1939.3	0.001729	0.00971	0.01144	-37.66	97.36	59.70
120	2513.0	0.001915	0.00608	0.00799	-22.42	76.63	54.21
125	3208.0	0.002355	0.00254	0.00490	-0.83	40.73	39.90
126.2	3397.8	0.003194	0	0.00319	18.94	0	18.94

TABLE B.8.2 SI *Superheated Nitrogen*

Temp. K	v m³/kg	h kJ/kg	s kJ/kg K	v m³/kg	h kJ/kg	s kJ/kg K	v m³/kg	h kJ/kg	s kJ/kg K
	100 kPa (77.24)			200 kPa (83.62)			500 kPa (93.98)		
Sat.	0.21903	76.61	5.4059	0.11520	81.05	5.2673	0.04834	86.15	5.0802
100	0.29103	101.94	5.6944	0.14252	100.24	5.4775	0.05306	94.46	5.1660
120	0.35208	123.15	5.8878	0.17397	121.93	5.6753	0.06701	118.12	5.3821
140	0.41253	144.20	6.0501	0.20476	143.28	5.8399	0.08007	140.44	5.5541
160	0.47263	165.17	6.1901	0.23519	164.44	5.9812	0.09272	162.22	5.6996
180	0.53254	186.09	6.3132	0.26542	185.49	6.1052	0.10515	183.70	5.8261
200	0.59231	206.97	6.4232	0.29551	206.48	6.2157	0.11744	205.00	5.9383
220	0.65199	227.83	6.5227	0.32552	227.41	6.3155	0.12964	226.18	6.0392
240	0.71161	248.67	6.6133	0.35546	248.32	6.4064	0.14177	247.27	6.1310
260	0.77118	269.51	6.6967	0.38535	269.21	6.4900	0.15385	268.31	6.2152
280	0.83072	290.33	6.7739	0.41520	290.08	6.5674	0.16590	289.31	6.2930

(Continued)

TABLE B.8.1 SI (Continued) *Saturated Nitrogen*							
Temp.	Press.	Enthalpy, kJ/kg			Entropy, kJ/kg K		
K	kPa	Sat. Liquid	Evap.	Sat. Vapor	Sat. Liquid	Evap.	Sat. Vapor
T	P	h_f	h_{fg}	h_g	s_f	s_{fg}	s_g
63.1	12.5	-150.91	215.39	64.48	2.4234	3.4109	5.8343
65	17.4	-147.17	213.38	66.21	2.4816	3.2828	5.7645
70	38.6	-137.09	207.79	70.70	2.6307	2.9684	5.5991
75	76.1	-126.95	201.82	74.87	2.7700	2.6909	5.4609
77.3	101.3	-122.15	198.84	76.69	2.8326	2.5707	5.4033
80	137.0	-116.69	195.32	78.63	2.9014	2.4415	5.3429
85	229.1	-106.25	188.15	81.90	3.0266	2.2135	5.2401
90	360.8	-95.58	180.13	84.55	3.1466	2.0015	5.1480
95	541.1	-84.59	171.07	86.47	3.2627	1.8007	5.0634
100	779.2	-73.20	160.68	87.48	3.3761	1.6068	4.9829
105	1084.6	-61.24	148.59	87.35	3.4883	1.4151	4.9034
110	1467.6	-48.45	134.15	85.71	3.6017	1.2196	4.8213
115	1939.3	-34.31	116.19	81.88	3.7204	1.0104	4.7307
120	2513.0	-17.61	91.91	74.30	3.8536	0.7659	4.6195
125	3208.0	6.73	48.88	55.60	4.0399	0.3910	4.4309
126.2	3397.8	29.79	0	29.79	4.2193	0	4.2193

TABLE B.8.2 SI (Continued) *Superheated Nitrogen*									
Temp. K	v m³/kg	h kJ/kg	s kJ/kg K	v m³/kg	h kJ/kg	s kJ/kg K	v m³/kg	h kJ/kg	s kJ/kg K
	100 kPa (77.24)			200 kPa (83.62)			500 kPa (93.98)		
300	0.89023	311.16	6.8457	0.44503	310.94	6.6393	0.17792	310.28	6.3653
350	1.03891	363.24	7.0063	0.51952	363.09	6.8001	0.20788	362.63	6.5267
400	1.18752	415.41	7.1456	0.59392	415.31	6.9396	0.23777	414.99	6.6666
450	1.33607	467.77	7.2690	0.66827	467.70	7.0630	0.26759	467.49	6.7902
500	1.48458	520.41	7.3799	0.74258	520.37	7.1740	0.29739	520.24	6.9014
600	1.78154	626.94	7.5741	0.89114	626.94	7.3682	0.35691	626.93	7.0959
700	2.07845	735.58	7.7415	1.03965	735.61	7.5357	0.41637	735.68	7.2635
800	2.37532	846.60	7.8897	1.18812	846.64	7.6839	0.47581	846.78	7.4118
900	2.67217	960.01	8.0232	1.33657	960.07	7.8175	0.53522	960.24	7.5454
1000	2.96900	1075.68	8.1451	1.48501	1075.75	7.9393	0.59462	1075.96	7.6673

(Continued)

TABLE B.8.2 SI (Continued) *Superheated Nitrogen*

Temp. K	v m³/kg	h kJ/kg	s kJ/kg K	v m³/kg	h kJ/kg	s kJ/kg K	v m³/kg	h kJ/kg	s kJ/kg K
	600 kPa (96.37)			800 kPa (100.38)			1000 kPa (103.73)		
Sat.	0.04046	86.85	5.0411	0.03038	87.52	4.9768	0.02416	87.51	4.9237
120	0.05510	116.79	5.3204	0.04017	114.02	5.2191	0.03117	111.08	5.1357
140	0.06620	139.47	5.4953	0.04886	137.50	5.4002	0.03845	135.47	5.3239
160	0.07689	161.47	5.6422	0.05710	159.95	5.5501	0.04522	158.42	5.4772
180	0.08734	183.10	5.7696	0.06509	181.89	5.6793	0.05173	180.67	5.6082
200	0.09766	204.50	5.8823	0.07293	203.51	5.7933	0.05809	202.52	5.7234
220	0.10788	225.76	5.9837	0.08067	224.94	5.8954	0.06436	224.11	5.8263
240	0.11803	246.92	6.0757	0.08835	246.23	5.9880	0.07055	245.53	5.9194
260	0.12813	268.01	6.1601	0.09599	267.42	6.0728	0.07670	266.83	6.0047
280	0.13820	289.05	6.2381	0.10358	288.54	6.1511	0.08281	288.04	6.0833
300	0.14824	310.06	6.3105	0.11115	309.62	6.2238	0.08889	309.18	6.1562
350	0.17326	362.48	6.4722	0.12998	362.17	6.3858	0.10401	361.87	6.3187
400	0.19819	414.89	6.6121	0.14873	414.68	6.5260	0.11905	414.47	6.4591
450	0.22308	467.42	6.7359	0.16743	467.28	6.6500	0.13404	467.15	6.5832
500	0.24792	520.20	6.8471	0.18609	520.12	6.7613	0.14899	520.04	6.6947
600	0.29755	626.93	7.0416	0.22335	626.93	6.9560	0.17883	626.92	6.8895
700	0.34712	735.70	7.2093	0.26056	735.76	7.1237	0.20862	735.81	7.0573
800	0.39666	846.82	7.3576	0.29773	846.91	7.2721	0.23837	847.00	7.2057
900	0.44618	960.30	7.4912	0.33488	960.42	7.4058	0.26810	960.54	7.3394
1000	0.49568	1076.02	7.6131	0.37202	1076.16	7.5277	0.29782	1076.30	7.4614
	1500 kPa (110.38)			2000 kPa (115.58)			3000 kPa (123.61)		
Sat.	0.01555	85.51	4.8148	0.01100	81.25	4.7193	0.00582	63.47	4.5032
120	0.01899	102.75	4.9650	0.01260	92.10	4.8116	—	—	—
140	0.02452	130.15	5.1767	0.01752	124.40	5.0618	0.01038	111.13	4.8706
160	0.02937	154.50	5.3394	0.02144	150.43	5.2358	0.01350	141.85	5.0763
180	0.03393	177.60	5.4755	0.02503	174.48	5.3775	0.01614	168.09	5.2310
200	0.03832	200.03	5.5937	0.02844	197.53	5.4989	0.01857	192.49	5.3596
220	0.04260	222.05	5.6987	0.03174	219.99	5.6060	0.02088	215.88	5.4711
240	0.04682	243.80	5.7933	0.03496	242.08	5.7021	0.02312	238.66	5.5702
260	0.05099	265.36	5.8796	0.03814	263.90	5.7894	0.02531	261.02	5.6597
280	0.05512	286.78	5.9590	0.04128	285.53	5.8696	0.02746	283.09	5.7414
300	0.05922	308.10	6.0325	0.04440	307.03	5.9438	0.02958	304.94	5.8168
350	0.06940	361.13	6.1960	0.05209	360.39	6.1083	0.03480	358.96	5.9834
400	0.07949	413.96	6.3371	0.05971	413.47	6.2500	0.03993	412.50	6.1264
450	0.08953	466.82	6.4616	0.06727	466.49	6.3750	0.04502	465.87	6.2521
500	0.09953	519.84	6.5733	0.07480	519.65	6.4870	0.05008	519.29	6.3647
600	0.11948	626.92	6.7685	0.08980	626.93	6.6825	0.06013	626.95	6.5609
700	0.13937	735.94	6.9365	0.10474	736.07	6.8507	0.07012	736.35	6.7295
800	0.15923	847.22	7.0851	0.11965	847.45	6.9994	0.08008	847.92	6.8785
900	0.17906	960.83	7.2189	0.13454	961.13	7.1333	0.09003	961.73	7.0125
1000	0.19889	1076.65	7.3409	0.14942	1077.01	7.2553	0.09996	1077.72	7.1347

TABLE B.8.2 SI *Superheated Nitrogen*

Temp. K	v m^3/kg	h kJ/kg	s kJ/kg K	v m^3/kg	h kJ/kg	s kJ/kg K	v m^3/kg	h kJ/kg	s kJ/kg K
	6000 kPa			8000 kPa			10000 kPa		
140	0.002941	47.44	4.2926	0.002224	27.78	4.1167	0.002003	20.87	4.0373
160	0.005556	112.16	4.7292	0.003748	91.80	4.5453	0.002908	76.52	4.4088
180	0.007309	148.02	4.9411	0.005193	134.69	4.7988	0.004021	122.65	4.6813
200	0.008771	177.29	5.0955	0.006387	167.47	4.9717	0.005014	158.35	4.8697
220	0.010095	203.77	5.2217	0.007449	196.07	5.1082	0.005902	188.88	5.0153
240	0.011337	228.73	5.3303	0.008433	222.48	5.2231	0.006721	216.64	5.1362
260	0.012526	252.73	5.4264	0.009367	247.55	5.3235	0.007495	242.72	5.2406
280	0.013678	276.09	5.5130	0.010264	271.74	5.4131	0.008235	267.69	5.3331
300	0.014803	298.99	5.5920	0.011135	295.32	5.4945	0.008952	291.90	5.4167
350	0.017532	354.95	5.7646	0.013236	352.51	5.6709	0.010670	350.26	5.5967
400	0.020187	409.83	5.9111	0.015264	408.24	5.8197	0.012320	406.79	5.7477
450	0.022794	464.19	6.0392	0.017248	463.22	5.9492	0.013927	462.36	5.8786
500	0.025370	518.37	6.1534	0.019202	517.88	6.0644	0.015507	517.48	5.9948
600	0.030463	627.12	6.3516	0.023053	627.32	6.2639	0.018611	627.58	6.1955
700	0.035506	737.27	6.5214	0.026856	737.94	6.4344	0.021669	738.65	6.3667
800	0.040519	849.37	6.6710	0.030631	850.38	6.5845	0.024700	851.43	6.5172
900	0.045514	963.59	6.8055	0.034388	964.86	6.7194	0.027714	966.15	6.6523
1000	0.050495	1079.88	6.9281	0.038132	1081.35	6.8421	0.030715	1082.84	6.7753
	15000 kPa			20000 kPa			50000 kPa		
140	0.001770	14.81	3.9273	0.001655	13.75	3.8587	0.001391	28.05	3.6405
160	0.002183	59.14	4.2232	0.001929	53.63	4.1250	0.001497	61.62	3.8647
180	0.002749	102.34	4.4778	0.002281	93.02	4.3570	0.001612	94.31	4.0573
200	0.003365	140.60	4.6796	0.002687	130.17	4.5529	0.001736	126.15	4.2250
220	0.003964	174.10	4.8394	0.003108	164.26	4.7154	0.001867	157.12	4.3726
240	0.004531	204.33	4.9710	0.003525	195.59	4.8518	0.002003	187.24	4.5037
260	0.005071	232.41	5.0834	0.003930	224.82	4.9689	0.002143	216.53	4.6209
280	0.005589	259.01	5.1820	0.004323	252.50	5.0714	0.002285	245.02	4.7266
300	0.006088	284.56	5.2702	0.004704	279.01	5.1629	0.002428	272.78	4.8223
350	0.007280	345.47	5.4581	0.005617	341.86	5.3568	0.002786	339.44	5.0280
400	0.008416	403.79	5.6139	0.006487	401.65	5.5166	0.003138	403.08	5.1980
450	0.009517	460.71	5.7480	0.007329	459.70	5.6534	0.003484	464.64	5.3431
500	0.010593	516.88	5.8664	0.008149	516.78	5.7737	0.003823	524.82	5.4699
600	0.012697	628.50	6.0699	0.009748	629.76	5.9797	0.004484	642.94	5.6853
700	0.014759	740.63	6.2427	0.011310	742.85	6.1540	0.005129	760.04	5.8658
800	0.016797	854.18	6.3943	0.012849	857.11	6.3065	0.005762	877.47	6.0226
900	0.018818	969.50	6.5301	0.014374	972.98	6.4430	0.006385	995.87	6.1621
1000	0.020828	1086.64	6.6535	0.015887	1090.55	6.5668	0.007001	1115.51	6.2881

TABLE B.9 SI *Thermodynamic Properties of Oxygen*

TABLE B.9.1 SI *Saturated Oxygen*

Temp. K T	Press. kPa P	Specific Volume, m³/kg Sat. Liquid v_f	Evap. v_{fg}	Sat. Vapor v_g	Internal Energy, kJ/kg Sat. Liquid u_f	Evap. u_{fg}	Sat. Vapor u_g
54.4	0.1	0.000766	96.54192	96.54268	-193.55	228.60	35.05
55	0.2	0.000767	80.00808	80.00885	-192.48	227.94	35.46
60	0.7	0.000780	21.46087	21.46165	-184.12	222.80	38.68
65	2.3	0.000794	7.21804	7.21884	-175.75	217.62	41.87
70	6.3	0.000808	2.89166	2.89247	-167.36	212.40	45.04
75	14.5	0.000824	1.32849	1.32931	-158.97	207.15	48.18
80	30.1	0.000840	0.68015	0.68099	-150.57	201.82	51.24
85	56.8	0.000857	0.37962	0.38047	-142.16	196.35	54.19
90	99.4	0.000876	0.22706	0.22794	-133.71	190.68	56.97
90.2	101.3	0.000876	0.22298	0.22386	-133.39	190.46	57.07
95	163.1	0.000895	0.14361	0.14450	-125.21	184.74	59.53
100	254.0	0.000917	0.09501	0.09592	-116.62	178.47	61.85
105	378.5	0.000940	0.06518	0.06612	-107.93	171.81	63.88
110	543.4	0.000966	0.04602	0.04699	-99.11	164.69	65.58
115	755.6	0.000994	0.03324	0.03424	-90.11	157.02	66.91
120	1022.3	0.001027	0.02441	0.02544	-80.90	148.70	67.81
125	1350.9	0.001064	0.01813	0.01919	-71.40	139.59	68.19
130	1749.1	0.001108	0.01352	0.01463	-61.54	129.48	67.94
135	2225.0	0.001161	0.01004	0.01120	-51.18	118.05	66.87
140	2787.8	0.001230	0.00734	0.00856	-40.07	104.72	64.66
145	3447.7	0.001324	0.00513	0.00646	-27.73	88.36	60.63
150	4218.6	0.001480	0.00317	0.00465	-12.86	65.85	52.99
154.6	5043.0	0.002293	0	0.00229	20.92	0	20.92

TABLE B.9.2 SI *Superheated Oxygen*

Temp. K	v m³/kg	h kJ/kg	s kJ/kg K	v m³/kg	h kJ/kg	s kJ/kg K	v m³/kg	h kJ/kg	s kJ/kg K
	100 kPa (90.06)			200 kPa (97.24)			400 kPa (105.73)		
Sat.	0.22658	79.66	5.3065	0.11970	84.54	5.1860	0.06278	89.26	5.0648
100	0.25373	89.05	5.4054	0.12365	87.25	5.2135	—	—	—
120	0.30746	107.67	5.5752	0.15150	106.40	5.3881	0.07343	103.73	5.1932
140	0.36054	126.17	5.7178	0.17863	125.21	5.5331	0.08764	123.25	5.3437
160	0.41325	144.56	5.8406	0.20537	143.81	5.6574	0.10142	142.29	5.4709
180	0.46575	162.90	5.9486	0.23189	162.29	5.7662	0.11496	161.06	5.5814
200	0.51811	181.20	6.0450	0.25828	180.70	5.8631	0.12836	179.68	5.6795
220	0.57039	199.49	6.1321	0.28457	199.06	5.9507	0.14166	198.20	5.7678
240	0.62260	217.78	6.2117	0.31081	217.41	6.0305	0.15491	216.68	5.8481
260	0.67477	236.08	6.2850	0.33699	235.76	6.1039	0.16810	235.12	5.9220

(Continued)

TABLE B.9.1 SI (Continued) *Saturated Oxygen*

Temp.	Press.	Enthalpy, kJ/kg			Entropy, kJ/kg K		
K	kPa	Sat. Liquid	Evap.	Sat. Vapor	Sat. Liquid	Evap.	Sat. Vapor
T	P	h_f	h_{fg}	h_g	s_f	s_{fg}	s_g
54.4	0.1	-193.55	242.72	49.17	2.0922	4.4650	6.5572
55	0.2	-192.48	242.23	49.75	2.1117	4.4042	6.5159
60	0.7	-184.12	238.38	54.25	2.2571	3.9729	6.2301
65	2.3	-175.74	234.47	58.73	2.3912	3.6072	5.9985
70	6.3	-167.35	230.51	63.16	2.5156	3.2930	5.8086
75	14.5	-158.96	226.48	67.52	2.6313	3.0197	5.6510
80	30.1	-150.55	222.31	71.76	2.7397	2.7788	5.5185
85	56.8	-142.11	217.92	75.81	2.8417	2.5638	5.4055
90	99.4	-133.63	213.24	79.61	2.9383	2.3693	5.3076
90.2	101.3	-133.31	213.06	79.75	2.9418	2.3624	5.3042
95	163.1	-125.06	208.16	83.10	3.0303	2.1912	5.2215
100	254.0	-116.39	202.60	86.22	3.1184	2.0260	5.1445
105	378.5	-107.58	196.48	88.91	3.2033	1.8713	5.0745
110	543.4	-98.58	189.70	91.11	3.2855	1.7245	5.0100
115	755.6	-89.36	182.13	92.78	3.3656	1.5838	4.9494
120	1022.3	-79.85	173.66	93.81	3.4443	1.4472	4.8914
125	1350.9	-69.96	164.08	94.11	3.5221	1.3126	4.8348
130	1749.1	-59.60	153.13	93.52	3.6000	1.1779	4.7779
135	2225.0	-48.59	140.39	91.80	3.6790	1.0400	4.7190
140	2787.8	-36.64	125.17	88.53	3.7611	0.8941	4.6552
145	3447.7	-23.16	106.05	82.89	3.8497	0.7314	4.5811
150	4218.6	-6.61	79.23	72.62	3.9545	0.5282	4.4828
154.6	5043.0	32.48	0	32.48	4.2008	0	4.2008

TABLE B.9.2 SI (Continued) *Superheated Oxygen*

Temp. K	v m³/kg	h kJ/kg	s kJ/kg K	v m³/kg	h kJ/kg	s kJ/kg K	v m³/kg	h kJ/kg	s kJ/kg K
	100 kPa (90.06)			200 kPa (97.24)			400 kPa (105.73)		
280	0.72691	254.41	6.3529	0.36314	254.13	6.1720	0.18126	253.57	5.9904
300	0.77902	272.78	6.4162	0.38927	272.54	6.2355	0.19439	272.05	6.0541
320	0.83110	291.21	6.4757	0.41537	290.99	6.2950	0.20751	290.56	6.1138
340	0.88317	309.71	6.5318	0.44146	309.52	6.3512	0.22060	309.14	6.1701
360	0.93523	328.29	6.5849	0.46753	328.12	6.4044	0.23369	327.78	6.2234
380	0.98728	346.96	6.6354	0.49360	346.81	6.4549	0.24676	346.51	6.2740
400	1.03931	365.74	6.6835	0.51965	365.61	6.5031	0.25982	365.34	6.3223
420	1.09134	384.63	6.7296	0.54569	384.51	6.5492	0.27287	384.28	6.3685
440	1.14337	403.64	6.7738	0.57173	403.53	6.5935	0.28592	403.32	6.4128

(Continued)

TABLE B.9.2 SI (Continued) *Superheated Oxygen*

Temp. *K*	*v* m³/kg	*h* kJ/kg	*s* kJ/kg K	*v* m³/kg	*h* kJ/kg	*s* kJ/kg K	*v* m³/kg	*h* kJ/kg	*s* kJ/kg K
	600 kPa (111.46)			800 kPa (115.91)			1000 kPa (119.62)		
Sat.	0.04275	91.66	4.9920	0.03239	93.02	4.9387	0.02600	93.76	4.8958
120	0.04731	100.86	5.0716	0.03415	97.74	4.9788	0.02615	94.24	4.8998
140	0.05727	121.21	5.2286	0.04206	119.09	5.1435	0.03291	116.88	5.0746
160	0.06675	140.74	5.3589	0.04941	139.14	5.2774	0.03899	137.52	5.2125
180	0.07598	159.82	5.4714	0.05648	158.56	5.3918	0.04478	157.28	5.3289
200	0.08505	178.65	5.5706	0.06339	177.62	5.4922	0.05040	176.58	5.4306
220	0.09403	197.34	5.6596	0.07021	196.47	5.5821	0.05592	195.60	5.5213
240	0.10294	215.94	5.7405	0.07696	215.20	5.6635	0.06137	214.45	5.6033
260	0.11181	234.49	5.8148	0.08366	233.85	5.7382	0.06677	233.21	5.6783
280	0.12064	253.02	5.8835	0.09032	252.46	5.8072	0.07214	251.91	5.7476
300	0.12944	271.56	5.9474	0.09696	271.07	5.8714	0.07748	270.59	5.8121
320	0.13822	290.13	6.0074	0.10358	289.70	5.9315	0.08279	289.28	5.8724
340	0.14699	308.76	6.0638	0.11018	308.38	5.9881	0.08809	308.00	5.9291
360	0.15574	327.45	6.1172	0.11676	327.11	6.0416	0.09338	326.77	5.9828
380	0.16448	346.21	6.1680	0.12334	345.92	6.0925	0.09866	345.62	6.0337
400	0.17321	365.08	6.2163	0.12990	364.81	6.1409	0.10392	364.55	6.0823
420	0.18193	384.04	6.2626	0.13646	383.80	6.1873	0.10918	383.57	6.1287
440	0.19065	403.11	6.3069	0.14301	402.90	6.2317	0.11443	402.70	6.1732
	1500 kPa (126.98)			2000 kPa (132.74)			3000 kPa (141.69)		
Sat.	0.01721	94.00	4.8124	0.01263	92.74	4.7460	0.00780	86.95	4.6316
140	0.02060	110.86	4.9390	0.01431	103.86	4.8277	—	—	—
160	0.02507	133.26	5.0887	0.01807	128.71	4.9939	0.01097	118.42	4.8417
180	0.02916	154.00	5.2109	0.02134	150.60	5.1229	0.01349	143.38	4.9890
200	0.03307	173.93	5.3160	0.02440	171.23	5.2316	0.01572	165.64	5.1063
220	0.03686	193.40	5.4087	0.02733	191.17	5.3267	0.01781	186.63	5.2064
240	0.04058	212.59	5.4922	0.03019	210.71	5.4117	0.01981	206.92	5.2947
260	0.04426	231.60	5.5683	0.03300	230.00	5.4889	0.02175	226.77	5.3741
280	0.04789	250.52	5.6384	0.03577	249.12	5.5598	0.02365	246.34	5.4467
300	0.05150	269.37	5.7035	0.03851	268.16	5.6254	0.02553	265.74	5.5136
320	0.05508	288.21	5.7642	0.04123	287.14	5.6867	0.02738	285.02	5.5758
340	0.05865	307.05	5.8214	0.04393	306.12	5.7442	0.02922	304.25	5.6341
360	0.06220	325.94	5.8753	0.04662	325.11	5.7985	0.03104	323.46	5.6890
380	0.06575	344.88	5.9265	0.04930	344.14	5.8500	0.03285	342.69	5.7409
400	0.06928	363.89	5.9753	0.05196	363.24	5.8989	0.03465	361.95	5.7903
420	0.07281	382.99	6.0219	0.05462	382.41	5.9457	0.03644	381.26	5.8375
440	0.07633	402.18	6.0665	0.05728	401.66	5.9905	0.03823	400.65	5.8826

(Continued)

TABLE B.9.2 SI (Continued) ***Superheated Oxygen***

Temp. K	v m³/kg	h kJ/kg	s kJ/kg K	v m³/kg	h kJ/kg	s kJ/kg K	v m³/kg	h kJ/kg	s kJ/kg K
	6000 kPa			8000 kPa			10000 kPa		
140	0.001167	-39.45	3.7136	0.001141	-40.21	3.6917	0.001120	-40.58	3.6729
160	0.002570	49.33	4.2932	0.001527	10.75	4.0300	0.001397	4.12	3.9703
180	0.005511	117.07	4.6992	0.003458	94.23	4.5235	0.002363	70.41	4.3596
200	0.007027	147.30	4.8588	0.004862	133.80	4.7329	0.003602	119.78	4.6210
220	0.008293	172.41	4.9786	0.005936	162.55	4.8701	0.004548	152.63	4.7779
240	0.009444	195.34	5.0784	0.006875	187.54	4.9789	0.005356	179.83	4.8963
260	0.010527	217.07	5.1654	0.007742	210.65	5.0715	0.006089	204.36	4.9946
280	0.011565	238.06	5.2432	0.008562	232.65	5.1530	0.006776	227.38	5.0799
300	0.012572	258.59	5.3140	0.009351	253.95	5.2265	0.007431	249.46	5.1561
320	0.013557	278.80	5.3792	0.010116	274.78	5.2937	0.008063	270.90	5.2253
340	0.014523	298.80	5.4398	0.010864	295.29	5.3559	0.008678	291.92	5.2890
360	0.015476	318.66	5.4966	0.011598	315.59	5.4139	0.009280	312.64	5.3482
380	0.016417	338.45	5.5501	0.012321	335.75	5.4684	0.009871	333.15	5.4037
400	0.017349	358.20	5.6008	0.013035	355.82	5.5199	0.010453	353.54	5.4559
420	0.018274	377.95	5.6489	0.013742	375.85	5.5687	0.011028	373.84	5.5055
440	0.019192	397.72	5.6949	0.014442	395.87	5.6153	0.011597	394.10	5.5526
	15000 kPa			20000 kPa			50000 kPa		
140	0.001081	-40.50	3.6342	0.001052	-39.57	3.6028	0.000954	-27.33	3.4768
160	0.001261	-1.97	3.8911	0.001192	-3.80	3.8414	0.001023	3.29	3.6812
180	0.001600	42.87	4.1547	0.001405	34.75	4.0683	0.001103	33.78	3.8608
200	0.002183	90.42	4.4054	0.001726	75.27	4.2817	0.001195	64.06	4.0203
220	0.002827	129.89	4.5938	0.002130	113.74	4.4652	0.001298	93.97	4.1629
240	0.003414	161.90	4.7333	0.002550	147.62	4.6127	0.001411	123.31	4.2905
260	0.003948	189.73	4.8447	0.002953	177.53	4.7325	0.001532	151.90	4.4050
280	0.004444	215.13	4.9389	0.003336	204.71	4.8333	0.001658	179.64	4.5078
300	0.004912	239.03	5.0214	0.003699	230.05	4.9207	0.001788	206.50	4.6005
320	0.005360	261.91	5.0952	0.004048	254.14	4.9985	0.001919	232.54	4.6845
340	0.005794	284.11	5.1625	0.004384	277.34	5.0688	0.002050	257.84	4.7612
360	0.006215	305.82	5.2246	0.004711	299.90	5.1333	0.002181	282.51	4.8317
380	0.006627	327.18	5.2823	0.005029	322.00	5.1930	0.002311	306.64	4.8969
400	0.007031	348.29	5.3365	0.005341	343.75	5.2488	0.002439	330.32	4.9577
420	0.007428	369.23	5.3876	0.005647	365.25	5.3013	0.002567	353.65	5.0146
440	0.007820	390.06	5.4360	0.005949	386.58	5.3509	0.002693	376.67	5.0682

TABLE B.10 SI *Thermodynamic Properties of Propane*
TABLE B.10.1 SI *Saturated Propane*

Temp.	Press.	Specific Volume, m³/kg			Internal Energy, kJ/kg		
K	kPa	Sat. Liquid	Evap.	Sat. Vapor	Sat. Liquid	Evap.	Sat. Vapor
T	P	v_f	v_{fg}	v_g	u_f	u_{fg}	u_g
85.5	1.65E-7	0.001364	9.57E7	9.57E7	-495.72	546.50	50.78
100	2.52E-5	0.001392	7.47E5	7.47E5	-467.82	529.12	61.30
120	2.95E-3	0.001432	7.66E3	7.66E3	-428.94	505.99	77.05
140	7.87E-2	0.001475	3.354E2	3.354E2	-389.60	483.70	94.10
160	0.847	0.001521	35.58281	35.58433	-349.68	461.99	112.31
180	5.0	0.001570	6.68856	6.69013	-308.95	440.52	131.56
200	20.1	0.001624	1.84775	1.84938	-267.18	418.86	151.68
220	60.5	0.001684	0.66584	0.66753	-224.06	396.52	172.47
231.1	101.3	0.001720	0.41170	0.41342	-199.50	383.69	184.18
240	147.9	0.001752	0.28900	0.29076	-179.27	372.99	193.72
260	310.5	0.001831	0.14292	0.14475	-132.45	347.64	215.19
280	581.5	0.001925	0.07727	0.07919	-83.17	319.69	236.52
300	997.4	0.002043	0.04414	0.04618	-30.84	287.94	257.10
320	1598.4	0.002200	0.02573	0.02793	25.38	250.39	275.78
340	2430.6	0.002433	0.01450	0.01693	87.18	202.82	290.00
360	3554.0	0.002894	0.00660	0.00949	160.40	130.66	291.07
369.9	4247.7	0.004535	0	0.00454	240.34	0	240.34

TABLE B.10.2 SI *Superheated Propane*

Temp. K	v m³/kg	h kJ/kg	s kJ/kg K	v m³/kg	h kJ/kg	s kJ/kg K	v m³/kg	h kJ/kg	s kJ/kg K
	50 kPa (216.22)			100 kPa (230.77)			200 kPa (247.72)		
Sat.	0.79617	208.31	5.7715	0.41851	225.72	5.7227	0.21917	245.83	5.6826
240	0.89026	241.87	5.9187	0.43749	239.29	5.7804	—	—	—
260	0.96824	271.64	6.0378	0.47787	269.57	5.9016	0.23242	265.25	5.7591
280	1.04557	302.97	6.1539	0.51755	301.27	6.0190	0.25338	297.76	5.8795
300	1.12246	335.96	6.2677	0.55677	334.53	6.1337	0.27382	331.61	5.9962
320	1.19904	370.68	6.3797	0.59566	369.45	6.2463	0.29391	366.96	6.1103
340	1.27540	407.16	6.4902	0.63433	406.09	6.3574	0.31375	403.93	6.2223
360	1.35160	445.41	6.5995	0.67283	444.47	6.4670	0.33341	442.57	6.3327
380	1.42767	485.45	6.7077	0.71120	484.61	6.5755	0.35294	482.91	6.4418
400	1.50365	527.25	6.8149	0.74947	526.49	6.6829	0.37237	524.96	6.5496

(Continued)

TABLE B.10.1 SI (Continued) ***Saturated Propane***

Temp. K T	Press. kPa P	Enthalpy, kJ/kg Sat. Liquid h_f	Evap. h_{fg}	Sat. Vapor h_g	Entropy, kJ/kg K Sat. Liquid s_f	Evap. s_{fg}	Sat. Vapor s_g
85.5	1.65E-7	-495.72	562.62	66.90	1.8813	6.5826	8.4639
100	2.52E-5	-467.82	547.98	80.15	2.1826	5.4798	7.6624
120	2.95E-3	-428.94	528.61	99.67	2.5370	4.4051	6.9421
140	7.87E-2	-389.60	510.09	120.49	2.8402	3.6435	6.4837
160	0.847	-349.68	492.12	142.44	3.1067	3.0757	6.1824
180	5.0	-308.95	474.30	165.35	3.3465	2.6350	5.9814
200	20.1	-267.14	456.07	188.92	3.5665	2.2803	5.8468
220	60.5	-223.95	436.79	212.83	3.7719	1.9854	5.7573
231.1	101.3	-199.33	425.40	226.07	3.8808	1.8411	5.7219
240	147.9	-179.01	415.72	236.71	3.9668	1.7322	5.6989
260	310.5	-131.88	392.02	260.14	4.1542	1.5078	5.6619
280	581.5	-82.05	364.62	282.57	4.3369	1.3022	5.6391
300	997.4	-28.81	331.97	303.16	4.5176	1.1066	5.6242
320	1598.4	28.90	291.52	320.42	4.6996	0.9110	5.6106
340	2430.6	93.09	238.06	331.15	4.8883	0.7002	5.5884
360	3554.0	170.69	154.11	324.79	5.1013	0.4281	5.5293
369.9	4247.7	259.61	0	259.61	5.3376	0	5.3376

TABLE B.10.2 SI (Continued) ***Superheated Propane***

Temp. K	v m³/kg	h kJ/kg	s kJ/kg K	v m³/kg	h kJ/kg	s kJ/kg K	v m³/kg	h kJ/kg	s kJ/kg K
	50 kPa (216.22)			100 kPa (230.77)			200 kPa (247.72)		
420	1.57956	570.79	6.9211	0.78767	570.10	6.7893	0.39172	568.72	6.6563
440	1.65540	616.06	7.0264	0.82581	615.43	6.8947	0.41101	614.16	6.7620
460	1.73119	663.01	7.1307	0.86389	662.43	6.9991	0.43024	661.27	6.8667
480	1.80694	711.61	7.2341	0.90194	711.08	7.1027	0.44943	710.00	6.9704
500	1.88266	761.83	7.3366	0.93995	761.33	7.2052	0.46859	760.34	7.0731
520	1.95835	813.62	7.4382	0.97793	813.16	7.3068	0.48772	812.23	7.1749
540	2.03402	866.95	7.5388	1.01589	866.52	7.4075	0.50683	865.65	7.2757
560	2.10966	921.77	7.6385	1.05382	921.37	7.5073	0.52591	920.56	7.3755
580	2.18529	978.05	7.7372	1.09174	977.67	7.6060	0.54497	976.91	7.4744
600	2.26090	1035.74	7.8350	1.12964	1035.38	7.7039	0.56402	1034.67	7.5722

(Continued)

TABLE B.10.2 SI (Continued) ***Superheated Propane***

Temp. K	v m³/kg	h kJ/kg	s kJ/kg K	v m³/kg	h kJ/kg	s kJ/kg K	v m³/kg	h kJ/kg	s kJ/kg K
	300 kPa (258.99)			400 kPa (267.69)			500 kPa (274.89)		
Sat.	0.14956	258.97	5.6634	0.11371	268.92	5.6518	0.09172	276.97	5.6439
280	0.16516	294.08	5.7937	0.12092	290.21	5.7296	0.09424	286.11	5.6768
300	0.17940	328.57	5.9127	0.13211	325.42	5.8510	0.10367	322.14	5.8011
320	0.19326	364.39	6.0282	0.14288	361.75	5.9682	0.11261	359.03	5.9202
340	0.20684	401.71	6.1413	0.15336	399.45	6.0825	0.12124	397.14	6.0356
360	0.22024	440.63	6.2525	0.16363	438.66	6.1945	0.12965	436.65	6.1485
380	0.23350	481.19	6.3622	0.17377	479.44	6.3047	0.13791	477.67	6.2594
400	0.24665	523.42	6.4705	0.18379	521.86	6.4135	0.14606	520.28	6.3687
420	0.25973	567.32	6.5775	0.19372	565.91	6.5210	0.15411	564.49	6.4765
440	0.27273	612.89	6.6835	0.20359	611.60	6.6272	0.16210	610.31	6.5831
460	0.28569	660.10	6.7884	0.21341	658.92	6.7324	0.17003	657.74	6.6885
480	0.29860	708.92	6.8923	0.22318	707.84	6.8365	0.17792	706.75	6.7928
500	0.31147	759.34	6.9952	0.23291	758.33	6.9395	0.18577	757.32	6.8960
520	0.32432	811.30	7.0971	0.24262	810.37	7.0415	0.19359	809.43	6.9982
540	0.33714	864.78	7.1980	0.25229	863.91	7.1426	0.20139	863.04	7.0993
560	0.34994	919.75	7.2979	0.26195	918.93	7.2426	0.20916	918.12	7.1994
580	0.36271	976.15	7.3969	0.27159	975.39	7.3417	0.21691	974.62	7.2986
600	0.37548	1033.95	7.4949	0.28121	1033.24	7.4397	0.22465	1032.52	7.3967
	600 kPa (281.08)			700 kPa (286.56)			800 kPa (291.48)		
Sat.	0.07680	283.74	5.6381	0.06598	289.58	5.6336	0.05776	294.69	5.6299
300	0.08463	318.71	5.7585	0.07097	315.12	5.7207	0.06066	311.33	5.6862
320	0.09239	356.22	5.8795	0.07791	353.32	5.8440	0.06701	350.31	5.8120
340	0.09980	394.77	5.9963	0.08446	392.33	5.9622	0.07294	389.84	5.9318
360	0.10698	434.60	6.1102	0.09077	432.51	6.0770	0.07860	430.39	6.0477
380	0.11400	475.88	6.2217	0.09691	474.05	6.1893	0.08408	472.20	6.1607
400	0.12090	518.68	6.3315	0.10292	517.06	6.2996	0.08943	515.43	6.2715
420	0.12770	563.05	6.4397	0.10884	561.60	6.4082	0.09468	560.14	6.3806
440	0.13444	609.01	6.5466	0.11468	607.69	6.5154	0.09985	606.37	6.4881
460	0.14112	656.54	6.6522	0.12046	655.35	6.6213	0.10497	654.14	6.5943
480	0.14775	705.65	6.7567	0.12620	704.55	6.7260	0.11004	703.45	6.6992
500	0.15435	756.31	6.8601	0.13190	755.30	6.8296	0.11507	754.28	6.8029
520	0.16091	808.49	6.9625	0.13757	807.55	6.9321	0.12006	806.61	6.9055
540	0.16745	862.17	7.0637	0.14321	861.29	7.0335	0.12503	860.41	7.0071
560	0.17397	917.30	7.1640	0.14883	916.48	7.1338	0.12998	915.66	7.1075
580	0.18046	973.86	7.2632	0.15443	973.09	7.2331	0.13490	972.33	7.2069
600	0.18694	1031.80	7.3614	0.16001	1031.08	7.3314	0.13981	1030.37	7.3053

(Continued)

TABLE B.10.2 SI (Continued) *Superheated Propane*

Temp. K	v m³/kg	h kJ/kg	s kJ/kg K	v m³/kg	h kJ/kg	s kJ/kg K	v m³/kg	h kJ/kg	s kJ/kg K
	1000 kPa (300.10)			2000 kPa (330.42)			4000 kPa (366.53)		
Sat.	0.04606	303.26	5.6241	0.02157	327.17	5.6011	0.00715	308.67	5.4748
340	0.05675	384.63	5.8787	0.02353	352.08	5.6754	—	—	—
360	0.06153	426.00	5.9969	0.02695	400.50	5.8139	—	—	—
380	0.06610	468.41	6.1115	0.02990	447.30	5.9404	0.01064	383.27	5.6754
400	0.07053	512.09	6.2236	0.03259	494.04	6.0602	0.01312	447.49	5.8402
420	0.07485	557.16	6.3335	0.03511	541.37	6.1757	0.01500	503.81	5.9777
440	0.07910	603.69	6.4417	0.03753	589.66	6.2880	0.01663	557.89	6.1034
460	0.08328	651.71	6.5484	0.03986	639.09	6.3979	0.01811	611.45	6.2225
480	0.08741	701.22	6.6538	0.04213	689.76	6.5057	0.01949	665.27	6.3370
500	0.09150	752.23	6.7579	0.04435	741.75	6.6118	0.02080	719.74	6.4482
520	0.09555	804.71	6.8608	0.04653	795.07	6.7163	0.02206	775.09	6.5567
540	0.09958	858.65	6.9626	0.04869	849.73	6.8195	0.02328	831.45	6.6631
560	0.10358	914.02	7.0632	0.05081	905.74	6.9213	0.02447	888.91	6.7676
580	0.10757	970.79	7.1628	0.05291	963.07	7.0219	0.02564	947.50	6.8703
600	0.11153	1028.93	7.2614	0.05499	1021.71	7.1213	0.02678	1007.24	6.9716
	10000 kPa			20000 kPa			50000 kPa		
380	0.002618	207.08	5.1520	0.002290	194.83	5.0560	0.001991	207.83	4.9236
400	0.002999	278.45	5.3350	0.002437	254.18	5.2082	0.002056	261.94	5.0623
420	0.003608	356.92	5.5263	0.002609	315.64	5.3581	0.002124	317.54	5.1979
440	0.004454	437.64	5.7141	0.002809	378.97	5.5054	0.002196	374.58	5.3306
460	0.005371	513.72	5.8833	0.003037	443.82	5.6495	0.002272	433.00	5.4604
480	0.006225	583.73	6.0323	0.003291	509.79	5.7899	0.002351	492.72	5.5875
500	0.006996	649.74	6.1670	0.003568	576.48	5.9260	0.002432	553.69	5.7119
520	0.007700	713.66	6.2924	0.003861	643.58	6.0576	0.002517	615.85	5.8338
540	0.008353	776.65	6.4113	0.004164	710.85	6.1846	0.002604	679.14	5.9532
560	0.008968	839.42	6.5254	0.004470	778.21	6.3070	0.002693	743.52	6.0703
580	0.009553	902.39	6.6359	0.004775	845.66	6.4254	0.002784	808.94	6.1851
600	0.010115	965.84	6.7434	0.005076	913.27	6.5400	0.002876	875.37	6.2977

TABLE B.11 SI *Thermodynamic Properties of R-11*
TABLE B.11.1 SI *Saturated R-11*

Temp.	Press.	Specific Volume, m^3/kg			Internal Energy, kJ/kg		
C	kPa	Sat. Liquid	Evap.	Sat. Vapor	Sat. Liquid	Evap.	Sat. Vapor
T	P	v_f	v_{fg}	v_g	u_f	u_{fg}	u_g
-70	0.6	0.000593	21.33962	21.34021	-25.38	201.41	176.03
-60	1.3	0.000601	9.95986	9.96046	-16.91	197.09	180.18
-50	2.7	0.000608	5.04550	5.04611	-8.45	192.88	184.43
-40	5.1	0.000616	2.74106	2.74168	0.00	188.76	188.76
-30	9.2	0.000625	1.58111	1.58174	8.46	184.71	193.17
-20	15.8	0.000633	0.96030	0.96093	16.94	180.70	197.64
-10	25.7	0.000642	0.60981	0.61045	25.44	176.71	202.16
0	40.2	0.000652	0.40247	0.40312	33.99	172.72	206.70
10	60.6	0.000662	0.27467	0.27534	42.57	168.70	211.27
20	88.4	0.000672	0.19299	0.19366	51.20	164.64	215.84
23.8	101.3	0.000676	0.16983	0.17051	54.50	163.09	217.59
30	125.4	0.000683	0.13908	0.13976	59.87	160.54	220.41
40	173.5	0.000695	0.10246	0.10315	68.59	156.37	224.96
50	234.6	0.000707	0.07693	0.07764	77.35	152.12	229.48
60	311.1	0.000720	0.05873	0.05945	86.18	147.78	233.96
70	405.2	0.000734	0.04548	0.04621	95.07	143.31	238.38
80	519.2	0.000749	0.03564	0.03639	104.05	138.70	242.75
90	655.8	0.000766	0.02821	0.02897	113.13	133.90	247.03
100	817.7	0.000784	0.02250	0.02329	122.33	128.87	251.21
110	1007.6	0.000804	0.01806	0.01886	131.70	123.56	255.26
120	1228.4	0.000826	0.01455	0.01537	141.27	117.89	259.17
130	1483.4	0.000850	0.01173	0.01258	151.11	111.76	262.87
140	1775.8	0.000879	0.00944	0.01032	161.29	105.02	266.32
150	2109.1	0.000913	0.00755	0.00846	171.93	97.49	269.41
160	2487.0	0.000954	0.00596	0.00691	183.17	88.85	272.02
170	2913.8	0.001007	0.00458	0.00559	195.27	78.61	273.88
180	3393.9	0.001081	0.00335	0.00443	208.73	65.75	274.48
190	3932.4	0.001208	0.00211	0.00332	224.91	47.33	272.24
198.0	4400.2	0.001806	0	0.00181	252.76	0	252.76

(Continued)

TABLE B.11.1 SI (Continued) *Saturated R-11*

Temp.	Press.	Enthalpy, kJ/kg			Entropy, kJ/kg K		
C T	kPa P	Sat. Liquid h_f	Evap. h_{fg}	Sat. Vapor h_g	Sat. Liquid s_f	Evap. s_{fg}	Sat. Vapor s_g
-70	0.6	-25.38	213.70	188.32	-0.1165	1.0519	0.9354
-60	1.3	-16.91	209.97	193.06	-0.0758	0.9851	0.9093
-50	2.7	-8.45	206.35	197.90	-0.0370	0.9247	0.8877
-40	5.1	0	202.81	202.81	0	0.8699	0.8699
-30	9.2	8.46	199.32	207.78	0.0355	0.8198	0.8553
-20	15.8	16.95	195.86	212.80	0.0697	0.7737	0.8434
-10	25.7	25.46	192.39	217.85	0.1027	0.7311	0.8338
0	40.2	34.01	188.89	222.90	0.1345	0.6915	0.8260
10	60.6	42.61	185.33	227.94	0.1654	0.6545	0.8199
20	88.4	51.26	181.70	232.96	0.1953	0.6198	0.8152
23.8	101.3	54.57	180.29	234.86	0.2065	0.6071	0.8137
30	125.4	59.95	177.98	237.93	0.2244	0.5871	0.8115
40	173.5	68.71	174.14	242.85	0.2527	0.5561	0.8088
50	234.6	77.52	170.18	247.70	0.2803	0.5266	0.8069
60	311.1	86.40	166.05	252.45	0.3072	0.4984	0.8056
70	405.2	95.37	161.73	257.11	0.3335	0.4713	0.8048
80	519.2	104.44	157.20	261.64	0.3593	0.4451	0.8045
90	655.8	113.63	152.40	266.03	0.3847	0.4197	0.8044
100	817.7	122.97	147.28	270.25	0.4097	0.3947	0.8044
110	1007.6	132.51	141.76	274.27	0.4346	0.3700	0.8045
120	1228.4	142.29	135.76	278.05	0.4593	0.3453	0.8046
130	1483.4	152.37	129.16	281.53	0.4841	0.3204	0.8045
140	1775.8	162.86	121.79	284.64	0.5091	0.2948	0.8039
150	2109.1	173.85	113.41	287.26	0.5347	0.2680	0.8027
160	2487.0	185.54	103.67	289.21	0.5612	0.2393	0.8005
170	2913.8	198.20	91.97	290.17	0.5891	0.2075	0.7967
180	3393.9	212.40	77.11	289.51	0.6197	0.1702	0.7899
190	3932.4	229.66	55.63	285.28	0.6560	0.1201	0.7761
198.0	4400.2	260.71	0	260.71	0.7210	0	0.7210

TABLE B.11.2 SI *Superheated R-11*

Temp. C	v m^3/kg	h kJ/kg	s kJ/kg K	v m^3/kg	h kJ/kg	s kJ/kg K	v m^3/kg	h kJ/kg	s kJ/kg K
	10 kPa (-28.58)			25 kPa (-10.60)			50 kPa (5.23)		
Sat.	1.46934	208.50	0.8534	0.62646	217.55	0.8343	0.32898	225.54	0.8227
-20	1.52216	213.01	0.8716	—	—	—	—	—	—
-10	1.58361	218.35	0.8923	0.62795	217.87	0.8355	—	—	—
0	1.64494	223.79	0.9125	0.65299	223.35	0.8559	—	—	—
10	1.70618	229.31	0.9324	0.67792	228.91	0.8759	0.33510	228.23	0.8323
20	1.76733	234.92	0.9519	0.70277	234.55	0.8955	0.34786	233.93	0.8520
30	1.82842	240.61	0.9710	0.72754	240.27	0.9147	0.36055	239.70	0.8714
40	1.88944	246.37	0.9897	0.75226	246.07	0.9335	0.37316	245.54	0.8904
50	1.95040	252.22	1.0080	0.77691	251.93	0.9519	0.38572	251.45	0.9089
60	2.01132	258.13	1.0260	0.80152	257.87	0.9700	0.39823	257.42	0.9271
70	2.07220	264.11	1.0437	0.82609	263.87	0.9878	0.41070	263.46	0.9450
80	2.13304	270.16	1.0611	0.85062	269.93	1.0052	0.42313	269.55	0.9625
90	2.19385	276.27	1.0782	0.87512	276.06	1.0223	0.43553	275.71	0.9797
100	2.25464	282.44	1.0949	0.89959	282.24	1.0391	0.44790	281.92	0.9965
110	2.31539	288.67	1.1114	0.92404	288.49	1.0556	0.46024	288.18	1.0131
120	2.37613	294.95	1.1276	0.94846	294.78	1.0718	0.47256	294.50	1.0294
130	2.43685	301.29	1.1435	0.97287	301.13	1.0878	0.48487	300.86	1.0454
	100 kPa (23.44)			200 kPa (44.62)			300 kPa (58.67)		
Sat.	0.17261	234.68	0.8138	0.09024	245.10	0.8079	0.06154	251.83	0.8058
30	0.17697	238.54	0.8267	—	—	—	—	—	—
40	0.18355	244.48	0.8459	—	—	—	—	—	—
50	0.19008	250.47	0.8648	0.09214	248.43	0.8182	—	—	—
60	0.19655	256.52	0.8832	0.09561	254.65	0.8372	0.06187	252.68	0.8083
70	0.20297	262.63	0.9013	0.09903	260.91	0.8557	0.06431	259.11	0.8273
80	0.20936	268.78	0.9190	0.10241	267.20	0.8738	0.06670	265.55	0.8458
90	0.21571	274.99	0.9363	0.10575	273.53	0.8914	0.06904	272.01	0.8639
100	0.22203	281.25	0.9533	0.10905	279.89	0.9087	0.07135	278.49	0.8815
110	0.22833	287.56	0.9700	0.11233	286.30	0.9257	0.07363	285.00	0.8987
120	0.23460	293.92	0.9864	0.11559	292.74	0.9423	0.07589	291.53	0.9155
130	0.24085	300.32	1.0025	0.11882	299.22	0.9586	0.07812	298.10	0.9320
140	0.24709	306.77	1.0183	0.12204	305.74	0.9745	0.08033	304.69	0.9482
150	0.25331	313.26	1.0338	0.12523	312.30	0.9902	0.08252	311.31	0.9640
160	0.25951	319.80	1.0490	0.12842	318.89	1.0056	0.08470	317.97	0.9796
170	0.26571	326.37	1.0640	0.13159	325.52	1.0207	0.08687	324.65	0.9948
180	0.27189	332.98	1.0788	0.13474	332.18	1.0356	0.08902	331.36	1.0098

(Continued)

TABLE B.11.2 SI (Continued) ***Superheated R-11***

Temp. C	v m³/kg	h kJ/kg	s kJ/kg K	v m³/kg	h kJ/kg	s kJ/kg K	v m³/kg	h kJ/kg	s kJ/kg K
	400 kPa (69.50)			600 kPa (86.12)			800 kPa (98.98)		
Sat.	0.04678	256.88	0.8049	0.03161	264.34	0.8044	0.02380	269.83	0.8044
90	0.05065	270.42	0.8433	0.03215	267.04	0.8118	—	—	—
100	0.05247	277.04	0.8612	0.03350	273.95	0.8306	0.02391	270.57	0.8064
110	0.05425	283.66	0.8787	0.03481	280.83	0.8488	0.02501	277.77	0.8254
120	0.05601	290.29	0.8958	0.03609	287.69	0.8665	0.02606	284.90	0.8438
130	0.05775	296.94	0.9125	0.03733	294.54	0.8837	0.02707	291.98	0.8616
140	0.05946	303.62	0.9289	0.03855	301.38	0.9004	0.02806	299.02	0.8789
150	0.06115	310.31	0.9449	0.03975	308.23	0.9168	0.02902	306.05	0.8956
160	0.06283	317.02	0.9606	0.04094	315.08	0.9328	0.02996	313.05	0.9120
170	0.06449	323.76	0.9760	0.04210	321.94	0.9485	0.03088	320.05	0.9280
180	0.06615	330.52	0.9911	0.04325	328.82	0.9638	0.03179	327.05	0.9436
190	0.06778	337.31	1.0059	0.04440	335.70	0.9789	0.03268	334.04	0.9589
200	0.06941	344.12	1.0204	0.04552	342.60	0.9936	0.03357	341.04	0.9738
210	0.07104	350.95	1.0347	0.04665	349.52	1.0081	0.03444	348.05	0.9885
220	0.07265	357.81	1.0488	0.04776	356.45	1.0223	0.03530	355.06	1.0028
230	0.07425	364.69	1.0626	0.04886	363.40	1.0362	0.03616	362.08	1.0169
240	0.07585	371.59	1.0762	0.04996	370.37	1.0499	0.03700	369.12	1.0308
	1000 kPa (109.63)			2000 kPa (146.87)			4000 kPa (191.18)		
Sat.	0.01901	274.13	0.8045	0.00901	286.50	0.8032	0.00318	284.27	0.7735
140	0.02172	296.52	0.8609	—	—	—	—	—	—
150	0.02255	303.75	0.8781	0.00922	289.34	0.8099	—	—	—
160	0.02335	310.93	0.8949	0.00983	298.09	0.8304	—	—	—
170	0.02413	318.08	0.9112	0.01040	306.49	0.8495	—	—	—
180	0.02489	325.21	0.9271	0.01092	314.63	0.8677	—	—	—
190	0.02564	332.33	0.9427	0.01141	322.60	0.8851	—	—	—
200	0.02638	339.43	0.9579	0.01189	330.44	0.9018	0.00398	301.03	0.8093
210	0.02710	346.53	0.9727	0.01234	338.17	0.9180	0.00451	313.64	0.8356
220	0.02782	353.63	0.9872	0.01278	345.82	0.9337	0.00494	324.41	0.8577
230	0.02853	360.73	1.0015	0.01320	353.41	0.9489	0.00530	334.28	0.8775
240	0.02922	367.83	1.0155	0.01361	360.95	0.9638	0.00563	343.60	0.8959
250	0.02992	374.94	1.0292	0.01402	368.44	0.9782	0.00593	352.54	0.9131
260	0.03060	382.06	1.0427	0.01441	375.91	0.9924	0.00621	361.21	0.9295

TABLE B.12 SI *Thermodynamic Properties of R-12*
TABLE B.12.1 SI *Saturated R-12*

Temp.	Press.	Specific Volume, m^3/kg			Internal Energy, kJ/kg		
C	kPa	Sat. Liquid	Evap.	Sat. Vapor	Sat. Liquid	Evap.	Sat. Vapor
T	P	v_f	v_{fg}	v_g	u_f	u_{fg}	u_g
-90	2.8	0.000608	4.41494	4.41555	-43.29	177.20	133.91
-80	6.2	0.000617	2.13773	2.13835	-34.73	172.54	137.82
-70	12.3	0.000627	1.12665	1.12728	-26.14	167.94	141.81
-60	22.6	0.000637	0.63727	0.63791	-17.50	163.36	145.86
-50	39.1	0.000648	0.38246	0.38310	-8.80	158.76	149.95
-45	50.4	0.000654	0.30203	0.30268	-4.43	156.44	152.01
-40	64.2	0.000659	0.24125	0.24191	-0.04	154.11	154.07
-35	80.7	0.000666	0.19473	0.19540	4.37	151.77	156.13
-30	100.4	0.000672	0.15870	0.15937	8.79	149.40	158.19
-29.8	101.3	0.000672	0.15736	0.15803	8.98	149.30	158.28
-25	123.7	0.000679	0.13049	0.13117	13.24	147.01	160.25
-20	150.9	0.000685	0.10816	0.10885	17.71	144.59	162.31
-15	182.6	0.000693	0.09033	0.09102	22.20	142.15	164.35
-10	219.1	0.000700	0.07595	0.07665	26.72	139.67	166.39
-5	261.0	0.000708	0.06426	0.06496	31.26	137.16	168.42
0	308.6	0.000716	0.05467	0.05539	35.83	134.61	170.44
5	362.6	0.000724	0.04676	0.04749	40.43	132.01	172.44
10	423.3	0.000733	0.04018	0.04091	45.06	129.36	174.42
15	491.4	0.000743	0.03467	0.03541	49.73	126.65	176.38
20	567.3	0.000752	0.03003	0.03078	54.45	123.87	178.32
25	651.6	0.000763	0.02609	0.02685	59.21	121.03	180.23
30	744.9	0.000774	0.02273	0.02351	64.02	118.09	182.11
35	847.7	0.000786	0.01986	0.02064	68.88	115.06	183.95
40	960.7	0.000798	0.01737	0.01817	73.82	111.92	185.74
45	1084.3	0.000811	0.01522	0.01603	78.83	108.66	187.49
50	1219.3	0.000826	0.01334	0.01417	83.93	105.24	189.17
55	1366.3	0.000841	0.01170	0.01254	89.12	101.66	190.78
60	1525.9	0.000858	0.01025	0.01111	94.43	97.88	192.31
65	1698.8	0.000877	0.00897	0.00985	99.87	93.86	193.73
70	1885.8	0.000897	0.00783	0.00873	105.46	89.56	195.03
75	2087.5	0.000920	0.00680	0.00772	111.23	84.94	196.17
80	2304.6	0.000946	0.00588	0.00682	117.21	79.90	197.11
85	2538.0	0.000976	0.00503	0.00600	123.45	74.34	197.80
90	2788.5	0.001012	0.00425	0.00526	130.02	68.12	198.14
95	3056.9	0.001056	0.00351	0.00456	137.01	60.98	197.99
100	3344.1	0.001113	0.00279	0.00390	144.59	52.48	197.07
105	3650.9	0.001197	0.00205	0.00324	153.15	41.58	194.73
110	3978.5	0.001364	0.00110	0.00246	164.12	24.08	188.20
112.0	4116.8	0.001792	0	0.00179	176.06	0	176.06

(Continued)

TABLE B.12.1 SI (Continued) *Saturated R-12*

Temp.	Press.	Enthalpy, kJ/kg			Entropy, kJ/kg K		
C	kPa	Sat. Liquid	Evap.	Sat. Vapor	Sat. Liquid	Evap.	Sat. Vapor
T	P	h_f	h_{fg}	h_g	s_f	s_{fg}	s_g
-90	2.8	-43.28	189.75	146.46	-0.2086	1.0359	0.8273
-80	6.2	-34.72	185.74	151.02	-0.1631	0.9616	0.7984
-70	12.3	-26.13	181.76	155.64	-0.1198	0.8947	0.7749
-60	22.6	-17.49	177.77	160.29	-0.0783	0.8340	0.7557
-50	39.1	-8.78	173.73	164.95	-0.0384	0.7785	0.7401
-45	50.4	-4.40	171.68	167.28	-0.0190	0.7524	0.7334
-40	64.2	0	169.59	169.59	0	0.7274	0.7274
-35	80.7	4.42	167.48	171.90	0.0187	0.7032	0.7219
-30	100.4	8.86	165.34	174.20	0.0371	0.6799	0.7170
-29.8	101.3	9.05	165.24	174.29	0.0379	0.6790	0.7168
-25	123.7	13.33	163.15	176.48	0.0552	0.6574	0.7126
-20	150.9	17.82	160.92	178.74	0.0731	0.6356	0.7087
-15	182.6	22.33	158.64	180.97	0.0906	0.6145	0.7051
-10	219.1	26.87	156.31	183.19	0.1080	0.5940	0.7019
-5	261.0	31.45	153.93	185.37	0.1251	0.5740	0.6991
0	308.6	36.05	151.48	187.53	0.1420	0.5545	0.6965
5	362.6	40.69	148.96	189.65	0.1587	0.5355	0.6942
10	423.3	45.37	146.37	191.74	0.1752	0.5169	0.6921
15	491.4	50.10	143.68	193.78	0.1915	0.4986	0.6902
20	567.3	54.87	140.91	195.78	0.2078	0.4806	0.6884
25	651.6	59.70	138.03	197.73	0.2239	0.4629	0.6868
30	744.9	64.59	135.03	199.62	0.2399	0.4454	0.6853
35	847.7	69.55	131.90	201.45	0.2559	0.4280	0.6839
40	960.7	74.59	128.61	203.20	0.2718	0.4107	0.6825
45	1084.3	79.71	125.16	204.87	0.2877	0.3934	0.6811
50	1219.3	84.94	121.51	206.45	0.3037	0.3760	0.6797
55	1366.3	90.27	117.65	207.92	0.3197	0.3585	0.6782
60	1525.9	95.74	113.52	209.26	0.3358	0.3407	0.6765
65	1698.8	101.36	109.10	210.46	0.3521	0.3226	0.6747
70	1885.8	107.15	104.33	211.48	0.3686	0.3040	0.6726
75	2087.5	113.15	99.14	212.29	0.3854	0.2847	0.6702
80	2304.6	119.39	93.44	212.83	0.4027	0.2646	0.6672
85	2538.0	125.93	87.11	213.04	0.4204	0.2432	0.6636
90	2788.5	132.84	79.96	212.80	0.4389	0.2202	0.6590
95	3056.9	140.23	71.71	211.94	0.4583	0.1948	0.6531
100	3344.1	148.31	61.81	210.12	0.4793	0.1656	0.6449
105	3650.9	157.52	49.05	206.57	0.5028	0.1297	0.6325
110	3978.5	169.55	28.44	197.99	0.5333	0.0742	0.6076
112.0	4116.8	183.43	0	183.43	0.5689	0	0.5689

TABLE B.12.2 SI *Superheated R-12*

Temp. C	v m³/kg	h kJ/kg	s kJ/kg K	v m³/kg	h kJ/kg	s kJ/kg K	v m³/kg	h kJ/kg	s kJ/kg K
	25 kPa (-58.26)			50 kPa (-45.18)			100 kPa (-30.10)		
Sat.	0.58130	161.10	0.7527	0.30515	167.19	0.7336	0.15999	174.15	0.7171
-30	0.66179	176.19	0.8187	0.32738	175.55	0.7691	0.16006	174.21	0.7174
-20	0.69001	181.74	0.8410	0.34186	181.17	0.7917	0.16770	179.99	0.7406
-10	0.71811	187.40	0.8630	0.35623	186.89	0.8139	0.17522	185.84	0.7633
0	0.74613	193.17	0.8844	0.37051	192.70	0.8356	0.18265	191.77	0.7854
10	0.77409	199.03	0.9055	0.38472	198.61	0.8568	0.18999	197.77	0.8070
20	0.80198	204.99	0.9262	0.39886	204.62	0.8776	0.19728	203.85	0.8281
30	0.82982	211.05	0.9465	0.41296	210.71	0.8981	0.20451	210.02	0.8488
40	0.85762	217.20	0.9665	0.42701	216.89	0.9181	0.21169	216.26	0.8691
50	0.88538	223.45	0.9861	0.44103	223.16	0.9378	0.21884	222.58	0.8889
60	0.91312	229.77	1.0054	0.45502	229.51	0.9572	0.22596	228.98	0.9084
70	0.94083	236.19	1.0244	0.46898	235.95	0.9762	0.23305	235.46	0.9276
80	0.96852	242.68	1.0430	0.48292	242.46	0.9949	0.24011	242.01	0.9464
90	0.99618	249.26	1.0614	0.49684	249.05	1.0133	0.24716	248.63	0.9649
100	1.02384	255.91	1.0795	0.51074	255.71	1.0314	0.25419	255.32	0.9831
110	1.05148	262.63	1.0972	0.52463	262.45	1.0493	0.26121	262.08	1.0009
120	1.07910	269.43	1.1148	0.53851	269.26	1.0668	0.26821	268.91	1.0185
	200 kPa (-12.53)			300 kPa (-0.86)			400 kPa (8.15)		
Sat.	0.08354	182.07	0.7035	0.05690	187.16	0.6969	0.04321	190.97	0.6928
0	0.08861	189.80	0.7325	0.05715	187.72	0.6989	—	—	—
10	0.09255	196.02	0.7548	0.05998	194.17	0.7222	0.04363	192.21	0.6972
20	0.09642	202.28	0.7766	0.06273	200.64	0.7446	0.04584	198.91	0.7204
30	0.10023	208.60	0.7978	0.06542	207.12	0.7663	0.04797	205.58	0.7428
40	0.10399	214.97	0.8184	0.06805	213.64	0.7875	0.05005	212.25	0.7645
50	0.10771	221.41	0.8387	0.07064	220.19	0.8081	0.05207	218.94	0.7855
60	0.11140	227.90	0.8585	0.07319	226.79	0.8282	0.05406	225.65	0.8060
70	0.11506	234.46	0.8779	0.07571	233.44	0.8479	0.05601	232.40	0.8259
80	0.11869	241.09	0.8969	0.07820	240.15	0.8671	0.05794	239.19	0.8454
90	0.12230	247.77	0.9156	0.08067	246.90	0.8860	0.05985	246.02	0.8645
100	0.12590	254.53	0.9339	0.08313	253.72	0.9045	0.06173	252.89	0.8831
110	0.12948	261.34	0.9519	0.08557	260.58	0.9226	0.06360	259.81	0.9015
120	0.13305	268.21	0.9696	0.08799	267.50	0.9405	0.06546	266.79	0.9194
130	0.13661	275.15	0.9870	0.09041	274.48	0.9580	0.06730	273.81	0.9370
140	0.14016	282.14	1.0042	0.09281	281.51	0.9752	0.06913	280.88	0.9544
150	0.14370	289.19	1.0210	0.09520	288.59	0.9922	0.07095	287.99	0.9714

(Continued)

TABLE B.12.2 SI (Continued) ***Superheated R-12***

Temp. C	v m³/kg	h kJ/kg	s kJ/kg K	v m³/kg	h kJ/kg	s kJ/kg K	v m³/kg	h kJ/kg	s kJ/kg K
	500 kPa (15.60)			750 kPa (30.26)			1000 kPa (41.64)		
Sat.	0.03482	194.03	0.6899	0.02335	199.72	0.6852	0.01744	203.76	0.6820
30	0.03746	203.96	0.7235	—	—	—	—	—	—
40	0.03921	210.81	0.7457	0.02467	206.91	0.7086	—	—	—
50	0.04091	217.64	0.7672	0.02595	214.18	0.7314	0.01837	210.32	0.7026
60	0.04257	224.48	0.7881	0.02718	221.37	0.7533	0.01941	217.97	0.7259
70	0.04418	231.33	0.8083	0.02837	228.52	0.7745	0.02040	225.49	0.7481
80	0.04577	238.21	0.8281	0.02952	235.65	0.7949	0.02134	232.91	0.7695
90	0.04734	245.11	0.8473	0.03064	242.76	0.8148	0.02225	240.28	0.7900
100	0.04889	252.05	0.8662	0.03174	249.89	0.8342	0.02313	247.61	0.8100
110	0.05041	259.03	0.8847	0.03282	257.03	0.8530	0.02399	254.93	0.8293
120	0.05193	266.06	0.9028	0.03388	264.19	0.8715	0.02483	262.25	0.8482
130	0.05343	273.12	0.9205	0.03493	271.38	0.8895	0.02566	269.57	0.8665
140	0.05492	280.23	0.9379	0.03596	278.59	0.9072	0.02647	276.90	0.8845
150	0.05640	287.39	0.9550	0.03699	285.84	0.9246	0.02728	284.26	0.9021
160	0.05788	294.59	0.9718	0.03801	293.13	0.9416	0.02807	291.63	0.9193
170	0.05934	301.83	0.9884	0.03902	300.45	0.9583	0.02885	299.04	0.9362
180	0.06080	309.12	1.0046	0.04002	307.81	0.9747	0.02963	306.47	0.9528
	1500 kPa (59.22)			2000 kPa (72.88)			4000 kPa (110.32)		
Sat.	0.01132	209.06	0.6768	0.00813	211.97	0.6713	0.00239	196.90	0.6046
80	0.01305	226.73	0.7284	0.00870	219.02	0.6914	—	—	—
90	0.01377	234.77	0.7508	0.00941	228.23	0.7171	—	—	—
100	0.01446	242.65	0.7722	0.01003	236.94	0.7408	—	—	—
110	0.01512	250.41	0.7928	0.01061	245.34	0.7630	—	—	—
120	0.01575	258.10	0.8126	0.01116	253.53	0.7841	0.00374	225.18	0.6777
130	0.01636	265.74	0.8318	0.01168	261.58	0.8043	0.00433	238.69	0.7116
140	0.01696	273.35	0.8504	0.01217	269.53	0.8238	0.00478	249.93	0.7392
150	0.01754	280.94	0.8686	0.01265	277.41	0.8426	0.00517	260.12	0.7636
160	0.01811	288.52	0.8863	0.01312	285.24	0.8609	0.00552	269.71	0.7860
170	0.01867	296.11	0.9036	0.01357	293.04	0.8787	0.00585	278.90	0.8069
180	0.01922	303.70	0.9205	0.01401	300.82	0.8961	0.00615	287.82	0.8269
190	0.01977	311.31	0.9371	0.01445	308.59	0.9131	0.00643	296.55	0.8459
200	0.02031	318.93	0.9534	0.01488	316.36	0.9297	0.00671	305.14	0.8642
210	0.02084	326.58	0.9694	0.01530	324.14	0.9459	0.00697	313.61	0.8820
220	0.02137	334.24	0.9851	0.01572	331.92	0.9619	0.00723	322.01	0.8992

TABLE B.13 SI ***Thermodynamic Properties of R-13***
TABLE B.13.1 SI ***Saturated R-13***

Temp.	Press.	Specific Volume, m³/kg			Internal Energy, kJ/kg		
C	kPa	Sat. Liquid	Evap.	Sat. Vapor	Sat. Liquid	Evap.	Sat. Vapor
T	P	v_f	v_{fg}	v_g	u_f	u_{fg}	u_g
-100	33.1	0.000629	0.40762	0.40825	-55.74	144.56	88.82
-90	62.5	0.000644	0.22562	0.22627	-47.26	139.63	92.37
-85	83.4	0.000651	0.17222	0.17287	-42.91	137.04	94.13
-81.5	101.3	0.000657	0.14345	0.14410	-39.76	135.13	95.38
-80	109.4	0.000660	0.13343	0.13409	-38.48	134.36	95.89
-75	141.5	0.000668	0.10476	0.10543	-33.97	131.59	97.62
-70	180.3	0.000677	0.08325	0.08393	-29.39	128.73	99.34
-65	226.7	0.000687	0.06688	0.06756	-24.73	125.76	101.03
-60	281.8	0.000697	0.05424	0.05494	-20.01	122.70	102.69
-55	346.4	0.000707	0.04438	0.04508	-15.21	119.53	104.32
-50	421.5	0.000718	0.03659	0.03730	-10.36	116.25	105.90
-45	508.0	0.000730	0.03036	0.03109	-5.44	112.87	107.43
-40	607.0	0.000742	0.02535	0.02609	-0.45	109.37	108.92
-35	719.4	0.000756	0.02126	0.02202	4.59	105.74	110.34
-30	846.4	0.000770	0.01791	0.01868	9.70	101.98	111.69
-25	988.8	0.000786	0.01513	0.01591	14.88	98.08	112.96
-20	1147.9	0.000803	0.01280	0.01361	20.12	94.01	114.14
-15	1324.6	0.000822	0.01084	0.01167	25.45	89.76	115.20
-10	1520.2	0.000843	0.00918	0.01002	30.86	85.28	116.14
-5	1735.9	0.000867	0.00775	0.00862	36.38	80.54	116.92
0	1972.9	0.000894	0.00651	0.00740	42.04	75.46	117.50
5	2232.9	0.000926	0.00542	0.00635	47.89	69.94	117.83
10	2517.6	0.000965	0.00445	0.00541	53.99	63.81	117.80
15	2829.2	0.001015	0.00356	0.00457	60.49	56.76	117.26
20	3170.8	0.001085	0.00270	0.00379	67.71	48.11	115.83
29	3870.0	0.001731	0	0.001731	96.28	0	96.28

(Continued)

TABLE B.13.1 SI (Continued) *Saturated R-13*

Temp.	Press.	Enthalpy, kJ/kg			Entropy, kJ/kg K		
C	kPa	Sat. Liquid	Evap.	Sat. Vapor	Sat. Liquid	Evap.	Sat. Vapor
T	P	h_f	h_{fg}	h_g	s_f	s_{fg}	s_g
-100	33.1	-55.72	158.06	102.34	-0.2730	0.9128	0.6398
-90	62.5	-47.22	153.72	106.50	-0.2254	0.8393	0.6139
-85	83.4	-42.85	151.40	108.55	-0.2020	0.8047	0.6027
-81.5	101.3	-39.69	149.67	109.98	-0.1854	0.7808	0.5954
-80	109.4	-38.40	148.96	110.56	-0.1787	0.7712	0.5925
-75	141.5	-33.87	146.41	112.54	-0.1557	0.7389	0.5832
-70	180.3	-29.26	143.73	114.47	-0.1328	0.7075	0.5747
-65	226.7	-24.58	140.93	116.35	-0.1102	0.6770	0.5669
-60	281.8	-19.81	137.98	118.17	-0.0877	0.6474	0.5596
-55	346.4	-14.97	134.90	119.93	-0.0655	0.6184	0.5529
-50	421.5	-10.05	131.67	121.62	-0.0435	0.5901	0.5466
-45	508.0	-5.06	128.29	123.23	-0.0216	0.5623	0.5407
-40	607.0	0	124.75	124.75	0	0.5351	0.5351
-35	719.4	5.14	121.04	126.18	0.0215	0.5082	0.5297
-30	846.4	10.36	117.14	127.50	0.0427	0.4818	0.5245
-25	988.8	15.66	113.04	128.69	0.0639	0.4555	0.5194
-20	1147.9	21.04	108.71	129.75	0.0848	0.4294	0.5143
-15	1324.6	26.53	104.12	130.66	0.1058	0.4033	0.5091
-10	1520.2	32.14	99.24	131.38	0.1266	0.3771	0.5038
-5	1735.9	37.89	93.99	131.88	0.1476	0.3505	0.4981
0	1972.9	43.81	88.31	132.11	0.1687	0.3233	0.4920
5	2232.9	49.95	82.05	132.00	0.1901	0.2950	0.4851
10	2517.6	56.42	75.02	131.43	0.2122	0.2649	0.4771
15	2829.2	63.37	66.83	130.19	0.2354	0.2319	0.4674
20	3170.8	71.16	56.67	127.83	0.2610	0.1933	0.4543
29	3870.0	102.98	0	102.98	0.3645	0	0.3645

TABLE B.13.2 SI *Superheated R-13*

Temp. C	v m³/kg	h kJ/kg	s kJ/kg K	v m³/kg	h kJ/kg	s kJ/kg K	v m³/kg	h kJ/kg	s kJ/kg K
	25 kPa (-104.04)			50 kPa (-93.65)			100 kPa (-81.69)		
Sat.	0.52995	100.64	0.6519	0.27830	104.99	0.6228	0.14589	109.88	0.5959
-80	0.60873	112.36	0.7166	0.30122	111.84	0.6596	0.14737	110.77	0.6005
-70	0.64126	117.48	0.7425	0.31784	117.02	0.6857	0.15607	116.07	0.6272
-60	0.67368	122.74	0.7677	0.33436	122.33	0.7112	0.16465	121.48	0.6532
-50	0.70602	128.14	0.7925	0.35079	127.77	0.7361	0.17314	127.01	0.6786
-40	0.73830	133.68	0.8168	0.36715	133.34	0.7606	0.18155	132.66	0.7033
-30	0.77051	139.35	0.8406	0.38345	139.05	0.7845	0.18990	138.43	0.7275
-20	0.80268	145.16	0.8640	0.39971	144.88	0.8080	0.19820	144.31	0.7513
-10	0.83481	151.09	0.8870	0.41592	150.84	0.8311	0.20646	150.32	0.7745
0	0.86690	157.15	0.9096	0.43210	156.92	0.8538	0.21468	156.45	0.7974
10	0.89897	163.34	0.9318	0.44825	163.12	0.8761	0.22288	162.69	0.8198
20	0.93101	169.65	0.9537	0.46437	169.45	0.8980	0.23104	169.05	0.8419
30	0.96303	176.07	0.9753	0.48047	175.89	0.9197	0.23919	175.52	0.8636
40	0.99503	182.62	0.9965	0.49656	182.44	0.9409	0.24731	182.10	0.8850
50	1.02702	189.27	1.0174	0.51263	189.11	0.9619	0.25543	188.79	0.9060
60	1.05899	196.04	1.0380	0.52868	195.89	0.9825	0.26352	195.59	0.9267
70	1.09095	202.91	1.0583	0.54472	202.77	1.0029	0.27161	202.49	0.9471
	200 kPa (-67.77)			300 kPa (-58.51)			400 kPa (-51.36)		
Sat.	0.07608	115.32	0.5711	0.05175	118.70	0.5576	0.03924	121.17	0.5483
-40	0.08869	131.25	0.6438	0.05768	129.77	0.6070	0.04213	128.23	0.5793
-30	0.09308	137.15	0.6686	0.06077	135.84	0.6325	0.04457	134.47	0.6055
-20	0.09742	143.16	0.6929	0.06379	141.97	0.6572	0.04695	140.75	0.6308
-10	0.10171	149.27	0.7165	0.06676	148.20	0.6813	0.04927	147.09	0.6554
0	0.10596	155.49	0.7397	0.06970	154.51	0.7049	0.05155	153.51	0.6793
10	0.11017	161.81	0.7624	0.07259	160.91	0.7279	0.05379	160.00	0.7027
20	0.11437	168.24	0.7847	0.07546	167.41	0.7504	0.05600	166.58	0.7255
30	0.11854	174.77	0.8067	0.07831	174.01	0.7726	0.05819	173.24	0.7478
40	0.12269	181.40	0.8282	0.08114	180.70	0.7943	0.06036	179.99	0.7697
50	0.12682	188.14	0.8494	0.08395	187.49	0.8156	0.06250	186.83	0.7912
60	0.13094	194.98	0.8702	0.08674	194.37	0.8366	0.06464	193.76	0.8124
70	0.13505	201.92	0.8907	0.08952	201.35	0.8572	0.06676	200.78	0.8331
80	0.13914	208.96	0.9109	0.09229	208.42	0.8776	0.06886	207.88	0.8535
90	0.14323	216.09	0.9309	0.09505	215.59	0.8976	0.07096	215.08	0.8736
100	0.14731	223.31	0.9505	0.09780	222.84	0.9173	0.07305	222.36	0.8934
110	0.15138	230.63	0.9698	0.10055	230.18	0.9367	0.07513	229.73	0.9129

(Continued)

TABLE B.13.2 SI (Continued) *Superheated R-13*

Temp. C	v m^3/kg	h kJ/kg	s kJ/kg K	v m^3/kg	h kJ/kg	s kJ/kg K	v m^3/kg	h kJ/kg	s kJ/kg K
	500 kPa (-45.43)			750 kPa (-33.74)			1000 kPa (-24.63)		
Sat.	0.03158	123.09	0.5412	0.02111	126.52	0.5284	0.01573	128.78	0.5190
-30	0.03483	133.04	0.5834	0.02171	129.15	0.5393	—	—	—
-20	0.03682	139.48	0.6094	0.02324	136.09	0.5672	0.01634	132.28	0.5330
-10	0.03876	145.96	0.6345	0.02469	142.95	0.5938	0.01758	139.67	0.5616
0	0.04065	152.48	0.6588	0.02607	149.80	0.6194	0.01873	146.91	0.5886
10	0.04250	159.07	0.6825	0.02741	156.65	0.6440	0.01983	154.08	0.6144
20	0.04432	165.72	0.7056	0.02872	163.52	0.6679	0.02089	161.21	0.6391
30	0.04611	172.45	0.7282	0.02999	170.44	0.6911	0.02191	168.35	0.6631
40	0.04788	179.27	0.7503	0.03124	177.42	0.7137	0.02290	175.51	0.6863
50	0.04964	186.16	0.7720	0.03247	184.45	0.7358	0.02387	182.70	0.7089
60	0.05137	193.14	0.7932	0.03368	191.55	0.7575	0.02482	189.93	0.7310
70	0.05309	200.20	0.8141	0.03487	198.73	0.7787	0.02575	197.22	0.7525
80	0.05481	207.34	0.8346	0.03606	205.97	0.7995	0.02668	204.57	0.7736
90	0.05651	214.57	0.8548	0.03723	213.29	0.8199	0.02759	211.98	0.7943
100	0.05820	221.89	0.8747	0.03839	220.68	0.8400	0.02849	219.45	0.8146
110	0.05988	229.28	0.8942	0.03955	228.14	0.8597	0.02938	226.99	0.8345
120	0.06156	236.76	0.9135	0.04070	235.68	0.8791	0.03026	234.60	0.8541
	1500 kPa (-10.49)			2000 kPa (0.54)			3000 kPa (17.56)		
Sat.	0.01017	131.32	0.5043	0.00728	132.12	0.4913	0.00417	129.17	0.4613
0	0.01124	140.28	0.5378	—	—	—	—	—	—
10	0.01215	148.35	0.5668	0.00816	141.42	0.5247	—	—	—
20	0.01299	156.18	0.5940	0.00894	150.37	0.5557	0.00445	133.20	0.4751
30	0.01377	163.85	0.6197	0.00964	158.83	0.5841	0.00529	145.95	0.5179
40	0.01452	171.45	0.6444	0.01029	167.02	0.6107	0.00592	156.42	0.5518
50	0.01524	179.01	0.6681	0.01090	175.04	0.6359	0.00647	165.95	0.5818
60	0.01594	186.55	0.6911	0.01148	182.97	0.6601	0.00696	174.98	0.6093
70	0.01662	194.11	0.7135	0.01204	190.83	0.6834	0.00741	183.69	0.6351
80	0.01728	201.69	0.7352	0.01257	198.68	0.7059	0.00783	192.22	0.6596
90	0.01793	209.30	0.7565	0.01310	206.51	0.7278	0.00824	200.62	0.6831
100	0.01857	216.95	0.7773	0.01361	214.36	0.7491	0.00863	208.94	0.7057
110	0.01920	224.64	0.7976	0.01411	222.23	0.7699	0.00900	217.21	0.7275
120	0.01982	232.39	0.8176	0.01460	230.13	0.7903	0.00937	225.46	0.7488
130	0.02044	240.19	0.8372	0.01508	238.07	0.8102	0.00972	233.71	0.7695
140	0.02105	248.04	0.8564	0.01556	246.04	0.8297	0.01007	241.95	0.7897
150	0.02165	255.95	0.8753	0.01603	254.06	0.8489	0.01041	250.21	0.8095

TABLE B.14 SI *Thermodynamic Properties of R-14*
TABLE B.14.1 SI *Saturated R-14*

Temp.	Press.	Specific Volume, m^3/kg			Internal Energy, kJ/kg		
C	kPa	Sat. Liquid	Evap.	Sat. Vapor	Sat. Liquid	Evap.	Sat. Vapor
T	P	v_f	v_{fg}	v_g	u_f	u_{fg}	u_g
-150	15.4	0.000572	0.74669	0.74726	94.61	134.10	228.71
-145	25.2	0.000580	0.47241	0.47299	98.84	131.42	230.26
-140	39.5	0.000589	0.31055	0.31114	103.19	128.61	231.80
-135	59.7	0.000598	0.21107	0.21166	107.66	125.64	233.30
-130	87.4	0.000608	0.14769	0.14830	112.26	122.52	234.78
-128.0	101.3	0.000612	0.12859	0.12920	114.18	121.19	235.37
-125	124.4	0.000618	0.10600	0.10662	116.99	119.23	236.21
-120	172.4	0.000630	0.07778	0.07841	121.82	115.77	237.59
-115	233.5	0.000641	0.05818	0.05883	126.76	112.14	238.91
-110	309.7	0.000654	0.04426	0.04491	131.81	108.35	240.16
-105	403.2	0.000668	0.03415	0.03482	136.94	104.39	241.33
-100	516.2	0.000683	0.02667	0.02735	142.15	100.26	242.41
-95	650.9	0.000699	0.02103	0.02173	147.43	95.96	243.39
-90	809.6	0.000717	0.01672	0.01744	152.79	91.47	244.26
-85	994.6	0.000737	0.01336	0.01410	158.22	86.78	245.00
-80	1208.5	0.000759	0.01071	0.01147	163.72	81.85	245.57
-75	1453.8	0.000785	0.00858	0.00936	169.33	76.62	245.95
-70	1733.6	0.000815	0.00684	0.00765	175.07	71.00	246.07
-65	2051.4	0.000852	0.00539	0.00624	181.01	64.83	245.84
-60	2411.5	0.000898	0.00415	0.00505	187.31	57.77	245.08
-55	2819.7	0.000961	0.00304	0.00400	194.28	49.12	243.41
-50	3284.6	0.001065	0.00195	0.00301	202.94	36.78	239.72
-45.6	3740.0	0.001598	0	0.00160	222.60	0	222.60

(Continued)

TABLE B.14.1 SI (Continued) *Saturated R-14*

Temp.	Press.	Enthalpy, kJ/kg			Entropy, kJ/kg K		
C	kPa	Sat. Liquid	Evap.	Sat. Vapor	Sat. Liquid	Evap.	Sat. Vapor
T	P	h_f	h_{fg}	h_g	s_f	s_{fg}	s_g
-150	15.4	94.62	145.59	240.21	1.4617	1.1821	2.6438
-145	25.2	98.85	143.31	242.16	1.4953	1.1182	2.6135
-140	39.5	103.21	140.87	244.08	1.5286	1.0579	2.5865
-135	59.7	107.70	138.25	245.95	1.5616	1.0006	2.5622
-130	87.4	112.32	135.43	247.75	1.5943	0.9460	2.5403
-128.0	101.3	114.24	134.22	248.46	1.6076	0.9244	2.5319
-125	124.4	117.06	132.41	249.47	1.6267	0.8937	2.5204
-120	172.4	121.93	129.18	251.11	1.6588	0.8434	2.5022
-115	233.5	126.91	125.73	252.64	1.6906	0.7949	2.4855
-110	309.7	132.01	122.06	254.07	1.7220	0.7481	2.4701
-105	403.2	137.21	118.16	255.36	1.7530	0.7027	2.4557
-100	516.2	142.50	114.03	256.53	1.7836	0.6585	2.4421
-95	650.9	147.89	109.65	257.54	1.8137	0.6154	2.4292
-90	809.6	153.37	105.01	258.38	1.8435	0.5733	2.4167
-85	994.6	158.95	100.07	259.02	1.8728	0.5318	2.4046
-80	1208.5	164.64	94.79	259.43	1.9018	0.4907	2.3925
-75	1453.8	170.47	89.09	259.56	1.9306	0.4496	2.3802
-70	1733.6	176.48	82.86	259.34	1.9595	0.4079	2.3673
-65	2051.4	182.76	75.88	258.65	1.9887	0.3645	2.3533
-60	2411.5	189.47	67.78	257.25	2.0191	0.3180	2.3370
-55	2819.7	196.99	57.70	254.69	2.0522	0.2645	2.3167
-50	3284.6	206.44	43.17	249.60	2.0928	0.1934	2.2863
-45.6	3740.0	228.58	0	228.58	2.1883	0	2.1883

TABLE B.14.2 SI *Superheated R-14*

Temp. C	v m³/kg	h kJ/kg	s kJ/kg K	v m³/kg	h kJ/kg	s kJ/kg K	v m³/kg	h kJ/kg	s kJ/kg K
	25 kPa (-145.07)			50 kPa (-137.20)			100 kPa (-128.14)		
Sat.	0.47587	242.14	2.6139	0.24975	245.13	2.5726	0.13080	248.40	2.5327
-100	0.65048	264.04	2.7599	0.32329	263.70	2.6930	0.15966	263.01	2.6247
-80	0.72698	274.72	2.8182	0.36201	274.46	2.7518	0.17951	273.94	2.6844
-60	0.80319	286.00	2.8738	0.40044	285.79	2.8076	0.19906	285.39	2.7408
-40	0.87923	297.87	2.9270	0.43869	297.71	2.8610	0.21842	297.38	2.7945
-20	0.95515	310.33	2.9782	0.47683	310.19	2.9124	0.23766	309.93	2.8461
0	1.03099	323.36	3.0278	0.51488	323.24	2.9620	0.25683	323.02	2.8959
20	1.10678	336.94	3.0758	0.55288	336.85	3.0100	0.27593	336.66	2.9441
40	1.18253	351.08	3.1224	0.59084	351.00	3.0567	0.29500	350.83	2.9909
60	1.25824	365.75	3.1678	0.62877	365.67	3.1021	0.31403	365.53	3.0363
80	1.33394	380.92	3.2120	0.66668	380.86	3.1464	0.33305	380.73	3.0806
100	1.40962	396.60	3.2552	0.70457	396.54	3.1896	0.35204	396.43	3.1239
120	1.48529	412.75	3.2973	0.74245	412.70	3.2318	0.37103	412.59	3.1661
140	1.56094	429.36	3.3385	0.78032	429.31	3.2730	0.39000	429.22	3.2073
160	1.63659	446.41	3.3788	0.81818	446.36	3.3133	0.40897	446.28	3.2476
180	1.71223	463.87	3.4183	0.85603	463.83	3.3527	0.42793	463.75	3.2871
200	1.78787	481.73	3.4568	0.89387	481.70	3.3913	0.44688	481.62	3.3256
	200 kPa (-117.60)			300 kPa (-110.58)			400 kPa (-105.16)		
Sat.	0.06814	251.86	2.4941	0.04631	253.91	2.4718	0.03509	255.33	2.4561
-70	0.09332	278.66	2.6442	0.06130	277.71	2.6025	0.04528	276.72	2.5718
-50	0.10333	290.58	2.7001	0.06818	289.84	2.6594	0.05060	289.08	2.6298
-30	0.11319	303.00	2.7534	0.07490	302.40	2.7133	0.05575	301.79	2.6843
-10	0.12294	315.92	2.8044	0.08151	315.43	2.7648	0.06078	314.93	2.7362
10	0.13263	329.36	2.8537	0.08804	328.95	2.8143	0.06575	328.53	2.7860
30	0.14227	343.33	2.9013	0.09453	342.98	2.8621	0.07067	342.62	2.8341
50	0.15187	357.81	2.9476	0.10099	357.50	2.9085	0.07555	357.19	2.8806
70	0.16145	372.80	2.9925	0.10742	372.53	2.9536	0.08040	372.25	2.9258
90	0.17101	388.28	3.0364	0.11383	388.04	2.9976	0.08524	387.79	2.9699
110	0.18055	404.24	3.0791	0.12023	404.02	3.0404	0.09006	403.80	3.0128
130	0.19009	420.65	3.1209	0.12661	420.46	3.0822	0.09487	420.26	3.0546
150	0.19961	437.52	3.1617	0.13298	437.34	3.1231	0.09967	437.16	3.0955
170	0.20912	454.80	3.2016	0.13935	454.64	3.1630	0.10446	454.47	3.1355
190	0.21863	472.49	3.2407	0.14571	472.34	3.2021	0.10925	472.19	3.1746
210	0.22813	490.56	3.2789	0.15206	490.43	3.2403	0.11403	490.29	3.2129
230	0.23763	509.00	3.3163	0.15841	508.87	3.2777	0.11880	508.74	3.2503

(Continued)

TABLE B.14.2 SI (Continued) ***Superheated R-14***

Temp. C	v m³/kg	h kJ/kg	s kJ/kg K	v m³/kg	h kJ/kg	s kJ/kg K	v m³/kg	h kJ/kg	s kJ/kg K
	500 kPa (-100.66)			600 kPa (-96.79)			800 kPa (-90.28)		
Sat.	0.02822	256.38	2.4439	0.02357	257.19	2.4337	0.01765	258.33	2.4174
-70	0.03565	275.72	2.5471	0.02922	274.68	2.5261	0.02115	272.51	2.4910
-50	0.04004	288.31	2.6062	0.03300	287.53	2.5864	0.02418	285.91	2.5539
-30	0.04425	301.18	2.6614	0.03659	300.56	2.6423	0.02700	299.30	2.6114
-10	0.04835	314.43	2.7138	0.04006	313.93	2.6951	0.02969	312.90	2.6651
10	0.05237	328.12	2.7639	0.04345	327.69	2.7456	0.03230	326.84	2.7162
30	0.05635	342.26	2.8121	0.04680	341.90	2.7940	0.03486	341.18	2.7651
50	0.06028	356.88	2.8588	0.05011	356.57	2.8409	0.03739	355.95	2.8123
70	0.06420	371.98	2.9042	0.05339	371.71	2.8863	0.03988	371.16	2.8579
90	0.06809	387.55	2.9483	0.05665	387.31	2.9305	0.04236	386.82	2.9023
110	0.07196	403.58	2.9912	0.05990	403.36	2.9735	0.04482	402.93	2.9454
130	0.07583	420.06	3.0331	0.06313	419.87	3.0155	0.04727	419.47	2.9875
150	0.07968	436.98	3.0741	0.06636	436.80	3.0565	0.04971	436.44	3.0286
170	0.08353	454.31	3.1141	0.06958	454.15	3.0966	0.05214	453.82	3.0687
190	0.08737	472.04	3.1532	0.07279	471.89	3.1357	0.05456	471.59	3.1080
210	0.09121	490.15	3.1915	0.07599	490.01	3.1740	0.05698	489.74	3.1463
230	0.09504	508.61	3.2290	0.07919	508.49	3.2115	0.05939	508.24	3.1838
	1000 kPa (-84.86)			2000 kPa (-65.77)			5000 kPa		
Sat.	0.01402	259.03	2.4043	0.00644	258.79	2.3555	—	—	—
-70	0.01626	270.18	2.4613	—	—	—	—	—	—
-50	0.01887	284.23	2.5273	0.00808	274.38	2.4281	0.00106	206.44	2.0928
-30	0.02123	298.00	2.5864	0.00964	290.90	2.4990	0.00213	253.88	2.2862
-10	0.02347	311.86	2.6411	0.01099	306.34	2.5601	0.00336	285.20	2.4106
10	0.02561	325.98	2.6929	0.01222	321.50	2.6156	0.00414	306.02	2.4869
30	0.02770	340.45	2.7422	0.01338	336.70	2.6675	0.00478	324.43	2.5498
50	0.02976	355.32	2.7897	0.01449	352.11	2.7167	0.00535	341.96	2.6058
70	0.03178	370.61	2.8356	0.01558	367.82	2.7638	0.00588	359.18	2.6575
90	0.03378	386.33	2.8801	0.01664	383.87	2.8093	0.00638	376.39	2.7062
110	0.03577	402.49	2.9234	0.01768	400.30	2.8533	0.00686	393.71	2.7527
130	0.03775	419.08	2.9656	0.01872	417.11	2.8961	0.00733	411.25	2.7973
150	0.03971	436.09	3.0068	0.01974	434.30	2.9377	0.00778	429.05	2.8404
170	0.04167	453.50	3.0470	0.02075	451.88	2.9783	0.00823	447.13	2.8821
190	0.04362	471.30	3.0863	0.02176	469.82	3.0179	0.00867	465.51	2.9227
210	0.04557	489.46	3.1247	0.02276	488.11	3.0565	0.00910	484.19	2.9622
230	0.04751	507.98	3.1622	0.02375	506.74	3.0943	0.00953	503.16	3.0006

TABLE B.15 SI *Thermodynamic Properties of R-21*
TABLE B.15.1 SI *Saturated R-21*

Temp.	Press.	Specific Volume, m³/kg			Internal Energy, kJ/kg		
C T	kPa P	Sat. Liquid v_f	Evap. v_{fg}	Sat. Vapor v_g	Sat. Liquid u_f	Evap. u_{fg}	Sat. Vapor u_g
-50	4.8	0.000651	3.75145	3.75210	-190.50	256.59	66.09
-40	9.2	0.000660	2.04428	2.04494	-180.21	250.46	70.25
-35	12.4	0.000665	1.54492	1.54558	-175.10	247.45	72.35
-30	16.5	0.000670	1.18412	1.18479	-169.98	244.46	74.48
-25	21.6	0.000674	0.91951	0.92018	-164.87	241.49	76.62
-20	28.0	0.000679	0.72272	0.72340	-159.76	238.53	78.77
-15	35.8	0.000684	0.57446	0.57515	-154.63	235.58	80.95
-10	45.4	0.000690	0.46139	0.46208	-149.49	232.62	83.13
-5	56.8	0.000695	0.37418	0.37487	-144.33	229.66	85.33
0	70.5	0.000700	0.30618	0.30688	-139.15	226.68	87.53
5	86.6	0.000706	0.25263	0.25334	-133.93	223.68	89.74
9.0	101.3	0.000711	0.21816	0.21887	-129.78	221.28	91.50
10	105.5	0.000712	0.21007	0.21078	-128.69	220.65	91.96
15	127.5	0.000718	0.17593	0.17665	-123.40	217.58	94.18
20	153.0	0.000724	0.14833	0.14905	-118.07	214.47	96.40
25	182.2	0.000730	0.12583	0.12656	-112.70	211.32	98.61
30	215.5	0.000737	0.10735	0.10809	-107.28	208.11	100.82
35	253.4	0.000744	0.09207	0.09281	-101.81	204.84	103.03
40	296.1	0.000751	0.07935	0.08010	-96.29	201.51	105.21
45	344.1	0.000758	0.06870	0.06945	-90.72	198.10	107.39
50	397.7	0.000766	0.05972	0.06048	-85.08	194.63	109.54
55	457.4	0.000774	0.05211	0.05288	-79.39	191.06	111.67
60	523.5	0.000782	0.04562	0.04641	-73.64	187.42	113.77
65	596.5	0.000791	0.04007	0.04086	-67.83	183.67	115.84
70	676.9	0.000799	0.03529	0.03609	-61.96	179.83	117.87
75	764.9	0.000809	0.03116	0.03197	-56.04	175.88	119.84
80	861.1	0.000819	0.02757	0.02839	-50.05	171.82	121.77
90	1079.8	0.000840	0.02169	0.02253	-37.91	163.32	125.41
100	1336.5	0.000863	0.01715	0.01801	-25.58	154.28	128.70
110	1635.0	0.000890	0.01358	0.01447	-13.13	144.65	131.52
120	1979.2	0.000921	0.01075	0.01167	-0.70	134.40	133.70
130	2373.3	0.000957	0.00848	0.00944	11.41	123.63	135.04
140	2821.9	0.001000	0.00667	0.00767	22.63	112.72	135.34
150	3330.0	0.001056	0.00526	0.00632	32.03	102.64	134.67
160	3903.7	0.001133	0.00422	0.00535	38.88	94.82	133.70
170	4550.7	0.001261	0.00345	0.00471	44.03	89.34	133.38
175	4905.1	0.001390	0.00309	0.00448	47.51	86.09	133.60
178.5	5164.6	0.001908	0	0.00191	40.02	0	40.02

(Continued)

TABLE B.15.1 SI (Continued) *Saturated R-21*

Temp.	Press.	Enthalpy, kJ/kg			Entropy, kJ/kg K		
C	kPa	Sat. Liquid	Evap.	Sat. Vapor	Sat. Liquid	Evap.	Sat. Vapor
T	P	h_f	h_{fg}	h_g	s_f	s_{fg}	s_g
-50	4.8	-190.49	274.57	84.08	0.4074	1.2307	1.6381
-40	9.2	-180.21	269.22	89.01	0.4525	1.1549	1.6074
-35	12.4	-175.09	266.58	91.49	0.4742	1.1196	1.5938
-30	16.5	-169.97	263.97	94.00	0.4955	1.0858	1.5813
-25	21.6	-164.86	261.37	96.51	0.5163	1.0534	1.5697
-20	28.0	-159.74	258.77	99.03	0.5367	1.0223	1.5591
-15	35.8	-154.61	256.16	101.56	0.5568	0.9925	1.5492
-10	45.4	-149.46	253.55	104.09	0.5765	0.9637	1.5402
-5	56.8	-144.29	250.92	106.62	0.5959	0.9359	1.5318
0	70.5	-139.10	248.25	109.15	0.6151	0.9090	1.5241
5	86.6	-133.87	245.55	111.68	0.6340	0.8829	1.5169
9.0	101.3	-129.71	243.39	113.68	0.6488	0.8629	1.5117
10	105.5	-128.61	242.81	114.20	0.6527	0.8577	1.5104
15	127.5	-123.31	240.02	116.71	0.6712	0.8331	1.5043
20	153.0	-117.96	237.16	119.20	0.6895	0.8091	1.4987
25	182.2	-112.57	234.24	121.67	0.7077	0.7857	1.4935
30	215.5	-107.12	231.25	124.12	0.7257	0.7629	1.4887
35	253.4	-101.63	228.17	126.54	0.7436	0.7405	1.4842
40	296.1	-96.07	225.00	128.93	0.7614	0.7186	1.4800
45	344.1	-90.45	221.74	131.29	0.7791	0.6971	1.4762
50	397.7	-84.78	218.38	133.60	0.7967	0.6759	1.4725
55	457.4	-79.04	214.90	135.86	0.8142	0.6550	1.4691
60	523.5	-73.23	211.30	138.07	0.8316	0.6343	1.4659
65	596.5	-67.36	207.58	140.22	0.8489	0.6139	1.4628
70	676.9	-61.42	203.72	142.30	0.8662	0.5937	1.4599
75	764.9	-55.42	199.71	144.30	0.8833	0.5737	1.4570
80	861.1	-49.34	195.56	146.21	0.9004	0.5538	1.4542
90	1079.8	-37.00	186.74	149.74	0.9344	0.5143	1.4487
100	1336.5	-24.42	177.20	152.77	0.9680	0.4749	1.4429
110	1635.0	-11.67	166.85	155.18	1.0010	0.4355	1.4365
120	1979.2	1.12	155.67	156.80	1.0332	0.3960	1.4292
130	2373.3	13.68	143.76	157.43	1.0638	0.3566	1.4204
140	2821.9	25.45	131.54	156.98	1.0915	0.3184	1.4099
150	3330.0	35.55	120.16	155.71	1.1145	0.2840	1.3985
160	3903.7	43.30	111.30	154.59	1.1311	0.2570	1.3881
170	4550.7	49.77	105.05	154.82	1.1441	0.2371	1.3812
175	4905.1	54.33	101.23	155.56	1.1533	0.2259	1.3792
178.5	5164.6	49.87	0	49.87	1.1306	0	1.1306

TABLE B.15.2 SI *Superheated R-21*

Temp. C	v m³/kg	h kJ/kg	s kJ/kg K	v m³/kg	h kJ/kg	s kJ/kg K	v m³/kg	h kJ/kg	s kJ/kg K
	25 kPa (-22.22)			50 kPa (-7.86)			75 kPa (1.49)		
Sat.	0.80375	97.91	1.5637	0.42206	105.17	1.5365	0.28956	109.91	1.5219
-10	0.84369	104.42	1.5890	—	—	—	—	—	—
0	0.87635	109.86	1.6093	0.43504	109.47	1.5525	—	—	—
10	0.90899	115.41	1.6293	0.45153	115.04	1.5725	0.29901	114.66	1.5389
20	0.94160	121.06	1.6489	0.46799	120.70	1.5921	0.31009	120.35	1.5586
30	0.97419	126.81	1.6682	0.48443	126.47	1.6115	0.32115	126.13	1.5780
40	1.00677	132.67	1.6872	0.50085	132.35	1.6306	0.33219	132.02	1.5972
50	1.03933	138.63	1.7059	0.51725	138.32	1.6493	0.34322	138.01	1.6160
60	1.07187	144.70	1.7244	0.53364	144.40	1.6679	0.35422	144.10	1.6346
70	1.10440	150.88	1.7427	0.55002	150.59	1.6862	0.36521	150.30	1.6529
80	1.13692	157.16	1.7607	0.56638	156.88	1.7042	0.37619	156.60	1.6710
90	1.16942	163.54	1.7786	0.58273	163.27	1.7221	0.38716	163.00	1.6889
100	1.20192	170.03	1.7962	0.59907	169.77	1.7397	0.39812	169.51	1.7065
110	1.23440	176.62	1.8136	0.61540	176.37	1.7572	0.40906	176.12	1.7240
120	1.26688	183.32	1.8309	0.63172	183.08	1.7745	0.42000	182.83	1.7413
130	1.29935	190.12	1.8480	0.64804	189.89	1.7916	0.43093	189.65	1.7585
140	1.33181	197.03	1.8649	0.66434	196.81	1.8085	0.44185	196.58	1.7754
	100 kPa (8.62)			150 kPa (19.45)			200 kPa (27.75)		
Sat.	0.22157	113.51	1.5121	0.15181	118.93	1.4993	0.11596	123.02	1.4908
20	0.23113	119.98	1.5346	0.15212	119.24	1.5003	—	—	—
30	0.23950	125.78	1.5541	0.15781	125.08	1.5199	0.11693	124.35	1.4952
40	0.24785	131.69	1.5733	0.16348	131.01	1.5392	0.12127	130.32	1.5145
50	0.25618	137.69	1.5921	0.16913	137.04	1.5581	0.12558	136.38	1.5336
60	0.26450	143.79	1.6107	0.17476	143.17	1.5768	0.12987	142.54	1.5524
70	0.27280	150.00	1.6291	0.18038	149.41	1.5953	0.13414	148.80	1.5709
80	0.28109	156.31	1.6472	0.18598	155.74	1.6134	0.13840	155.16	1.5892
90	0.28937	162.73	1.6651	0.19156	162.18	1.6314	0.14265	161.62	1.6072
100	0.29763	169.24	1.6828	0.19714	168.71	1.6492	0.14688	168.17	1.6250
110	0.30589	175.86	1.7004	0.20271	175.35	1.6667	0.15111	174.83	1.6426
120	0.31413	182.59	1.7177	0.20826	182.09	1.6841	0.15532	181.59	1.6600
130	0.32237	189.42	1.7348	0.21381	188.94	1.7013	0.15952	188.46	1.6773
140	0.33060	196.35	1.7518	0.21935	195.89	1.7183	0.16371	195.42	1.6943
150	0.33882	203.38	1.7686	0.22488	202.94	1.7352	0.16790	202.49	1.7112
160	0.34704	210.52	1.7853	0.23040	210.09	1.7519	0.17208	209.66	1.7280
170	0.35524	217.77	1.8019	0.23592	217.35	1.7685	0.17625	216.93	1.7446

(Continued)

TABLE B.15.2 SI (Continued) *Superheated R-21*

Temp. C	v m³/kg	h kJ/kg	s kJ/kg K	v m³/kg	h kJ/kg	s kJ/kg K	v m³/kg	h kJ/kg	s kJ/kg K
	300 kPa (40.43)			400 kPa (50.20)			600 kPa (65.22)		
Sat.	0.07911	129.14	1.4797	0.06015	133.69	1.4724	0.04063	140.31	1.4627
50	0.08197	135.01	1.4981	—	—	—	—	—	—
60	0.08493	141.23	1.5171	0.06241	139.86	1.4912	—	—	—
70	0.08787	147.55	1.5358	0.06469	146.24	1.5101	0.04142	143.45	1.4719
80	0.09080	153.96	1.5542	0.06696	152.71	1.5287	0.04304	150.07	1.4909
90	0.09371	160.47	1.5724	0.06921	159.28	1.5470	0.04464	156.76	1.5096
100	0.09660	167.07	1.5903	0.07144	165.93	1.5651	0.04622	163.54	1.5280
110	0.09949	173.77	1.6080	0.07366	172.68	1.5829	0.04778	170.40	1.5462
120	0.10236	180.57	1.6256	0.07586	179.53	1.6006	0.04933	177.34	1.5641
130	0.10522	187.47	1.6429	0.07805	186.47	1.6180	0.05086	184.38	1.5817
140	0.10807	194.47	1.6601	0.08024	193.50	1.6352	0.05238	191.50	1.5992
150	0.11092	201.57	1.6770	0.08241	200.64	1.6523	0.05389	198.71	1.6164
160	0.11375	208.77	1.6939	0.08458	207.87	1.6692	0.05540	206.02	1.6335
170	0.11658	216.07	1.7105	0.08674	215.21	1.6859	0.05689	213.42	1.6504
180	0.11940	223.48	1.7270	0.08889	222.64	1.7025	0.05837	220.92	1.6671
190	0.12222	230.98	1.7434	0.09104	230.17	1.7190	0.05985	228.51	1.6837
200	0.12503	238.59	1.7597	0.09318	237.80	1.7353	0.06132	236.20	1.7001
	800 kPa (76.87)			1000 kPa (86.54)			2000 kPa (120.56)		
Sat.	0.03057	145.03	1.4560	0.02439	148.57	1.4506	0.01153	156.86	1.4287
100	0.03354	160.96	1.5001	0.02588	158.17	1.4768	—	—	—
110	0.03479	167.96	1.5186	0.02695	165.34	1.4958	—	—	—
120	0.03602	175.03	1.5368	0.02800	172.56	1.5144	—	—	—
130	0.03723	182.17	1.5547	0.02902	179.84	1.5327	0.01228	165.21	1.4497
140	0.03843	189.40	1.5724	0.03003	187.18	1.5507	0.01301	173.78	1.4707
150	0.03961	196.70	1.5899	0.03103	194.60	1.5684	0.01369	182.20	1.4908
160	0.04079	204.10	1.6072	0.03201	202.09	1.5859	0.01434	190.54	1.5103
170	0.04195	211.58	1.6243	0.03298	209.66	1.6032	0.01496	198.83	1.5292
180	0.04311	219.15	1.6411	0.03394	217.31	1.6202	0.01556	207.11	1.5477
190	0.04425	226.81	1.6579	0.03489	225.05	1.6371	0.01615	215.40	1.5658
200	0.04539	234.56	1.6744	0.03583	232.87	1.6539	0.01671	223.72	1.5836
210	0.04652	242.41	1.6908	0.03677	240.79	1.6704	0.01727	232.07	1.6010
220	0.04765	250.35	1.7071	0.03769	248.79	1.6868	0.01781	240.48	1.6183
230	0.04877	258.39	1.7233	0.03861	256.88	1.7030	0.01835	248.94	1.6353
240	0.04988	266.52	1.7393	0.03953	265.07	1.7192	0.01887	257.47	1.6520
250	0.05099	274.75	1.7551	0.04044	273.35	1.7351	0.01939	266.06	1.6686

TABLE B.16 SI *Thermodynamic Properties of R-22*
TABLE B.16.1 SI *Saturated R-22*

Temp.	Press.	Specific Volume, m³/kg			Internal Energy, kJ/kg		
C	kPa	Sat. Liquid	Evap.	Sat. Vapor	Sat. Liquid	Evap.	Sat. Vapor
T	P	v_f	v_{fg}	v_g	u_f	u_{fg}	u_g
-70	20.5	0.000670	0.94027	0.94094	-30.62	230.13	199.51
-65	28.0	0.000676	0.70480	0.70547	-25.68	227.21	201.54
-60	37.5	0.000682	0.53647	0.53715	-20.68	224.25	203.57
-55	49.5	0.000689	0.41414	0.41483	-15.62	221.21	205.59
-50	64.4	0.000695	0.32386	0.32456	-10.50	218.11	207.61
-45	82.7	0.000702	0.25629	0.25699	-5.32	214.94	209.62
-40.8	101.3	0.000708	0.21191	0.21261	-0.87	212.18	211.31
-40	104.9	0.000709	0.20504	0.20575	-0.07	211.68	211.60
-35	131.7	0.000717	0.16568	0.16640	5.23	208.34	213.57
-30	163.5	0.000725	0.13512	0.13584	10.61	204.91	215.52
-25	201.0	0.000733	0.11113	0.11186	16.04	201.39	217.44
-20	244.8	0.000741	0.09210	0.09284	21.55	197.78	219.32
-15	295.7	0.000750	0.07688	0.07763	27.11	194.07	221.18
-10	354.3	0.000759	0.06458	0.06534	32.74	190.25	222.99
-5	421.3	0.000768	0.05457	0.05534	38.44	186.33	224.77
0	497.6	0.000778	0.04636	0.04714	44.20	182.30	226.50
5	583.8	0.000789	0.03957	0.04036	50.03	178.15	228.17
10	680.7	0.000800	0.03391	0.03471	55.92	173.87	229.79
15	789.1	0.000812	0.02918	0.02999	61.88	169.47	231.35
20	909.9	0.000824	0.02518	0.02600	67.92	164.92	232.85
25	1043.9	0.000838	0.02179	0.02262	74.04	160.22	234.26
30	1191.9	0.000852	0.01889	0.01974	80.23	155.35	235.59
35	1354.8	0.000867	0.01640	0.01727	86.53	150.30	236.82
40	1533.5	0.000884	0.01425	0.01514	92.92	145.02	237.94
45	1729.0	0.000902	0.01238	0.01328	99.42	139.50	238.93
50	1942.3	0.000922	0.01075	0.01167	106.06	133.70	239.76
55	2174.4	0.000944	0.00931	0.01025	112.85	127.56	240.41
60	2426.6	0.000969	0.00803	0.00900	119.83	121.01	240.84
65	2699.9	0.000997	0.00689	0.00789	127.04	113.94	240.98
70	2995.9	0.001030	0.00586	0.00689	134.54	106.22	240.76
75	3316.1	0.001069	0.00491	0.00598	142.44	97.61	240.05
80	3662.3	0.001118	0.00403	0.00515	150.92	87.71	238.63
85	4036.8	0.001183	0.00317	0.00436	160.32	75.78	236.10
90	4442.5	0.001282	0.00228	0.00356	171.51	59.90	231.41
95	4883.5	0.001521	0.00103	0.00255	188.93	29.89	218.83
96.0	4969.0	0.001906	0	0.00191	203.07	0	203.07

(Continued)

TABLE B.16.1 SI (Continued) *Saturated R-22*

Temp.	Press.	Enthalpy, kJ/kg			Entropy, kJ/kg K		
C	kPa	Sat. Liquid	Evap.	Sat. Vapor	Sat. Liquid	Evap.	Sat. Vapor
T	P	h_f	h_{fg}	h_g	s_f	s_{fg}	s_g
-70	20.5	-30.61	249.43	218.82	-0.1401	1.2277	1.0876
-65	28.0	-25.66	246.93	221.27	-0.1161	1.1862	1.0701
-60	37.5	-20.65	244.35	223.70	-0.0924	1.1463	1.0540
-55	49.5	-15.59	241.70	226.12	-0.0689	1.1079	1.0390
-50	64.4	-10.46	238.96	228.51	-0.0457	1.0708	1.0251
-45	82.7	-5.26	236.13	230.87	-0.0227	1.0349	1.0122
-40.8	101.3	-0.80	233.65	232.85	-0.0034	1.0053	1.0019
-40	104.9	0	233.20	233.20	0	1.0002	1.0002
-35	131.7	5.33	230.16	235.48	0.0225	0.9664	0.9889
-30	163.5	10.73	227.00	237.73	0.0449	0.9335	0.9784
-25	201.0	16.19	223.73	239.92	0.0670	0.9015	0.9685
-20	244.8	21.73	220.33	242.06	0.0890	0.8703	0.9593
-15	295.7	27.33	216.80	244.13	0.1107	0.8398	0.9505
-10	354.3	33.01	213.13	246.14	0.1324	0.8099	0.9422
-5	421.3	38.76	209.32	248.09	0.1538	0.7806	0.9344
0	497.6	44.59	205.36	249.95	0.1751	0.7518	0.9269
5	583.8	50.49	201.25	251.73	0.1963	0.7235	0.9197
10	680.7	56.46	196.96	253.42	0.2173	0.6956	0.9129
15	789.1	62.52	192.49	255.02	0.2382	0.6680	0.9062
20	909.9	68.67	187.84	256.51	0.2590	0.6407	0.8997
25	1043.9	74.91	182.97	257.88	0.2797	0.6137	0.8934
30	1191.9	81.25	177.87	259.12	0.3004	0.5867	0.8871
35	1354.8	87.70	172.52	260.22	0.3210	0.5598	0.8809
40	1533.5	94.27	166.88	261.15	0.3417	0.5329	0.8746
45	1729.0	100.98	160.91	261.90	0.3624	0.5058	0.8682
50	1942.3	107.85	154.58	262.43	0.3832	0.4783	0.8615
55	2174.4	114.91	147.80	262.71	0.4042	0.4504	0.8546
60	2426.6	122.18	140.50	262.68	0.4255	0.4217	0.8472
65	2699.9	129.73	132.55	262.28	0.4472	0.3920	0.8391
70	2995.9	137.63	123.77	261.40	0.4695	0.3607	0.8302
75	3316.1	145.99	113.90	259.89	0.4927	0.3272	0.8198
80	3662.3	155.01	102.47	257.49	0.5173	0.2902	0.8075
85	4036.8	165.09	88.60	253.69	0.5445	0.2474	0.7918
90	4442.5	177.20	70.04	247.24	0.5767	0.1929	0.7695
95	4883.5	196.36	34.93	231.28	0.6273	0.0949	0.7222
96.0	4969.0	212.54	0	212.54	0.6708	0	0.6708

TABLE B.16.2 SI *Superheated R-22*

Temp. C	v m³/kg	h kJ/kg	s kJ/kg K	v m³/kg	h kJ/kg	s kJ/kg K	v m³/kg	h kJ/kg	s kJ/kg K
	50 kPa (-54.80)			100 kPa (-41.03)			150 kPa (-32.02)		
Sat.	0.41077	226.21	1.0384	0.21525	232.72	1.0026	0.14727	236.83	0.9826
-40	0.44063	234.72	1.0762	0.21633	233.34	1.0052	—	—	—
-30	0.46064	240.60	1.1008	0.22675	239.36	1.0305	0.14872	238.08	0.9877
-20	0.48054	246.59	1.1250	0.23706	245.47	1.0551	0.15585	244.32	1.0129
-10	0.50036	252.68	1.1485	0.24728	251.67	1.0791	0.16288	250.63	1.0373
0	0.52010	258.87	1.1717	0.25742	257.96	1.1026	0.16982	257.02	1.0612
10	0.53977	265.18	1.1943	0.26749	264.35	1.1256	0.17670	263.50	1.0844
20	0.55939	271.59	1.2166	0.27750	270.83	1.1481	0.18352	270.06	1.1072
30	0.57897	278.12	1.2385	0.28747	277.42	1.1702	0.19028	276.71	1.1295
40	0.59851	284.74	1.2600	0.29739	284.10	1.1919	0.19701	283.45	1.1514
50	0.61801	291.48	1.2811	0.30729	290.89	1.2132	0.20370	290.29	1.1729
60	0.63749	298.32	1.3020	0.31715	297.77	1.2342	0.21036	297.22	1.1940
70	0.65694	305.26	1.3225	0.32699	304.76	1.2548	0.21700	304.25	1.2148
80	0.67636	312.31	1.3428	0.33680	311.84	1.2752	0.22361	311.37	1.2353
90	0.69577	319.47	1.3627	0.34660	319.03	1.2952	0.23020	318.58	1.2554
100	0.71516	326.72	1.3824	0.35637	326.31	1.3150	0.23678	325.90	1.2753
110	0.73454	334.07	1.4019	0.36614	333.69	1.3345	0.24333	333.30	1.2948
	200 kPa (-25.12)			250 kPa (-19.46)			300 kPa (-14.61)		
Sat.	0.11237	239.87	0.9688	0.09102	242.28	0.9583	0.07657	244.29	0.9499
-20	0.11520	243.14	0.9818	—	—	—	—	—	—
-10	0.12065	249.57	1.0068	0.09528	248.49	0.9823	0.07834	247.38	0.9617
0	0.12600	256.07	1.0310	0.09969	255.10	1.0069	0.08213	254.10	0.9868
10	0.13129	262.63	1.0546	0.10402	261.76	1.0309	0.08583	260.86	1.0111
20	0.13651	269.27	1.0776	0.10829	268.48	1.0542	0.08947	267.67	1.0347
30	0.14168	275.99	1.1002	0.11251	275.27	1.0770	0.09305	274.53	1.0577
40	0.14681	282.80	1.1222	0.11668	282.13	1.0993	0.09659	281.46	1.0802
50	0.15190	289.69	1.1439	0.12082	289.08	1.1211	0.10009	288.46	1.1022
60	0.15696	296.66	1.1652	0.12492	296.10	1.1425	0.10355	295.54	1.1238
70	0.16200	303.73	1.1861	0.12899	303.21	1.1635	0.10699	302.69	1.1449
80	0.16701	310.89	1.2066	0.13304	310.41	1.1842	0.11040	309.92	1.1657
90	0.17200	318.14	1.2269	0.13708	317.69	1.2045	0.11379	317.24	1.1861
100	0.17697	325.48	1.2468	0.14109	325.06	1.2246	0.11716	324.64	1.2062
110	0.18193	332.91	1.2665	0.14509	332.52	1.2443	0.12052	332.13	1.2260
120	0.18688	340.44	1.2858	0.14907	340.07	1.2637	0.12387	339.70	1.2455
130	0.19181	348.05	1.3050	0.15304	347.70	1.2829	0.12720	347.36	1.2648

(Continued)

TABLE B.16.2 SI (Continued) *Superheated R-22*

Temp. C	*v* m³/kg	*h* kJ/kg	*s* kJ/kg K	*v* m³/kg	*h* kJ/kg	*s* kJ/kg K	*v* m³/kg	*h* kJ/kg	*s* kJ/kg K
	400 kPa (-6.52)			500 kPa (0.15)			600 kPa (5.88)		
Sat.	0.05817	247.50	0.9367	0.04692	250.00	0.9267	0.03929	252.04	0.9185
0	0.06013	252.05	0.9536	—	—	—	—	—	—
10	0.06306	259.02	0.9787	0.04936	257.11	0.9522	0.04018	255.11	0.9295
20	0.06591	266.01	1.0029	0.05175	264.30	0.9772	0.04228	262.52	0.9552
30	0.06871	273.03	1.0265	0.05408	271.48	1.0013	0.04431	269.89	0.9799
40	0.07146	280.09	1.0494	0.05636	278.69	1.0247	0.04628	277.25	1.0038
50	0.07416	287.21	1.0717	0.05859	285.93	1.0474	0.04820	284.62	1.0270
60	0.07683	294.39	1.0936	0.06079	293.22	1.0696	0.05008	292.02	1.0495
70	0.07947	301.63	1.1150	0.06295	300.55	1.0913	0.05193	299.46	1.0715
80	0.08209	308.94	1.1361	0.06509	307.95	1.1126	0.05375	306.94	1.0930
90	0.08468	316.33	1.1567	0.06721	315.41	1.1334	0.05555	314.48	1.1140
100	0.08725	323.80	1.1770	0.06930	322.94	1.1539	0.05733	322.07	1.1347
110	0.08981	331.34	1.1969	0.07138	330.54	1.1740	0.05909	329.73	1.1549
120	0.09236	338.96	1.2165	0.07345	338.21	1.1937	0.06084	337.46	1.1748
130	0.09489	346.66	1.2359	0.07550	345.96	1.2132	0.06258	345.26	1.1944
140	0.09741	354.45	1.2550	0.07755	353.79	1.2324	0.06430	353.12	1.2137
150	0.09992	362.31	1.2738	0.07958	361.69	1.2513	0.06601	361.07	1.2327
	700 kPa (10.93)			800 kPa (15.47)			900 kPa (19.61)		
Sat.	0.03377	253.73	0.9116	0.02958	255.16	0.9056	0.02629	256.39	0.9002
20	0.03549	260.67	0.9357	0.03037	258.74	0.9179	0.02636	256.71	0.9013
30	0.03731	268.24	0.9611	0.03203	266.53	0.9440	0.02792	264.76	0.9283
40	0.03906	275.77	0.9855	0.03363	274.24	0.9690	0.02940	272.67	0.9540
50	0.04076	283.28	1.0091	0.03518	281.91	0.9931	0.03082	280.50	0.9786
60	0.04242	290.80	1.0320	0.03667	289.55	1.0164	0.03219	288.28	1.0023
70	0.04405	298.34	1.0543	0.03814	297.20	1.0391	0.03353	296.04	1.0253
80	0.04565	305.91	1.0761	0.03957	304.87	1.0611	0.03483	303.81	1.0476
90	0.04722	313.53	1.0973	0.04097	312.57	1.0826	0.03611	311.59	1.0693
100	0.04878	321.19	1.1181	0.04236	320.30	1.1036	0.03736	319.40	1.0905
110	0.05031	328.91	1.1386	0.04373	328.09	1.1242	0.03860	327.25	1.1113
120	0.05183	336.70	1.1586	0.04508	335.93	1.1444	0.03982	335.15	1.1316
130	0.05334	344.54	1.1783	0.04641	343.82	1.1642	0.04102	343.09	1.1516
140	0.05484	352.45	1.1977	0.04774	351.78	1.1837	0.04222	351.10	1.1712
150	0.05632	360.44	1.2168	0.04905	359.80	1.2029	0.04340	359.16	1.1905
160	0.05780	368.49	1.2356	0.05036	367.89	1.2218	0.04457	367.28	1.2094
170	0.05927	376.61	1.2541	0.05166	376.04	1.2404	0.04574	375.47	1.2281

(Continued)

TABLE B.16.2 SI (Continued) *Superheated R-22*

Temp. C	v m³/kg	h kJ/kg	s kJ/kg K	v m³/kg	h kJ/kg	s kJ/kg K	v m³/kg	h kJ/kg	s kJ/kg K
	1000 kPa (23.42)			1200 kPa (30.26)			1400 kPa (36.31)		
Sat.	0.02364	257.46	0.8954	0.01960	259.18	0.8868	0.01668	260.48	0.8792
30	0.02460	262.91	0.9136	—	—	—	—	—	—
40	0.02599	271.04	0.9400	0.02085	267.60	0.9141	0.01712	263.86	0.8901
50	0.02732	279.05	0.9651	0.02205	276.01	0.9405	0.01825	272.77	0.9181
60	0.02860	286.97	0.9893	0.02319	284.26	0.9657	0.01930	281.40	0.9444
70	0.02984	294.86	1.0126	0.02428	292.42	0.9898	0.02029	289.86	0.9694
80	0.03104	302.73	1.0352	0.02534	300.51	1.0131	0.02125	298.20	0.9934
90	0.03221	310.60	1.0572	0.02636	308.57	1.0356	0.02217	306.47	1.0165
100	0.03336	318.49	1.0786	0.02736	316.62	1.0574	0.02306	314.70	1.0388
110	0.03449	326.41	1.0996	0.02833	324.68	1.0788	0.02393	322.92	1.0606
120	0.03561	334.36	1.1200	0.02929	332.76	1.0996	0.02477	331.13	1.0817
130	0.03671	342.36	1.1401	0.03024	340.87	1.1199	0.02561	339.35	1.1024
140	0.03780	350.41	1.1599	0.03117	349.02	1.1399	0.02643	347.60	1.1226
150	0.03887	358.51	1.1792	0.03208	357.21	1.1595	0.02723	355.89	1.1424
160	0.03994	366.68	1.1983	0.03299	365.45	1.1787	0.02803	364.21	1.1618
170	0.04100	374.90	1.2171	0.03389	373.74	1.1977	0.02882	372.57	1.1809
180	0.04205	383.19	1.2356	0.03479	382.09	1.2163	0.02960	380.99	1.1997
	1600 kPa (41.75)			1800 kPa (46.71)			2000 kPa (51.28)		
Sat.	0.01446	261.43	0.8724	0.01271	262.10	0.8659	0.01129	262.53	0.8598
50	0.01535	269.26	0.8969	0.01305	265.42	0.8763	—	—	—
60	0.01635	278.36	0.9246	0.01403	275.10	0.9057	0.01213	271.56	0.8873
70	0.01728	287.17	0.9507	0.01492	284.33	0.9330	0.01301	281.31	0.9161
80	0.01817	295.80	0.9755	0.01576	293.28	0.9588	0.01381	290.64	0.9429
90	0.01901	304.30	0.9992	0.01655	302.05	0.9832	0.01456	299.70	0.9682
100	0.01983	312.73	1.0221	0.01730	310.68	1.0067	0.01528	308.57	0.9923
110	0.02061	321.10	1.0442	0.01803	319.24	1.0293	0.01596	317.32	1.0155
120	0.02138	329.46	1.0658	0.01874	327.75	1.0512	0.01662	325.99	1.0378
130	0.02213	337.81	1.0867	0.01943	336.22	1.0725	0.01726	334.61	1.0594
140	0.02287	346.16	1.1072	0.02010	344.70	1.0933	0.01788	343.20	1.0805
150	0.02359	354.54	1.1272	0.02076	353.17	1.1136	0.01849	351.78	1.1010
160	0.02430	362.95	1.1469	0.02141	361.67	1.1334	0.01909	360.37	1.1211
170	0.02501	371.39	1.1661	0.02204	370.19	1.1528	0.01967	368.97	1.1407
180	0.02570	379.87	1.1851	0.02267	378.74	1.1719	0.02025	377.60	1.1600
190	0.02639	388.40	1.2037	0.02330	387.33	1.1907	0.02082	386.25	1.1788
200	0.02707	396.97	1.2220	0.02391	395.96	1.2091	0.02138	394.94	1.1974

(Continued)

TABLE B.16.2 SI (Continued) *Superheated R-22*

Temp. C	v m³/kg	h kJ/kg	s kJ/kg K	v m³/kg	h kJ/kg	s kJ/kg K	v m³/kg	h kJ/kg	s kJ/kg K
	2500 kPa (61.38)			3000 kPa (70.07)			3500 kPa (77.70)		
Sat.	0.00868	262.61	0.8450	0.00688	261.38	0.8300	0.00552	258.72	0.8135
70	0.00946	272.68	0.8748	—	—	—	—	—	—
80	0.01024	283.33	0.9054	0.00775	274.53	0.8678	0.00576	262.74	0.8249
90	0.01095	293.34	0.9333	0.00847	286.04	0.9000	0.00660	277.27	0.8655
100	0.01160	302.94	0.9594	0.00910	296.66	0.9288	0.00726	289.50	0.8987
110	0.01221	312.26	0.9841	0.00967	306.74	0.9555	0.00783	300.64	0.9282
120	0.01279	321.40	1.0076	0.01021	316.47	0.9805	0.00835	311.13	0.9552
130	0.01334	330.41	1.0302	0.01072	325.96	1.0044	0.00883	321.20	0.9805
140	0.01388	339.34	1.0521	0.01120	335.27	1.0272	0.00928	330.98	1.0045
150	0.01440	348.21	1.0733	0.01166	344.47	1.0492	0.00970	340.55	1.0274
160	0.01491	357.04	1.0939	0.01211	353.58	1.0705	0.01011	349.99	1.0494
170	0.01540	365.86	1.1141	0.01255	362.65	1.0912	0.01051	359.32	1.0707
180	0.01589	374.68	1.1338	0.01298	371.68	1.1113	0.01089	368.59	1.0914
190	0.01636	383.51	1.1530	0.01339	380.70	1.1310	0.01127	377.81	1.1115
200	0.01683	392.35	1.1719	0.01380	389.71	1.1502	0.01163	387.00	1.1311
210	0.01730	401.23	1.1905	0.01420	398.73	1.1691	0.01199	396.19	1.1503
220	0.01776	410.13	1.2087	0.01460	407.77	1.1876	0.01234	405.37	1.1692
	4000 kPa (84.53)			5000 kPa			6000 kPa		
Sat.	0.00443	254.13	0.7935	—	—	—	—	—	—
110	0.00641	293.75	0.9009	0.00425	275.92	0.8406	0.00243	243.28	0.7467
120	0.00692	305.27	0.9306	0.00485	291.36	0.8804	0.00333	272.39	0.8218
130	0.00739	316.08	0.9578	0.00533	304.47	0.9134	0.00390	290.25	0.8668
140	0.00782	326.42	0.9831	0.00576	316.38	0.9426	0.00435	304.76	0.9023
150	0.00823	336.45	1.0071	0.00614	327.56	0.9693	0.00473	317.63	0.9331
160	0.00861	346.25	1.0300	0.00649	338.27	0.9943	0.00507	329.55	0.9609
170	0.00898	355.89	1.0520	0.00683	348.63	1.0180	0.00539	340.85	0.9867
180	0.00933	365.41	1.0732	0.00714	358.76	1.0406	0.00568	351.72	1.0110
190	0.00968	374.85	1.0939	0.00744	368.71	1.0623	0.00596	362.27	1.0340
200	0.01001	384.24	1.1139	0.00774	378.54	1.0833	0.00622	372.60	1.0561
210	0.01033	393.59	1.1335	0.00802	388.27	1.1036	0.00648	382.76	1.0773
220	0.01065	402.93	1.1526	0.00829	397.93	1.1234	0.00672	392.80	1.0979
230	0.01097	412.25	1.1713	0.00856	407.55	1.1427	0.00696	402.75	1.1179
240	0.01127	421.58	1.1897	0.00882	417.14	1.1616	0.00719	412.62	1.1373
250	0.01158	430.91	1.2077	0.00908	426.71	1.1801	0.00742	422.45	1.1563
260	0.01188	440.27	1.2254	0.00933	436.28	1.1982	0.00764	432.25	1.1748

TABLE B.17 SI ***Thermodynamic Properties of R-23***
TABLE B.17.1 SI ***Saturated R-23***

Temp.	Press.	Specific Volume, m³/kg			Internal Energy, kJ/kg		
C	kPa	Sat. Liquid	Evap.	Sat. Vapor	Sat. Liquid	Evap.	Sat. Vapor
T	P	v_f	v_{fg}	v_g	u_f	u_{fg}	u_g
-120	6.0	0.000647	3.02595	3.02660	-195.43	248.84	53.40
-115	9.5	0.000652	1.95919	1.95984	-188.98	244.30	55.32
-110	14.6	0.000657	1.30950	1.31015	-182.79	240.01	57.21
-105	21.8	0.000663	0.90017	0.90083	-176.80	235.88	59.08
-100	31.7	0.000669	0.63438	0.63505	-170.94	231.85	60.91
-95	45.0	0.000676	0.45709	0.45776	-165.15	227.84	62.69
-90	62.5	0.000683	0.33593	0.33662	-159.39	223.81	64.42
-85	85.1	0.000690	0.25132	0.25201	-153.61	219.71	66.10
-82.0	101.3	0.000695	0.21323	0.21392	-150.17	217.23	67.06
-80	113.8	0.000698	0.19105	0.19175	-147.78	215.49	67.71
-75	149.8	0.000707	0.14734	0.14805	-141.87	211.13	69.26
-70	194.1	0.000716	0.11512	0.11584	-135.87	206.61	70.73
-65	248.1	0.000725	0.09101	0.09173	-129.76	201.90	72.14
-60	313.0	0.000736	0.07271	0.07345	-123.53	197.00	73.46
-55	390.3	0.000747	0.05865	0.05940	-117.19	191.90	74.71
-50	481.4	0.000759	0.04771	0.04847	-110.74	186.61	75.87
-45	587.7	0.000772	0.03911	0.03988	-104.17	181.12	76.95
-40	710.6	0.000786	0.03227	0.03305	-97.51	175.45	77.93
-35	851.9	0.000801	0.02677	0.02757	-90.77	169.59	78.82
-30	1012.9	0.000818	0.02231	0.02313	-83.94	163.55	79.60
-25	1195.5	0.000836	0.01866	0.01950	-77.04	157.31	80.26
-20	1401.5	0.000857	0.01564	0.01650	-70.07	150.85	80.78
-15	1632.6	0.000880	0.01312	0.01400	-63.00	144.13	81.13
-10	1891.1	0.000906	0.01099	0.01189	-55.82	137.09	81.27
-5	2179.3	0.000936	0.00917	0.01011	-48.46	129.59	81.13
0	2500.1	0.000971	0.00760	0.00858	-40.81	121.45	80.64
5	2856.7	0.001013	0.00623	0.00724	-32.72	112.36	79.64
10	3252.8	0.001066	0.00499	0.00606	-23.89	101.78	77.89
15	3693.1	0.001135	0.00385	0.00498	-13.77	88.71	74.94
20	4183.1	0.001240	0.00271	0.00395	-1.11	70.71	69.60
25	4729.4	0.001493	0.00119	0.00268	19.89	35.38	55.27
25.9	4824.1	0.001905	0	0.00190	36.79	0	36.79

(Continued)

TABLE B.17.1 SI (Continued) ***Saturated R-23***

Temp.	Press.	Enthalpy, kJ/kg			Entropy, kJ/kg K		
C T	kPa P	Sat. Liquid h_f	Evap. h_{fg}	Sat. Vapor h_g	Sat. Liquid s_f	Evap. s_{fg}	Sat. Vapor s_g
-120	6.0	-195.43	266.91	71.47	0.5575	1.7426	2.3002
-115	9.5	-188.97	262.90	73.93	0.5990	1.6623	2.2613
-110	14.6	-182.78	259.13	76.35	0.6375	1.5882	2.2257
-105	21.8	-176.79	255.51	78.73	0.6737	1.5194	2.1931
-100	31.7	-170.92	251.96	81.04	0.7080	1.4551	2.1631
-95	45.0	-165.12	248.41	83.29	0.7410	1.3943	2.1353
-90	62.5	-159.35	244.81	85.46	0.7729	1.3366	2.1095
-85	85.1	-153.55	241.10	87.54	0.8040	1.2813	2.0854
-82.0	101.3	-150.09	238.83	88.74	0.8222	1.2496	2.0718
-80	113.8	-147.70	237.24	89.54	0.8346	1.2282	2.0628
-75	149.8	-141.77	233.20	91.43	0.8648	1.1768	2.0416
-70	194.1	-135.73	228.95	93.22	0.8947	1.1269	2.0217
-65	248.1	-129.58	224.47	94.89	0.9244	1.0784	2.0028
-60	313.0	-123.30	219.76	96.45	0.9540	1.0309	1.9850
-55	390.3	-116.90	214.79	97.89	0.9834	0.9846	1.9680
-50	481.4	-110.37	209.57	99.20	1.0127	0.9391	1.9518
-45	587.7	-103.72	204.10	100.38	1.0418	0.8946	1.9364
-40	710.6	-96.96	198.38	101.42	1.0707	0.8508	1.9216
-35	851.9	-90.09	192.39	102.31	1.0994	0.8078	1.9073
-30	1012.9	-83.11	186.15	103.03	1.1279	0.7655	1.8934
-25	1195.5	-76.04	179.62	103.57	1.1560	0.7238	1.8798
-20	1401.5	-68.87	172.77	103.90	1.1840	0.6824	1.8664
-15	1632.6	-61.57	165.55	103.98	1.2117	0.6413	1.8530
-10	1891.1	-54.11	157.87	103.76	1.2395	0.5999	1.8393
-5	2179.3	-46.42	149.58	103.16	1.2674	0.5578	1.8252
0	2500.1	-38.39	140.46	102.08	1.2959	0.5142	1.8102
5	2856.7	-29.83	130.15	100.32	1.3257	0.4679	1.7936
10	3252.8	-20.42	118.02	97.60	1.3578	0.4168	1.7746
15	3693.1	-9.57	102.92	93.34	1.3940	0.3571	1.7512
20	4183.1	4.07	82.04	86.11	1.4390	0.2798	1.7188
25	4729.4	26.95	41.02	67.96	1.5138	0.1376	1.6513
25.9	4824.1	45.98	0	45.98	1.5770	0	1.5770

TABLE B.17.2 SI *Superheated R-23*

Temp. C	v m³/kg	h kJ/kg	s kJ/kg K	v m³/kg	h kJ/kg	s kJ/kg K	v m³/kg	h kJ/kg	s kJ/kg K
	25 kPa (-103.21)			50 kPa (-93.43)			75 kPa (-87.08)		
Sat.	0.79293	79.56	2.1821	0.41478	83.98	2.1270	0.28373	86.69	2.0952
-90	0.85879	87.28	2.2259	0.42365	86.07	2.1385	—	—	—
-80	0.90807	93.22	2.2575	0.44923	92.21	2.1711	0.29622	91.18	2.1189
-70	0.95698	99.25	2.2879	0.47442	98.40	2.2024	0.31353	97.54	2.1510
-60	1.00562	105.40	2.3174	0.49933	104.68	2.2325	0.33054	103.94	2.1818
-50	1.05405	111.66	2.3461	0.52402	111.04	2.2617	0.34732	110.42	2.2114
-40	1.10230	118.04	2.3741	0.54853	117.51	2.2901	0.36393	116.97	2.2402
-30	1.15042	124.55	2.4014	0.57291	124.09	2.3177	0.38040	123.63	2.2681
-20	1.19844	131.20	2.4282	0.59718	130.80	2.3447	0.39675	130.39	2.2954
-10	1.24637	137.98	2.4545	0.62137	137.63	2.3712	0.41302	137.27	2.3220
0	1.29424	144.90	2.4803	0.64548	144.59	2.3972	0.42923	144.28	2.3482
10	1.34205	151.97	2.5057	0.66954	151.69	2.4227	0.44537	151.41	2.3738
20	1.38981	159.18	2.5308	0.69356	158.93	2.4478	0.46147	158.68	2.3991
30	1.43754	166.54	2.5554	0.71754	166.32	2.4726	0.47754	166.09	2.4239
40	1.48524	174.04	2.5798	0.74149	173.84	2.4970	0.49357	173.64	2.4484
50	1.53292	181.70	2.6038	0.76542	181.51	2.5211	0.50958	181.32	2.4725
60	1.58057	189.49	2.6276	0.78932	189.32	2.5449	0.52557	189.15	2.4964
	100 kPa (-82.26)			150 kPa (-74.97)			200 kPa (-69.40)		
Sat.	0.21658	88.65	2.0728	0.14783	91.44	2.0415	0.11259	93.42	2.0194
-70	0.23305	96.66	2.1135	0.15250	94.86	2.0585	—	—	—
-60	0.24612	103.20	2.1450	0.16166	101.69	2.0914	0.11938	100.14	2.0516
-50	0.25896	109.79	2.1751	0.17057	108.51	2.1226	0.12634	107.20	2.0840
-40	0.27162	116.43	2.2043	0.17928	115.34	2.1526	0.13309	114.23	2.1148
-30	0.28413	123.16	2.2325	0.18785	122.22	2.1814	0.13969	121.26	2.1443
-20	0.29653	129.99	2.2600	0.19630	129.17	2.2095	0.14618	128.34	2.1728
-10	0.30885	136.92	2.2869	0.20467	136.20	2.2367	0.15257	135.48	2.2005
0	0.32110	143.97	2.3132	0.21296	143.33	2.2633	0.15888	142.70	2.2274
10	0.33328	151.14	2.3389	0.22119	150.57	2.2893	0.16514	150.01	2.2537
20	0.34543	158.43	2.3643	0.22938	157.93	2.3149	0.17135	157.43	2.2794
30	0.35754	165.87	2.3892	0.23753	165.41	2.3400	0.17752	164.96	2.3047
40	0.36961	173.43	2.4137	0.24565	173.02	2.3647	0.18367	172.61	2.3295
50	0.38166	181.14	2.4380	0.25374	180.76	2.3890	0.18978	180.39	2.3540
60	0.39370	188.98	2.4619	0.26182	188.64	2.4130	0.19588	188.29	2.3781
70	0.40571	196.96	2.4855	0.26988	196.65	2.4367	0.20196	196.33	2.4018
80	0.41771	205.09	2.5088	0.27792	204.79	2.4601	0.20802	204.50	2.4253

(Continued)

TABLE B.17.2 SI (Continued) ***Superheated R-23***

Temp. C	v m³/kg	h kJ/kg	s kJ/kg K	v m³/kg	h kJ/kg	s kJ/kg K	v m³/kg	h kJ/kg	s kJ/kg K
	400 kPa (-54.43)			600 kPa (-44.46)			800 kPa (-36.76)		
Sat.	0.05801	98.05	1.9661	0.03907	100.50	1.9348	0.02937	102.01	1.9122
-40	0.06368	109.56	2.0171	0.04038	104.47	1.9520	—	—	—
-30	0.06737	117.29	2.0496	0.04316	113.04	1.9880	0.03095	108.44	1.9390
-20	0.07092	124.92	2.0803	0.04576	121.31	2.0213	0.03312	117.47	1.9754
-10	0.07437	132.51	2.1097	0.04825	129.40	2.0526	0.03515	126.14	2.0090
0	0.07774	140.09	2.1380	0.05065	137.39	2.0824	0.03707	134.57	2.0405
10	0.08104	147.71	2.1654	0.05298	145.33	2.1110	0.03892	142.87	2.0703
20	0.08429	155.37	2.1920	0.05525	153.26	2.1385	0.04071	151.09	2.0989
30	0.08750	163.12	2.2180	0.05748	161.23	2.1652	0.04245	159.30	2.1264
40	0.09068	170.95	2.2434	0.05967	169.24	2.1912	0.04416	167.51	2.1530
50	0.09383	178.87	2.2683	0.06184	177.33	2.2167	0.04583	175.75	2.1789
60	0.09696	186.91	2.2928	0.06398	185.49	2.2415	0.04749	184.06	2.2043
70	0.10007	195.05	2.3168	0.06611	193.75	2.2660	0.04912	192.44	2.2290
80	0.10317	203.31	2.3406	0.06822	202.11	2.2900	0.05074	200.90	2.2533
90	0.10626	211.70	2.3640	0.07031	210.58	2.3136	0.05234	209.45	2.2772
100	0.10933	220.20	2.3871	0.07240	219.16	2.3369	0.05393	218.11	2.3007
110	0.11239	228.83	2.4099	0.07448	227.85	2.3599	0.05552	226.87	2.3239
	1000 kPa (-30.38)			2000 kPa (-8.05)			4000 kPa (18.19)		
Sat.	0.02343	102.98	1.8944	0.01116	103.58	1.8339	0.00432	89.24	1.7321
0	0.02890	131.63	2.0058	0.01222	114.14	1.8732	—	—	—
10	0.03047	140.32	2.0370	0.01334	125.76	1.9150	—	—	—
20	0.03197	148.86	2.0667	0.01435	136.42	1.9520	0.00463	95.11	1.7522
30	0.03343	157.31	2.0950	0.01527	146.49	1.9858	0.00574	116.43	1.8238
40	0.03484	165.73	2.1223	0.01613	156.18	2.0172	0.00650	131.80	1.8737
50	0.03623	174.15	2.1488	0.01696	165.63	2.0469	0.00714	144.97	1.9151
60	0.03759	182.60	2.1746	0.01774	174.91	2.0752	0.00770	156.95	1.9516
70	0.03892	191.10	2.1997	0.01851	184.11	2.1024	0.00822	168.20	1.9849
80	0.04025	199.67	2.2243	0.01925	193.27	2.1287	0.00869	178.99	2.0159
90	0.04156	208.31	2.2484	0.01997	202.41	2.1542	0.00915	189.46	2.0451
100	0.04285	217.05	2.2722	0.02068	211.57	2.1791	0.00958	199.72	2.0730
110	0.04414	225.87	2.2955	0.02138	220.77	2.2034	0.01000	209.85	2.0998
120	0.04542	234.80	2.3185	0.02207	230.02	2.2273	0.01040	219.89	2.1257
130	0.04669	243.82	2.3412	0.02275	239.33	2.2506	0.01079	229.88	2.1508
140	0.04795	252.95	2.3635	0.02343	248.71	2.2736	0.01118	239.86	2.1752
150	0.04921	262.18	2.3856	0.02410	258.16	2.2962	0.01155	249.84	2.1991

TABLE B.18 SI *Thermodynamic Properties of R-113*
TABLE B.18.1 SI *Saturated R-113*

Temp.	Press.	Specific Volume, m³/kg			Internal Energy, kJ/kg		
C	kPa	Sat. Liquid	Evap.	Sat. Vapor	Sat. Liquid	Evap.	Sat. Vapor
T	P	v_f	v_{fg}	v_g	u_f	u_{fg}	u_g
-10	9.1	0.000608	1.27293	1.27353	26.46	150.02	176.48
0	15.1	0.000617	0.79415	0.79477	35.18	146.93	182.10
10	24.1	0.000626	0.51552	0.51614	43.97	143.82	187.79
20	36.9	0.000635	0.34646	0.34709	52.86	140.68	193.54
25	45.1	0.000640	0.28734	0.28798	57.35	139.09	196.43
30	54.7	0.000645	0.24001	0.24066	61.86	137.48	199.34
35	65.8	0.000650	0.20181	0.20246	66.40	135.85	202.26
40	78.7	0.000655	0.17075	0.17141	70.98	134.21	205.18
45	93.5	0.000660	0.14531	0.14597	75.58	132.54	208.12
47.4	101.3	0.000663	0.13478	0.13544	77.80	131.73	209.52
50	110.4	0.000665	0.12433	0.12500	80.22	130.84	211.06
55	129.6	0.000671	0.10692	0.10759	84.89	129.12	214.01
60	151.2	0.000677	0.09238	0.09306	89.59	127.37	216.96
65	175.6	0.000683	0.08017	0.08086	94.33	125.58	219.92
70	202.8	0.000689	0.06986	0.07055	99.11	123.77	222.88
75	233.2	0.000695	0.06111	0.06180	103.92	121.92	225.83
80	266.9	0.000702	0.05364	0.05434	108.76	120.03	228.79
85	304.2	0.000709	0.04724	0.04795	113.64	118.10	231.74
90	345.3	0.000716	0.04172	0.04244	118.56	116.13	234.69
95	390.4	0.000723	0.03695	0.03768	123.52	114.11	237.63
100	439.9	0.000731	0.03281	0.03354	128.51	112.05	240.55
105	493.8	0.000739	0.02919	0.02993	133.54	109.93	243.47
110	552.6	0.000747	0.02603	0.02677	138.60	107.76	246.36
115	616.4	0.000756	0.02324	0.02400	143.71	105.53	249.24
120	685.6	0.000765	0.02079	0.02155	148.85	103.24	252.09
125	760.3	0.000774	0.01861	0.01939	154.03	100.88	254.92
130	840.9	0.000784	0.01668	0.01746	159.25	98.45	257.71
135	927.7	0.000795	0.01496	0.01575	164.52	95.94	260.46
140	1020.9	0.000806	0.01341	0.01422	169.83	93.33	263.16
145	1120.9	0.000818	0.01203	0.01285	175.18	90.63	265.81
150	1228.0	0.000831	0.01078	0.01161	180.58	87.81	268.39
160	1464.8	0.000860	0.00863	0.00949	191.51	81.79	273.31
170	1734.3	0.000895	0.00685	0.00774	202.66	75.13	277.78
180	2039.6	0.000938	0.00534	0.00628	214.00	67.61	281.61
190	2384.8	0.000995	0.00404	0.00503	225.49	58.94	284.43
200	2774.5	0.001079	0.00291	0.00399	236.88	48.90	285.78
210	3215.4	0.001254	0.00195	0.00320	248.00	38.00	286.00
214.1	3412.8	0.001755	0	0.00175	257.32	0	257.32

(Continued)

TABLE B.18.1 SI (Continued) *Saturated R-113*

Temp.	Press.	Enthalpy, kJ/kg			Entropy, kJ/kg K		
C T	kPa P	Sat. Liquid h_f	Evap. h_{fg}	Sat. Vapor h_g	Sat. Liquid s_f	Evap. s_{fg}	Sat. Vapor s_g
-10	9.1	26.46	161.64	188.10	0.1052	0.6143	0.7195
0	15.1	35.18	158.95	194.13	0.1377	0.5820	0.7197
10	24.1	43.99	156.23	200.22	0.1694	0.5518	0.7212
20	36.9	52.89	153.46	206.34	0.2002	0.5235	0.7238
25	45.1	57.38	152.04	209.42	0.2154	0.5100	0.7254
30	54.7	61.90	150.60	212.50	0.2304	0.4969	0.7273
35	65.8	66.45	149.14	215.59	0.2453	0.4840	0.7293
40	78.7	71.03	147.65	218.68	0.2600	0.4716	0.7316
45	93.5	75.64	146.13	221.77	0.2746	0.4594	0.7340
47.4	101.3	77.86	145.39	223.25	0.2815	0.4536	0.7352
50	110.4	80.29	144.57	224.86	0.2891	0.4474	0.7365
55	129.6	84.98	142.97	227.95	0.3034	0.4357	0.7392
60	151.2	89.70	141.34	231.03	0.3177	0.4243	0.7420
65	175.6	94.45	139.66	234.11	0.3318	0.4131	0.7448
70	202.8	99.25	137.94	237.18	0.3458	0.4020	0.7478
75	233.2	104.08	136.17	240.25	0.3597	0.3912	0.7509
80	266.9	108.95	134.34	243.29	0.3735	0.3805	0.7540
85	304.2	113.86	132.47	246.33	0.3873	0.3699	0.7572
90	345.3	118.81	130.53	249.34	0.4009	0.3595	0.7604
95	390.4	123.80	128.54	252.33	0.4145	0.3492	0.7637
100	439.9	128.83	126.48	255.30	0.4280	0.3390	0.7669
105	493.8	133.90	124.35	258.25	0.4414	0.3289	0.7702
110	552.6	139.01	122.14	261.16	0.4547	0.3188	0.7735
115	616.4	144.17	119.86	264.03	0.4679	0.3088	0.7768
120	685.6	149.37	117.49	266.87	0.4811	0.2989	0.7800
125	760.3	154.62	115.04	269.66	0.4942	0.2890	0.7832
130	840.9	159.91	112.48	272.39	0.5073	0.2790	0.7863
135	927.7	165.26	109.81	275.07	0.5203	0.2691	0.7894
140	1020.9	170.65	107.03	277.68	0.5332	0.2591	0.7923
145	1120.9	176.10	104.11	280.21	0.5461	0.2490	0.7952
150	1228.0	181.60	101.05	282.65	0.5590	0.2388	0.7979
160	1464.8	192.77	94.44	287.21	0.5847	0.2180	0.8027
170	1734.3	204.21	87.00	291.21	0.6102	0.1963	0.8066
180	2039.6	215.91	78.49	294.41	0.6357	0.1732	0.8089
190	2384.8	227.86	68.57	296.44	0.6611	0.1481	0.8091
200	2774.5	239.88	56.97	296.85	0.6859	0.1204	0.8063
210	3215.4	252.04	44.26	296.30	0.7102	0.0916	0.8019
214.1	3412.8	263.31	0	263.31	0.7323	0	0.7323

TABLE B.18.2 SI *Superheated R-113*

Temp. C	v m³/kg	h kJ/kg	s kJ/kg K	v m³/kg	h kJ/kg	s kJ/kg K	v m³/kg	h kJ/kg	s kJ/kg K
	25 kPa (10.86)			50 kPa (27.65)			75 kPa (38.62)		
Sat.	0.49827	200.74	0.7214	0.26158	211.05	0.7264	0.17934	217.83	0.7309
20	0.51478	206.51	0.7414	—	—	—	—	—	—
30	0.53281	212.91	0.7629	0.26372	212.57	0.7314	—	—	—
40	0.55082	219.39	0.7839	0.27287	219.06	0.7525	0.18019	218.73	0.7338
50	0.56882	225.96	0.8045	0.28200	225.64	0.7732	0.18636	225.32	0.7546
60	0.58681	232.61	0.8248	0.29111	232.31	0.7935	0.19252	232.00	0.7749
70	0.60478	239.36	0.8448	0.30021	239.07	0.8135	0.19867	238.77	0.7949
80	0.62273	246.18	0.8644	0.30930	245.90	0.8331	0.20480	245.62	0.8146
90	0.64068	253.10	0.8837	0.31837	252.83	0.8525	0.21092	252.56	0.8340
100	0.65861	260.10	0.9027	0.32743	259.84	0.8715	0.21702	259.58	0.8531
110	0.67654	267.19	0.9215	0.33648	266.94	0.8903	0.22312	266.69	0.8719
120	0.69445	274.37	0.9400	0.34552	274.12	0.9088	0.22920	273.88	0.8904
130	0.71236	281.63	0.9582	0.35456	281.40	0.9271	0.23528	281.16	0.9087
140	0.73026	288.98	0.9762	0.36358	288.75	0.9451	0.24135	288.52	0.9267
150	0.74815	296.42	0.9940	0.37260	296.20	0.9629	0.24741	295.98	0.9446
160	0.76604	303.94	1.0116	0.38161	303.73	0.9805	0.25347	303.51	0.9622
170	0.78392	311.55	1.0289	0.39062	311.34	0.9979	0.25951	311.14	0.9796
	100 kPa (47.00)			150 kPa (59.73)			200 kPa (69.51)		
Sat.	0.13711	223.00	0.7350	0.09377	230.87	0.7418	0.07149	236.88	0.7475
50	0.13853	225.00	0.7412	—	—	—	—	—	—
60	0.14321	231.69	0.7616	0.09386	231.05	0.7423	—	—	—
70	0.14788	238.47	0.7816	0.09706	237.86	0.7625	0.07161	237.22	0.7485
80	0.15254	245.33	0.8013	0.10025	244.74	0.7823	0.07407	244.14	0.7684
90	0.15718	252.28	0.8207	0.10342	251.71	0.8017	0.07651	251.13	0.7879
100	0.16181	259.31	0.8398	0.10657	258.77	0.8209	0.07893	258.21	0.8071
110	0.16643	266.43	0.8587	0.10972	265.91	0.8398	0.08134	265.37	0.8261
120	0.17104	273.63	0.8772	0.11285	273.13	0.8584	0.08374	272.61	0.8447
130	0.17564	280.92	0.8955	0.11598	280.43	0.8767	0.08613	279.94	0.8631
140	0.18023	288.29	0.9136	0.11909	287.82	0.8948	0.08851	287.35	0.8813
150	0.18481	295.75	0.9314	0.12220	295.30	0.9127	0.09089	294.84	0.8992
160	0.18939	303.30	0.9491	0.12530	302.86	0.9304	0.09325	302.42	0.9169
170	0.19396	310.93	0.9665	0.12839	310.51	0.9478	0.09560	310.08	0.9344
180	0.19852	318.65	0.9837	0.13148	318.24	0.9651	0.09795	317.82	0.9517
190	0.20308	326.45	1.0007	0.13456	326.05	0.9821	0.10030	325.65	0.9688
200	0.20763	334.34	1.0176	0.13764	333.95	0.9990	0.10263	333.57	0.9857

(Continued)

TABLE B.18.2 SI (Continued) *Superheated R-113*

Temp. C	v m³/kg	h kJ/kg	s kJ/kg K	v m³/kg	h kJ/kg	s kJ/kg K	v m³/kg	h kJ/kg	s kJ/kg K
	400 kPa (96.00)			600 kPa (113.75)			800 kPa (127.51)		
Sat.	0.03680	252.93	0.7643	0.02466	263.32	0.7759	0.01839	271.04	0.7848
110	0.03867	263.09	0.7913	—	—	—	—	—	—
120	0.03998	270.43	0.8102	0.02527	268.00	0.7880	—	—	—
130	0.04129	277.85	0.8289	0.02623	275.55	0.8069	0.01859	272.97	0.7896
140	0.04257	285.35	0.8472	0.02717	283.16	0.8256	0.01939	280.74	0.8086
150	0.04385	292.92	0.8653	0.02810	290.84	0.8439	0.02016	288.56	0.8273
160	0.04512	300.57	0.8832	0.02902	298.58	0.8620	0.02091	296.43	0.8457
170	0.04638	308.30	0.9009	0.02992	306.40	0.8799	0.02164	304.36	0.8638
180	0.04763	316.11	0.9183	0.03081	314.29	0.8975	0.02236	312.35	0.8816
190	0.04887	324.00	0.9355	0.03169	322.25	0.9148	0.02307	320.40	0.8992
200	0.05010	331.97	0.9525	0.03257	330.29	0.9320	0.02377	328.52	0.9166
210	0.05133	340.02	0.9694	0.03343	338.41	0.9490	0.02446	336.72	0.9337
220	0.05256	348.16	0.9861	0.03429	346.60	0.9658	0.02514	344.98	0.9506
230	0.05377	356.38	1.0026	0.03515	354.88	0.9824	0.02582	353.32	0.9674
240	0.05499	364.68	1.0189	0.03599	363.23	0.9988	0.02649	361.73	0.9839
250	0.05619	373.06	1.0351	0.03684	371.67	1.0151	0.02715	370.22	1.0003
260	0.05740	381.53	1.0511	0.03767	380.18	1.0312	0.02781	378.79	1.0165
	1000 kPa (138.91)			2000 kPa (178.77)			3000 kPa (205.27)		
Sat.	0.01454	277.11	0.7917	0.00644	294.07	0.8087	0.00353	296.49	0.8038
150	0.01532	286.02	0.8130	—	—	—	—	—	—
160	0.01598	294.06	0.8318	—	—	—	—	—	—
170	0.01663	302.14	0.8502	—	—	—	—	—	—
180	0.01726	310.26	0.8683	0.00652	295.34	0.8116	—	—	—
190	0.01787	318.43	0.8862	0.00711	305.30	0.8333	—	—	—
200	0.01847	326.65	0.9037	0.00762	314.80	0.8536	—	—	—
210	0.01906	334.93	0.9211	0.00808	324.07	0.8730	0.00395	304.88	0.8212
220	0.01964	343.28	0.9382	0.00850	333.22	0.8917	0.00454	318.10	0.8483
230	0.02021	351.69	0.9551	0.00891	342.30	0.9100	0.00499	329.43	0.8711
240	0.02078	360.18	0.9718	0.00929	351.35	0.9278	0.00538	340.00	0.8919
250	0.02133	368.73	0.9883	0.00965	360.40	0.9453	0.00572	350.17	0.9115
260	0.02188	377.36	1.0046	0.01001	369.47	0.9624	0.00603	360.11	0.9303
270	0.02243	386.06	1.0208	0.01035	378.56	0.9793	0.00633	369.91	0.9485
280	0.02297	394.83	1.0368	0.01069	387.69	0.9960	0.00661	379.64	0.9663
290	0.02350	403.69	1.0526	0.01101	396.86	1.0124	0.00688	389.32	0.9836
300	0.02403	412.61	1.0684	0.01133	406.09	1.0286	0.00713	398.98	1.0006

TABLE B.19 SI *Thermodynamic Properties of R-114*
TABLE B.19.1 SI *Saturated R-114*

Temp.	Press.	Specific Volume, m^3/kg			Internal Energy, kJ/kg		
C	kPa	Sat. Liquid	Evap.	Sat. Vapor	Sat. Liquid	Evap.	Sat. Vapor
T	P	v_f	v_{fg}	v_g	u_f	u_{fg}	u_g
-60	3.7	0.000594	2.77503	2.77563	-16.65	143.60	126.94
-50	7.2	0.000602	1.49354	1.49414	-8.44	140.72	132.29
-40	13.1	0.000612	0.85503	0.85564	-0.01	137.77	137.77
-30	22.6	0.000621	0.51614	0.51676	8.65	134.72	143.37
-20	37.0	0.000631	0.32612	0.32675	17.55	131.53	149.08
-10	58.0	0.000642	0.21432	0.21496	26.69	128.18	154.88
-5	71.6	0.000648	0.17602	0.17667	31.36	126.44	157.80
0	87.5	0.000654	0.14571	0.14637	36.09	124.66	160.75
3.8	101.3	0.000658	0.12697	0.12763	39.70	123.28	162.98
5	106.2	0.000660	0.12150	0.12216	40.88	122.83	163.71
10	127.8	0.000666	0.10200	0.10267	45.73	120.95	166.68
15	152.7	0.000673	0.08617	0.08685	50.65	119.02	169.66
20	181.1	0.000680	0.07322	0.07390	55.62	117.03	172.65
25	213.5	0.000687	0.06255	0.06324	60.64	115.00	175.64
30	250.0	0.000694	0.05370	0.05439	65.73	112.91	178.64
35	291.2	0.000702	0.04631	0.04701	70.87	110.76	181.63
40	337.2	0.000710	0.04010	0.04081	76.06	108.56	184.62
45	388.5	0.000718	0.03486	0.03557	81.30	106.31	187.61
50	445.4	0.000727	0.03040	0.03112	86.59	103.99	190.58
55	508.3	0.000736	0.02659	0.02733	91.93	101.62	193.55
60	577.5	0.000746	0.02332	0.02406	97.31	99.18	196.50
65	653.4	0.000757	0.02050	0.02125	102.74	96.68	199.42
70	736.4	0.000768	0.01805	0.01881	108.22	94.11	202.32
75	827.0	0.000779	0.01591	0.01669	113.73	91.47	205.20
80	925.4	0.000792	0.01404	0.01484	119.29	88.74	208.03
85	1032.2	0.000806	0.01240	0.01320	124.89	85.93	210.82
90	1147.9	0.000820	0.01094	0.01176	130.54	83.01	213.55
95	1272.9	0.000836	0.00965	0.01049	136.24	79.98	216.22
100	1407.8	0.000854	0.00849	0.00935	142.00	76.81	218.81
105	1553.1	0.000873	0.00746	0.00833	147.82	73.48	221.30
110	1709.7	0.000895	0.00651	0.00741	153.73	69.94	223.67
115	1878.3	0.000920	0.00565	0.00657	159.74	66.14	225.88
120	2059.8	0.000949	0.00486	0.00581	165.89	62.00	227.89
125	2255.5	0.000984	0.00412	0.00510	172.24	57.36	229.60
130	2466.9	0.001027	0.00341	0.00443	178.91	52.00	230.90
135	2696.0	0.001085	0.00270	0.00379	186.13	45.39	231.52
140	2945.8	0.001174	0.00194	0.00311	194.52	36.24	230.76
145.7	3263.2	0.001719	0	0.00172	216.35	0	216.35

(Continued)

TABLE B.19.1 SI (Continued) *Saturated R-114*

Temp. C T	Press. kPa P	Enthalpy, kJ/kg Sat. Liquid h_f	Evap. h_{fg}	Sat. Vapor h_g	Entropy, kJ/kg K Sat. Liquid s_f	Evap. s_{fg}	Sat. Vapor s_g
-60	3.7	-16.65	153.93	137.28	-0.0746	0.7221	0.6475
-50	7.2	-8.43	151.52	143.08	-0.0370	0.6790	0.6420
-40	13.1	0	149.01	149.01	0	0.6391	0.6391
-30	22.6	8.66	146.38	155.05	0.0364	0.6020	0.6383
-20	37.0	17.57	143.59	161.16	0.0722	0.5672	0.6394
-10	58.0	26.73	140.61	167.34	0.1076	0.5343	0.6420
-5	71.6	31.41	139.04	170.45	0.1252	0.5185	0.6437
0	87.5	36.15	137.41	173.56	0.1427	0.5030	0.6457
3.8	101.3	39.76	136.15	175.91	0.1558	0.4916	0.6474
5	106.2	40.95	135.73	176.68	0.1601	0.4879	0.6480
10	127.8	45.82	133.98	179.80	0.1774	0.4732	0.6505
15	152.7	50.75	132.17	182.92	0.1946	0.4587	0.6532
20	181.1	55.74	130.29	186.03	0.2117	0.4444	0.6561
25	213.5	60.79	128.35	189.14	0.2287	0.4305	0.6591
30	250.0	65.90	126.33	192.24	0.2456	0.4167	0.6623
35	291.2	71.07	124.25	195.32	0.2624	0.4032	0.6656
40	337.2	76.30	122.09	198.39	0.2791	0.3899	0.6690
45	388.5	81.58	119.85	201.43	0.2958	0.3767	0.6724
50	445.4	86.92	117.53	204.45	0.3123	0.3637	0.6760
55	508.3	92.30	115.13	207.44	0.3287	0.3508	0.6795
60	577.5	97.75	112.65	210.39	0.3450	0.3381	0.6831
65	653.4	103.24	110.07	213.31	0.3612	0.3255	0.6867
70	736.4	108.78	107.40	216.18	0.3772	0.3130	0.6902
75	827.0	114.38	104.62	219.00	0.3932	0.3005	0.6937
80	925.4	120.02	101.74	221.76	0.4091	0.2881	0.6972
85	1032.2	125.72	98.72	224.45	0.4249	0.2756	0.7005
90	1147.9	131.48	95.57	227.06	0.4406	0.2632	0.7038
95	1272.9	137.31	92.26	229.57	0.4563	0.2506	0.7069
100	1407.8	143.20	88.77	231.97	0.4719	0.2379	0.7097
105	1553.1	149.18	85.06	234.24	0.4874	0.2249	0.7124
110	1709.7	155.26	81.08	236.34	0.5030	0.2116	0.7146
115	1878.3	161.47	76.76	238.23	0.5187	0.1978	0.7165
120	2059.8	167.84	72.01	239.85	0.5346	0.1832	0.7178
125	2255.5	174.46	66.65	241.11	0.5509	0.1674	0.7183
130	2466.9	181.44	60.40	241.84	0.5678	0.1498	0.7176
135	2696.0	189.05	52.68	241.73	0.5859	0.1291	0.7150
140	2945.8	197.98	41.95	239.93	0.6070	0.1015	0.7085
145.7	3263.2	221.95	0	221.95	0.6635	0	0.6635

TABLE B.19.2 SI *Superheated R-114*

Temp. C	v m³/kg	h kJ/kg	s kJ/kg K	v m³/kg	h kJ/kg	s kJ/kg K	v m³/kg	h kJ/kg	s kJ/kg K
	25 kPa (-28.02)			50 kPa (-13.39)			75 kPa (-3.85)		
Sat.	0.47037	156.25	0.6384	0.24678	165.24	0.6409	0.16909	171.16	0.6442
-10	0.50630	167.91	0.6843	0.25023	167.48	0.6495	—	—	—
0	0.52616	174.58	0.7092	0.26038	174.18	0.6745	0.17175	173.77	0.6538
10	0.54599	181.39	0.7337	0.27049	181.01	0.6991	0.17862	180.63	0.6784
20	0.56578	188.33	0.7578	0.28056	187.98	0.7232	0.18546	187.62	0.7027
30	0.58554	195.40	0.7815	0.29059	195.07	0.7470	0.19226	194.73	0.7265
40	0.60527	202.59	0.8048	0.30061	202.28	0.7704	0.19903	201.96	0.7500
50	0.62499	209.90	0.8278	0.31059	209.61	0.7935	0.20578	209.31	0.7731
60	0.64467	217.33	0.8504	0.32056	217.05	0.8161	0.21251	216.77	0.7958
70	0.66434	224.86	0.8727	0.33050	224.60	0.8385	0.21921	224.34	0.8182
80	0.68400	232.51	0.8947	0.34043	232.26	0.8605	0.22590	232.01	0.8403
90	0.70364	240.25	0.9163	0.35034	240.02	0.8821	0.23257	239.79	0.8620
100	0.72326	248.10	0.9376	0.36024	247.88	0.9035	0.23923	247.65	0.8833
110	0.74287	256.03	0.9586	0.37012	255.82	0.9245	0.24587	255.61	0.9044
120	0.76247	264.06	0.9793	0.38000	263.86	0.9452	0.25250	263.66	0.9251
130	0.78207	272.17	0.9996	0.38986	271.98	0.9656	0.25912	271.79	0.9455
140	0.80165	280.36	1.0197	0.39972	280.18	0.9857	0.26574	279.99	0.9656
	100 kPa (3.43)			200 kPa (22.99)			300 kPa (36.00)		
Sat.	0.12921	175.70	0.6473	0.06727	187.90	0.6579	0.04568	195.94	0.6663
30	0.14308	194.39	0.7118	0.06922	192.98	0.6749	—	—	—
40	0.14823	201.65	0.7353	0.07197	200.33	0.6987	0.04646	198.93	0.6759
50	0.15336	209.01	0.7585	0.07468	207.78	0.7222	0.04839	206.48	0.6996
60	0.15847	216.49	0.7813	0.07737	215.34	0.7452	0.05029	214.12	0.7229
70	0.16356	224.08	0.8037	0.08004	222.99	0.7678	0.05216	221.86	0.7458
80	0.16863	231.76	0.8258	0.08269	230.74	0.7901	0.05400	229.68	0.7683
90	0.17368	239.55	0.8475	0.08531	238.58	0.8120	0.05583	237.58	0.7903
100	0.17871	247.43	0.8689	0.08793	246.52	0.8335	0.05764	245.58	0.8120
110	0.18374	255.40	0.8900	0.09052	254.54	0.8547	0.05943	253.65	0.8334
120	0.18875	263.46	0.9108	0.09311	262.64	0.8756	0.06121	261.80	0.8544
130	0.19375	271.59	0.9312	0.09568	270.82	0.8961	0.06298	270.02	0.8750
140	0.19874	279.81	0.9513	0.09825	279.07	0.9164	0.06473	278.32	0.8954
150	0.20373	288.10	0.9712	0.10080	287.40	0.9363	0.06648	286.68	0.9154
160	0.20870	296.46	0.9907	0.10335	295.79	0.9559	0.06822	295.10	0.9350
170	0.21367	304.88	1.0099	0.10588	304.24	0.9752	0.06995	303.59	0.9544
180	0.21863	313.37	1.0288	0.10842	312.75	0.9942	0.07167	312.13	0.9735

(Continued)

TABLE B.19.2 SI (Continued) ***Superheated R-114***

Temp. C	*v* m³/kg	*h* kJ/kg	*s* kJ/kg K	*v* m³/kg	*h* kJ/kg	*s* kJ/kg K	*v* m³/kg	*h* kJ/kg	*s* kJ/kg K
	400 kPa (46.05)			600 kPa (61.53)			800 kPa (73.56)		
Sat.	0.03458	202.07	0.6732	0.02316	211.29	0.6842	0.01728	218.19	0.6927
60	0.03670	212.85	0.7062	—	—	—	—	—	—
70	0.03818	220.67	0.7294	0.02410	218.11	0.7043	—	—	—
80	0.03963	228.57	0.7521	0.02518	226.21	0.7276	0.01787	223.58	0.7081
90	0.04106	236.55	0.7743	0.02623	234.36	0.7503	0.01875	231.96	0.7315
100	0.04247	244.60	0.7962	0.02726	242.56	0.7726	0.01960	240.35	0.7543
110	0.04387	252.73	0.8177	0.02826	250.82	0.7944	0.02042	248.77	0.7766
120	0.04525	260.94	0.8388	0.02925	259.14	0.8159	0.02122	257.24	0.7984
130	0.04661	269.21	0.8596	0.03022	267.52	0.8369	0.02200	265.74	0.8197
140	0.04797	277.54	0.8801	0.03118	275.95	0.8576	0.02276	274.28	0.8407
150	0.04931	285.95	0.9001	0.03213	284.44	0.8779	0.02351	282.87	0.8612
160	0.05065	294.41	0.9199	0.03306	292.98	0.8978	0.02425	291.50	0.8814
170	0.05197	302.93	0.9394	0.03399	301.57	0.9174	0.02498	300.17	0.9012
180	0.05329	311.50	0.9585	0.03491	310.21	0.9367	0.02570	308.89	0.9206
190	0.05461	320.12	0.9773	0.03582	318.90	0.9557	0.02641	317.64	0.9397
200	0.05591	328.79	0.9958	0.03672	327.62	0.9743	0.02712	326.43	0.9585
210	0.05721	337.50	1.0140	0.03762	336.39	0.9926	0.02782	335.25	0.9769
	1000 kPa (83.53)			2000 kPa (118.39)			3000 kPa (141.03)		
Sat.	0.01366	223.67	0.6996	0.00605	239.37	0.7174	0.00296	239.16	0.7062
100	0.01494	237.93	0.7387	—	—	—	—	—	—
110	0.01566	246.56	0.7615	—	—	—	—	—	—
120	0.01636	255.20	0.7837	0.00616	241.15	0.7220	—	—	—
130	0.01704	263.85	0.8055	0.00678	251.71	0.7485	—	—	—
140	0.01769	272.53	0.8267	0.00731	261.74	0.7731	—	—	—
150	0.01833	281.23	0.8475	0.00778	271.48	0.7964	0.00384	255.62	0.7456
160	0.01896	289.96	0.8679	0.00823	281.05	0.8187	0.00439	268.34	0.7753
170	0.01957	298.72	0.8879	0.00864	290.51	0.8403	0.00483	279.66	0.8012
180	0.02017	307.52	0.9076	0.00903	299.89	0.8613	0.00521	290.34	0.8250
190	0.02077	316.34	0.9268	0.00941	309.23	0.8816	0.00555	300.66	0.8475
200	0.02135	325.20	0.9457	0.00978	318.53	0.9015	0.00586	310.73	0.8691
210	0.02193	334.08	0.9643	0.01013	327.81	0.9209	0.00616	320.65	0.8898
220	0.02250	342.99	0.9826	0.01047	337.07	0.9399	0.00644	330.44	0.9098
230	0.02307	351.91	1.0005	0.01081	346.31	0.9585	0.00670	340.14	0.9293
240	0.02363	360.86	1.0181	0.01113	355.55	0.9766	0.00696	349.77	0.9483
250	0.02419	369.83	1.0354	0.01146	364.78	0.9944	0.00721	359.35	0.9668

TABLE B.20 SI *Thermodynamic Properties of R-123*
TABLE B.20.1 SI *Saturated R-123*

Temp.	Press.	Specific Volume, m^3/kg			Internal Energy, kJ/kg		
C	kPa	Sat. Liquid	Evap.	Sat. Vapor	Sat. Liquid	Evap.	Sat. Vapor
T	P	v_f	v_{fg}	v_g	u_f	u_{fg}	u_g
-50	1.8	0.000609	6.84542	6.84602	-9.44	188.29	178.86
-40	3.6	0.000617	3.53134	3.53196	0.00	184.01	184.00
-30	6.7	0.000626	1.94639	1.94702	9.53	179.74	189.27
-20	12.0	0.000635	1.13579	1.13642	19.15	175.48	194.63
-10	20.2	0.000645	0.69626	0.69691	28.89	171.20	200.09
0	32.6	0.000655	0.44544	0.44609	38.73	166.90	205.63
5	40.8	0.000661	0.36137	0.36203	43.69	164.73	208.42
10	50.6	0.000666	0.29571	0.29637	48.68	162.54	211.22
15	62.1	0.000672	0.24392	0.24459	53.70	160.34	214.04
20	75.6	0.000677	0.20271	0.20338	58.75	158.12	216.87
25	91.4	0.000683	0.16963	0.17031	63.83	155.88	219.71
27.8	101.3	0.000687	0.15384	0.15453	66.71	154.60	221.31
30	109.6	0.000689	0.14287	0.14356	68.93	153.61	222.55
35	130.5	0.000695	0.12106	0.12175	74.07	151.33	225.40
40	154.5	0.000702	0.10315	0.10385	79.23	149.01	228.25
45	181.7	0.000709	0.08834	0.08905	84.43	146.67	231.10
50	212.5	0.000715	0.07603	0.07674	89.66	144.29	233.95
55	247.1	0.000723	0.06572	0.06644	94.92	141.88	236.79
60	285.9	0.000730	0.05704	0.05777	100.21	139.43	239.64
65	329.2	0.000738	0.04969	0.05043	105.53	136.93	242.47
70	377.2	0.000746	0.04344	0.04418	110.89	134.39	245.29
75	430.4	0.000754	0.03809	0.03885	116.29	131.81	248.10
80	489.1	0.000763	0.03350	0.03426	121.73	129.16	250.89
85	553.6	0.000772	0.02953	0.03030	127.20	126.46	253.66
90	624.2	0.000781	0.02609	0.02687	132.72	123.69	256.40
95	701.4	0.000791	0.02310	0.02389	138.28	120.84	259.12
100	785.5	0.000802	0.02048	0.02128	143.89	117.92	261.81
105	876.9	0.000813	0.01818	0.01899	149.55	114.91	264.46
110	976.0	0.000825	0.01615	0.01697	155.26	111.80	267.06
115	1083.2	0.000838	0.01435	0.01519	161.04	108.58	269.61
120	1199.0	0.000852	0.01276	0.01361	166.88	105.23	272.11
125	1323.7	0.000866	0.01134	0.01220	172.79	101.74	274.53
130	1457.8	0.000882	0.01006	0.01094	178.78	98.09	276.87
140	1756.3	0.000919	0.00788	0.00879	191.06	90.19	281.25
150	2098.7	0.000965	0.00607	0.00703	203.83	81.22	285.05
160	2490.1	0.001025	0.00452	0.00555	217.33	70.62	287.95
170	2937.2	0.001115	0.00313	0.00425	232.09	57.09	289.17
180	3450.6	0.001306	0.00162	0.00292	250.47	35.01	285.48
183.7	3661.8	0.001818	0	0.00182	269.49	0	269.49

(Continued)

TABLE B.20.1 SI (Continued) ***Saturated R-123***

Temp.	Press.	Enthalpy, kJ/kg			Entropy, kJ/kg K		
C	kPa	Sat. Liquid	Evap.	Sat. Vapor	Sat. Liquid	Evap.	Sat. Vapor
T	P	h_f	h_{fg}	h_g	s_f	s_{fg}	s_g
-50	1.8	-9.44	200.40	190.96	-0.0414	0.8980	0.8567
-40	3.6	0	196.63	196.63	0	0.8434	0.8434
-30	6.7	9.53	192.87	202.40	0.0400	0.7932	0.8332
-20	12.0	19.16	189.11	208.27	0.0788	0.7470	0.8258
-10	20.2	28.90	185.30	214.20	0.1165	0.7042	0.8207
0	32.6	38.75	181.44	220.19	0.1532	0.6642	0.8175
5	40.8	43.72	179.48	223.20	0.1712	0.6453	0.8165
10	50.6	48.72	177.49	226.21	0.1890	0.6269	0.8159
15	62.1	53.74	175.48	229.23	0.2066	0.6090	0.8156
20	75.6	58.80	173.44	232.25	0.2240	0.5917	0.8156
25	91.4	63.89	171.37	235.26	0.2411	0.5748	0.8159
27.8	101.3	66.78	170.19	236.97	0.2508	0.5655	0.8162
30	109.6	69.01	169.27	238.28	0.2581	0.5584	0.8165
35	130.5	74.16	167.13	241.29	0.2749	0.5424	0.8173
40	154.5	79.34	164.95	244.29	0.2916	0.5267	0.8183
45	181.7	84.56	162.72	247.28	0.3080	0.5115	0.8195
50	212.5	89.81	160.44	250.25	0.3243	0.4965	0.8208
55	247.1	95.10	158.12	253.21	0.3405	0.4818	0.8223
60	285.9	100.42	155.73	256.15	0.3565	0.4675	0.8240
65	329.2	105.78	153.29	259.07	0.3724	0.4533	0.8257
70	377.2	111.18	150.78	261.96	0.3881	0.4394	0.8275
75	430.4	116.62	148.20	264.82	0.4037	0.4257	0.8294
80	489.1	122.10	145.54	267.64	0.4192	0.4121	0.8314
85	553.6	127.63	142.80	270.43	0.4347	0.3987	0.8334
90	624.2	133.21	139.97	273.18	0.4500	0.3854	0.8354
95	701.4	138.83	137.04	275.88	0.4652	0.3723	0.8374
100	785.5	144.52	134.01	278.52	0.4803	0.3591	0.8395
105	876.9	150.26	130.85	281.11	0.4954	0.3460	0.8415
110	976.0	156.07	127.56	283.63	0.5105	0.3329	0.8434
115	1083.2	161.94	124.13	286.07	0.5255	0.3198	0.8453
120	1199.0	167.90	120.53	288.43	0.5405	0.3066	0.8471
125	1323.7	173.94	116.75	290.68	0.5555	0.2932	0.8487
130	1457.8	180.07	112.76	292.83	0.5705	0.2797	0.8502
140	1756.3	192.68	104.02	296.69	0.6007	0.2518	0.8525
150	2098.7	205.86	93.95	299.81	0.6315	0.2220	0.8535
160	2490.1	219.88	81.89	301.77	0.6633	0.1890	0.8524
170	2937.2	235.36	66.28	301.64	0.6975	0.1496	0.8471
180	3450.6	254.98	40.60	295.57	0.7399	0.0896	0.8295
183.7	3661.8	276.15	0	276.15	0.7857	0	0.7857

TABLE B.20.2 SI *Superheated R-123*

Temp. C	v m³/kg	h kJ/kg	s kJ/kg K	v m³/kg	h kJ/kg	s kJ/kg K	v m³/kg	h kJ/kg	s kJ/kg K
	25 kPa (-5.70)			50 kPa (9.73)			100 kPa (27.46)		
Sat.	0.57233	216.77	0.8191	0.29952	226.05	0.8159	0.15645	236.75	0.8162
10	0.60783	226.98	0.8562	0.29983	226.23	0.8165	—	—	—
20	0.63029	233.64	0.8793	0.31144	232.96	0.8399	—	—	—
30	0.65266	240.41	0.9020	0.32296	239.79	0.8628	0.15800	238.53	0.8221
40	0.67495	247.29	0.9243	0.33439	246.73	0.8853	0.16402	245.59	0.8450
50	0.69717	254.29	0.9463	0.34575	253.78	0.9075	0.16997	252.74	0.8675
60	0.71934	261.39	0.9680	0.35705	260.93	0.9293	0.17585	259.98	0.8895
70	0.74146	268.61	0.9893	0.36831	268.18	0.9507	0.18168	267.31	0.9112
80	0.76355	275.93	1.0103	0.37952	275.54	0.9719	0.18747	274.73	0.9325
90	0.78560	283.36	1.0311	0.39070	282.99	0.9927	0.19323	282.25	0.9535
100	0.80762	290.88	1.0515	0.40185	290.54	1.0132	0.19894	289.85	0.9742
120	0.85158	306.23	1.0916	0.42408	305.93	1.0534	0.21031	305.33	1.0146
140	0.89547	321.96	1.1306	0.44622	321.70	1.0925	0.22159	321.17	1.0539
160	0.93930	338.06	1.1686	0.46831	337.82	1.1306	0.23280	337.34	1.0921
180	0.98309	354.50	1.2057	0.49035	354.28	1.1677	0.24397	353.86	1.1294
200	1.02683	371.27	1.2420	0.51235	371.07	1.2040	0.25510	370.69	1.1657
220	1.07055	388.36	1.2773	0.53432	388.18	1.2394	0.26620	387.83	1.2012
	150 kPa (39.11)			200 kPa (48.05)			300 kPa (61.69)		
Sat.	0.10677	243.76	0.8181	0.08129	249.09	0.8203	0.05515	257.14	0.8245
50	0.11131	251.66	0.8430	0.08192	250.54	0.8248	—	—	—
60	0.11540	259.00	0.8653	0.08513	257.98	0.8475	—	—	—
70	0.11944	266.41	0.8873	0.08828	265.49	0.8697	0.05704	263.55	0.8434
80	0.12342	273.91	0.9088	0.09137	273.06	0.8914	0.05926	271.30	0.8657
90	0.12737	281.49	0.9300	0.09442	280.71	0.9128	0.06143	279.09	0.8875
100	0.13129	289.15	0.9508	0.09744	288.43	0.9337	0.06355	286.94	0.9088
120	0.13903	304.72	0.9914	0.10339	304.10	0.9746	0.06771	302.82	0.9502
140	0.14670	320.63	1.0309	0.10924	320.08	1.0143	0.07177	318.97	0.9903
160	0.15429	336.87	1.0693	0.11503	336.38	1.0528	0.07576	335.40	1.0291
180	0.16184	353.42	1.1066	0.12077	352.99	1.0903	0.07969	352.11	1.0668
200	0.16934	370.30	1.1431	0.12646	369.90	1.1268	0.08358	369.11	1.1036
220	0.17682	387.47	1.1786	0.13213	387.11	1.1624	0.08743	386.39	1.1393

(Continued)

TABLE B.20.2 SI (Continued) ***Superheated R-123***

Temp. C	v m^3/kg	h kJ/kg	s kJ/kg K	v m^3/kg	h kJ/kg	s kJ/kg K	v m^3/kg	h kJ/kg	s kJ/kg K
	400 kPa (72.20)			600 kPa (88.34)			800 kPa (100.82)		
Sat.	0.04173	263.22	0.8283	0.02796	272.27	0.8347	0.02088	278.95	0.8398
90	0.04487	277.40	0.8684	0.02818	273.67	0.8386	—	—	—
100	0.04657	285.39	0.8901	0.02948	282.03	0.8613	—	—	—
110	0.04822	293.42	0.9113	0.03072	290.37	0.8834	0.02186	286.99	0.8610
120	0.04985	301.50	0.9322	0.03192	298.71	0.9049	0.02288	295.66	0.8834
130	0.05145	309.64	0.9526	0.03309	307.06	0.9258	0.02384	304.29	0.9051
140	0.05302	317.83	0.9726	0.03422	315.44	0.9464	0.02478	312.90	0.9262
150	0.05457	326.08	0.9924	0.03534	323.86	0.9665	0.02568	321.52	0.9468
160	0.05611	334.39	1.0118	0.03643	332.32	0.9863	0.02656	330.14	0.9669
170	0.05763	342.77	1.0309	0.03751	340.83	1.0057	0.02743	338.80	0.9867
180	0.05914	351.22	1.0498	0.03858	349.39	1.0248	0.02827	347.48	1.0060
190	0.06064	359.73	1.0683	0.03963	358.00	1.0436	0.02910	356.21	1.0251
200	0.06213	368.31	1.0867	0.04067	366.67	1.0621	0.02992	364.98	1.0438
210	0.06361	376.95	1.1048	0.04170	375.40	1.0803	0.03073	373.80	1.0623
220	0.06508	385.66	1.1226	0.04272	384.18	1.0983	0.03153	382.67	1.0804
	1000 kPa (111.15)			1200 kPa (120.04)			1400 kPa (127.89)		
Sat.	0.01654	284.20	0.8438	0.01360	288.45	0.8471	0.01146	291.94	0.8496
120	0.01737	292.28	0.8646	—	—	—	—	—	—
130	0.01824	301.27	0.8872	0.01443	297.90	0.8708	0.01163	294.06	0.8549
140	0.01906	310.17	0.9090	0.01520	307.19	0.8936	0.01239	303.88	0.8789
150	0.01985	319.02	0.9302	0.01593	316.34	0.9155	0.01309	313.42	0.9017
160	0.02061	327.85	0.9508	0.01662	325.41	0.9366	0.01374	322.80	0.9236
170	0.02135	336.67	0.9710	0.01728	334.43	0.9572	0.01435	332.06	0.9448
180	0.02207	345.50	0.9907	0.01792	343.44	0.9773	0.01494	341.27	0.9653
190	0.02277	354.36	1.0100	0.01854	352.44	0.9970	0.01550	350.43	0.9853
200	0.02346	363.24	1.0290	0.01914	361.45	1.0162	0.01605	359.59	1.0049
210	0.02414	372.16	1.0476	0.01974	370.47	1.0351	0.01658	368.74	1.0240
220	0.02481	381.12	1.0660	0.02032	379.53	1.0537	0.01710	377.90	1.0428
	1600 kPa (134.94)			1800 kPa (141.36)			2000 kPa (147.25)		
Sat.	0.00983	294.81	0.8514	0.00854	297.17	0.8527	0.00748	299.04	0.8534
140	0.01021	300.11	0.8643	—	—	—	—	—	—
150	0.01091	310.20	0.8885	0.00916	306.56	0.8751	0.00769	302.28	0.8611
160	0.01154	319.97	0.9113	0.00980	316.87	0.8992	0.00837	313.39	0.8870
170	0.01213	329.54	0.9331	0.01038	326.82	0.9219	0.00895	323.85	0.9109
180	0.01268	338.98	0.9542	0.01091	336.55	0.9436	0.00948	333.95	0.9334
190	0.01321	348.34	0.9746	0.01142	346.14	0.9646	0.00997	343.82	0.9550
200	0.01372	357.65	0.9945	0.01190	355.64	0.9849	0.01043	353.53	0.9757
210	0.01421	366.94	1.0140	0.01236	365.08	1.0046	0.01087	363.15	0.9958
220	0.01469	376.22	1.0330	0.01280	374.49	1.0239	0.01129	372.71	1.0154

TABLE B.21 SI *Thermodynamic Properties of R-134a*
TABLE B.21.1 SI *Saturated R-134a*

Temp.	Press.	Specific Volume, m^3/kg			Internal Energy, kJ/kg		
C	kPa	Sat. Liquid	Evap.	Sat. Vapor	Sat. Liquid	Evap.	Sat. Vapor
T	P	v_f	v_{fg}	v_g	u_f	u_{fg}	u_g
-70	8.3	0.000675	1.97207	1.97274	119.46	218.74	338.20
-65	11.7	0.000679	1.42915	1.42983	123.18	217.76	340.94
-60	16.3	0.000684	1.05199	1.05268	127.52	216.19	343.71
-55	22.2	0.000689	0.78609	0.78678	132.36	214.14	346.50
-50	29.9	0.000695	0.59587	0.59657	137.60	211.71	349.31
-45	39.6	0.000701	0.45783	0.45853	143.15	208.99	352.15
-40	51.8	0.000708	0.35625	0.35696	148.95	206.05	355.00
-35	66.8	0.000715	0.28051	0.28122	154.93	202.93	357.86
-30	85.1	0.000722	0.22330	0.22402	161.06	199.67	360.73
-26.3	101.3	0.000728	0.18947	0.19020	165.73	197.16	362.89
-25	107.2	0.000730	0.17957	0.18030	167.30	196.31	363.61
-20	133.7	0.000738	0.14576	0.14649	173.65	192.85	366.50
-15	165.0	0.000746	0.11932	0.12007	180.07	189.32	369.39
-10	201.7	0.000755	0.09845	0.09921	186.57	185.70	372.27
-5	244.5	0.000764	0.08181	0.08257	193.14	182.01	375.15
0	294.0	0.000773	0.06842	0.06919	199.77	178.24	378.01
5	350.9	0.000783	0.05755	0.05833	206.48	174.38	380.85
10	415.8	0.000794	0.04866	0.04945	213.25	170.42	383.67
15	489.5	0.000805	0.04133	0.04213	220.10	166.35	386.45
20	572.8	0.000817	0.03524	0.03606	227.03	162.16	389.19
25	666.3	0.000829	0.03015	0.03098	234.04	157.83	391.87
30	771.0	0.000843	0.02587	0.02671	241.14	153.34	394.48
35	887.6	0.000857	0.02224	0.02310	248.34	148.68	397.02
40	1017.0	0.000873	0.01915	0.02002	255.65	143.81	399.46
45	1160.2	0.000890	0.01650	0.01739	263.08	138.71	401.79
50	1318.1	0.000908	0.01422	0.01512	270.63	133.35	403.98
55	1491.6	0.000928	0.01224	0.01316	278.33	127.68	406.01
60	1681.8	0.000951	0.01051	0.01146	286.19	121.66	407.85
65	1889.9	0.000976	0.00899	0.00997	294.24	115.22	409.46
70	2117.0	0.001005	0.00765	0.00866	302.51	108.27	410.78
75	2364.4	0.001038	0.00645	0.00749	311.06	100.68	411.74
80	2633.6	0.001078	0.00537	0.00645	319.96	92.26	412.22
85	2926.2	0.001128	0.00437	0.00550	329.35	82.67	412.01
90	3244.5	0.001195	0.00341	0.00461	339.51	71.24	410.75
95	3591.5	0.001297	0.00243	0.00373	351.17	56.25	407.42
100	3973.2	0.001557	0.00108	0.00264	368.55	28.19	396.74
101.2	4064.0	0.001969	0	0.00197	382.97	0	382.97

(Continued)

TABLE B.21.1 SI (Continued) ***Saturated R-134a***

Temp.	Press.	Enthalpy, kJ/kg			Entropy, kJ/kg K		
C T	kPa P	Sat. Liquid h_f	Evap. h_{fg}	Sat. Vapor h_g	Sat. Liquid s_f	Evap. s_{fg}	Sat. Vapor s_g
-70	8.3	119.47	235.15	354.62	0.6645	1.1575	1.8220
-65	11.7	123.18	234.55	357.73	0.6825	1.1268	1.8094
-60	16.3	127.53	233.33	360.86	0.7031	1.0947	1.7978
-55	22.2	132.37	231.63	364.00	0.7256	1.0618	1.7874
-50	29.9	137.62	229.54	367.16	0.7493	1.0286	1.7780
-45	39.6	143.18	227.14	370.32	0.7740	0.9956	1.7695
-40	51.8	148.98	224.50	373.48	0.7991	0.9629	1.7620
-35	66.8	154.98	221.67	376.64	0.8245	0.9308	1.7553
-30	85.1	161.12	218.68	379.80	0.8499	0.8994	1.7493
-26.3	101.3	165.80	216.36	382.16	0.8690	0.8763	1.7453
-25	107.2	167.38	215.57	382.95	0.8754	0.8687	1.7441
-20	133.7	173.74	212.34	386.08	0.9007	0.8388	1.7395
-15	165.0	180.19	209.00	389.20	0.9258	0.8096	1.7354
-10	201.7	186.72	205.56	392.28	0.9507	0.7812	1.7319
-5	244.5	193.32	202.02	395.34	0.9755	0.7534	1.7288
0	294.0	200.00	198.36	398.36	1.0000	0.7262	1.7262
5	350.9	206.75	194.57	401.32	1.0243	0.6995	1.7239
10	415.8	213.58	190.65	404.23	1.0485	0.6733	1.7218
15	489.5	220.49	186.58	407.07	1.0725	0.6475	1.7200
20	572.8	227.49	182.35	409.84	1.0963	0.6220	1.7183
25	666.3	234.59	177.92	412.51	1.1201	0.5967	1.7168
30	771.0	241.79	173.29	415.08	1.1437	0.5716	1.7153
35	887.6	249.10	168.42	417.52	1.1673	0.5465	1.7139
40	1017.0	256.54	163.28	419.82	1.1909	0.5214	1.7123
45	1160.2	264.11	157.85	421.96	1.2145	0.4962	1.7106
50	1318.1	271.83	152.08	423.91	1.2381	0.4706	1.7088
55	1491.6	279.72	145.93	425.65	1.2619	0.4447	1.7066
60	1681.8	287.79	139.33	427.13	1.2857	0.4182	1.7040
65	1889.9	296.09	132.21	428.30	1.3099	0.3910	1.7008
70	2117.0	304.64	124.47	429.11	1.3343	0.3627	1.6970
75	2364.4	313.51	115.94	429.45	1.3592	0.3330	1.6923
80	2633.6	322.79	106.40	429.19	1.3849	0.3013	1.6862
85	2926.2	332.65	95.45	428.10	1.4117	0.2665	1.6782
90	3244.5	343.38	82.31	425.70	1.4404	0.2267	1.6671
95	3591.5	355.83	64.98	420.81	1.4733	0.1765	1.6498
100	3973.2	374.74	32.47	407.21	1.5228	0.0870	1.6098
101.2	4064.0	390.98	0	390.98	1.5658	0	1.5658

TABLE B.21.2 SI *Superheated R-134a*

Temp. C	v m³/kg	h kJ/kg	s kJ/kg K	v m³/kg	h kJ/kg	s kJ/kg K	v m³/kg	h kJ/kg	s kJ/kg K
	50 kPa (-40.67)			100 kPa (-26.54)			150 kPa (-17.29)		
Sat.	0.36889	373.06	1.7629	0.19257	381.98	1.7456	0.13139	387.77	1.7372
-20	0.40507	388.82	1.8279	0.19860	387.22	1.7665	—	—	—
-10	0.42222	396.64	1.8582	0.20765	395.27	1.7978	0.13602	393.84	1.7606
0	0.43921	404.59	1.8878	0.21652	403.41	1.8281	0.14222	402.19	1.7917
10	0.45608	412.70	1.9170	0.22527	411.67	1.8578	0.14828	410.60	1.8220
20	0.47287	420.96	1.9456	0.23392	420.05	1.8869	0.15424	419.11	1.8515
30	0.48958	429.38	1.9739	0.24250	428.56	1.9155	0.16011	427.73	1.8804
40	0.50623	437.96	2.0017	0.25101	437.22	1.9436	0.16592	436.47	1.9088
50	0.52284	446.70	2.0292	0.25948	446.03	1.9712	0.17168	445.35	1.9367
60	0.53941	455.60	2.0563	0.26791	454.99	1.9985	0.17740	454.37	1.9642
70	0.55595	464.66	2.0831	0.27631	464.10	2.0255	0.18308	463.53	1.9913
80	0.57247	473.88	2.1096	0.28468	473.36	2.0521	0.18874	472.83	2.0180
90	0.58896	483.26	2.1358	0.29302	482.78	2.0784	0.19437	482.28	2.0444
100	0.60544	492.81	2.1617	0.30135	492.35	2.1044	0.19999	491.89	2.0705
110	0.62190	502.50	2.1874	0.30967	502.07	2.1301	0.20559	501.64	2.0963
120	0.63835	512.36	2.2128	0.31797	511.95	2.1555	0.21117	511.54	2.1218
130	0.65479	522.37	2.2379	0.32626	521.98	2.1807	0.21675	521.60	2.1470
	200 kPa (-10.22)			300 kPa (0.56)			400 kPa (8.84)		
Sat.	0.10002	392.15	1.7320	0.06787	398.69	1.7259	0.05136	403.56	1.7223
0	0.10501	400.91	1.7647	0.00077	—	—	—	—	—
10	0.10974	409.50	1.7956	0.07111	407.17	1.7564	0.05168	404.65	1.7261
20	0.11436	418.15	1.8256	0.07441	416.12	1.7874	0.05436	413.97	1.7584
30	0.11889	426.87	1.8549	0.07762	425.10	1.8175	0.05693	423.22	1.7895
40	0.12335	435.71	1.8836	0.08075	434.12	1.8468	0.05940	432.46	1.8195
50	0.12776	444.66	1.9117	0.08382	443.23	1.8755	0.06181	441.75	1.8487
60	0.13213	453.74	1.9394	0.08684	452.44	1.9035	0.06417	451.10	1.8772
70	0.13646	462.95	1.9666	0.08982	461.76	1.9311	0.06648	460.55	1.9051
80	0.14076	472.30	1.9935	0.09277	471.21	1.9582	0.06877	470.09	1.9325
90	0.14504	481.79	2.0200	0.09570	480.78	1.9850	0.07102	479.75	1.9595
100	0.14930	491.42	2.0461	0.09861	490.48	2.0113	0.07325	489.52	1.9860
110	0.15355	501.21	2.0720	0.10150	500.32	2.0373	0.07547	499.43	2.0122
120	0.15777	511.13	2.0976	0.10437	510.30	2.0631	0.07767	509.46	2.0381
130	0.16199	521.21	2.1229	0.10723	520.43	2.0885	0.07985	519.63	2.0636
140	0.16620	531.43	2.1479	0.11008	530.69	2.1136	0.08202	529.94	2.0889
150	0.17039	541.80	2.1727	0.11292	541.09	2.1385	0.08418	540.38	2.1139

(Continued)

TABLE B.21.2 SI (Continued) *Superheated R-134a*

Temp. C	*v* m³/kg	*h* kJ/kg	*s* kJ/kg K	*v* m³/kg	*h* kJ/kg	*s* kJ/kg K	*v* m³/kg	*h* kJ/kg	*s* kJ/kg K
	500 kPa (15.66)			600 kPa (21.52)			700 kPa (26.67)		
Sat.	0.04126	407.45	1.7198	0.03442	410.66	1.7179	0.02947	413.38	1.7163
20	0.04226	411.65	1.7342	—	—	—	—	—	—
30	0.04446	421.22	1.7663	0.03609	419.09	1.7461	0.03007	416.81	1.7277
40	0.04656	430.72	1.7971	0.03796	428.88	1.7779	0.03178	426.93	1.7606
50	0.04858	440.20	1.8270	0.03974	438.59	1.8084	0.03339	436.89	1.7919
60	0.05055	449.72	1.8560	0.04145	448.28	1.8379	0.03493	446.78	1.8220
70	0.05247	459.29	1.8843	0.04311	457.99	1.8666	0.03641	456.66	1.8512
80	0.05435	468.94	1.9120	0.04473	467.76	1.8947	0.03785	466.55	1.8796
90	0.05620	478.69	1.9392	0.04632	477.61	1.9222	0.03925	476.51	1.9074
100	0.05804	488.55	1.9660	0.04788	487.55	1.9492	0.04063	486.53	1.9347
110	0.05985	498.52	1.9924	0.04943	497.59	1.9758	0.04198	496.65	1.9614
120	0.06164	508.61	2.0184	0.05095	507.75	2.0019	0.04331	506.88	1.9878
130	0.06342	518.83	2.0440	0.05246	518.03	2.0277	0.04463	517.21	2.0137
140	0.06518	529.19	2.0694	0.05396	528.43	2.0532	0.04594	527.66	2.0393
150	0.06694	539.67	2.0945	0.05544	538.95	2.0784	0.04723	538.23	2.0646
160	0.06869	550.29	2.1193	0.05692	549.61	2.1033	0.04851	548.92	2.0896
170	0.07043	561.04	2.1438	0.05839	560.40	2.1279	0.04979	559.75	2.1143
	800 kPa (31.30)			900 kPa (35.50)			1000 kPa (39.37)		
Sat.	0.02571	415.72	1.7150	0.02276	417.76	1.7137	0.02038	419.54	1.7125
40	0.02711	424.86	1.7446	0.02345	422.64	1.7294	0.02047	420.25	1.7148
50	0.02861	435.11	1.7768	0.02487	433.23	1.7627	0.02185	431.24	1.7494
60	0.03002	445.22	1.8076	0.02619	443.60	1.7943	0.02311	441.89	1.7818
70	0.03137	455.27	1.8373	0.02745	453.83	1.8246	0.02429	452.34	1.8127
80	0.03268	465.31	1.8662	0.02865	464.03	1.8539	0.02542	462.70	1.8425
90	0.03394	475.38	1.8943	0.02981	474.22	1.8823	0.02650	473.03	1.8713
100	0.03518	485.50	1.9218	0.03094	484.44	1.9101	0.02754	483.36	1.8994
110	0.03639	495.70	1.9487	0.03204	494.73	1.9373	0.02856	493.74	1.9268
120	0.03758	505.99	1.9753	0.03313	505.09	1.9640	0.02956	504.17	1.9537
130	0.03876	516.38	2.0014	0.03419	515.54	1.9902	0.03053	514.69	1.9801
140	0.03992	526.88	2.0271	0.03524	526.10	2.0161	0.03150	525.30	2.0061
150	0.04107	537.50	2.0525	0.03628	536.76	2.0416	0.03244	536.02	2.0318
160	0.04221	548.23	2.0775	0.03731	547.54	2.0668	0.03338	546.84	2.0570
170	0.04334	559.09	2.1023	0.03832	558.44	2.0917	0.03431	557.77	2.0820
180	0.04446	570.08	2.1268	0.03933	569.45	2.1162	0.03523	568.83	2.1067

(Continued)

TABLE B.21.2 SI (Continued) *Superheated R-134a*

Temp. C	v m³/kg	h kJ/kg	s kJ/kg K	v m³/kg	h kJ/kg	s kJ/kg K	v m³/kg	h kJ/kg	s kJ/kg K
	1200 kPa (46.31)			1400 kPa (52.42)			1600 kPa (57.90)		
Sat.	0.01676	422.49	1.7102	0.01414	424.78	1.7077	0.01215	426.54	1.7051
50	0.01724	426.84	1.7237	—	—	—	—	—	—
60	0.01844	438.21	1.7584	0.01503	434.08	1.7360	0.01239	429.32	1.7135
70	0.01953	449.18	1.7908	0.01608	445.72	1.7704	0.01345	441.89	1.7507
80	0.02055	459.92	1.8217	0.01704	456.94	1.8026	0.01438	453.72	1.7847
90	0.02151	470.55	1.8514	0.01793	467.93	1.8333	0.01522	465.15	1.8166
100	0.02244	481.13	1.8801	0.01878	478.79	1.8628	0.01601	476.33	1.8469
110	0.02333	491.70	1.9081	0.01958	489.59	1.8914	0.01676	487.39	1.8762
120	0.02420	502.31	1.9354	0.02036	500.38	1.9192	0.01748	498.39	1.9045
130	0.02504	512.97	1.9621	0.02112	511.19	1.9463	0.01817	509.37	1.9321
140	0.02587	523.70	1.9884	0.02186	522.05	1.9730	0.01884	520.38	1.9591
150	0.02669	534.51	2.0143	0.02258	532.98	1.9991	0.01949	531.43	1.9855
160	0.02750	545.43	2.0398	0.02329	543.99	2.0248	0.02013	542.54	2.0115
170	0.02829	556.44	2.0649	0.02399	555.10	2.0502	0.02076	553.73	2.0370
180	0.02907	567.57	2.0898	0.02468	566.30	2.0752	0.02138	565.02	2.0622
	1800 kPa (62.89)			2000 kPa (67.48)			2500 kPa (77.57)		
Sat.	0.01057	427.85	1.7022	0.00930	428.75	1.6991	0.00694	429.41	1.6893
70	0.01134	437.56	1.7309	0.00958	432.53	1.7101	—	—	—
80	0.01227	450.20	1.7672	0.01055	446.30	1.7497	0.00722	433.80	1.7018
90	0.01310	462.16	1.8006	0.01137	458.95	1.7850	0.00816	449.50	1.7457
100	0.01385	473.74	1.8320	0.01211	471.00	1.8177	0.00891	463.28	1.7831
110	0.01456	485.09	1.8620	0.01279	482.69	1.8487	0.00956	476.13	1.8171
120	0.01523	496.32	1.8910	0.01342	494.19	1.8783	0.01015	488.46	1.8489
130	0.01587	507.50	1.9190	0.01403	505.57	1.9069	0.01069	500.47	1.8790
140	0.01649	518.66	1.9464	0.01461	516.90	1.9346	0.01121	512.31	1.9080
150	0.01709	529.84	1.9731	0.01517	528.22	1.9617	0.01170	524.04	1.9361
160	0.01768	541.07	1.9994	0.01571	539.57	1.9882	0.01217	535.72	1.9634
170	0.01825	552.36	2.0251	0.01624	550.96	2.0142	0.01262	547.40	1.9900
180	0.01881	563.72	2.0505	0.01676	562.42	2.0398	0.01307	559.10	2.0161

(Continued)

TABLE B.21.2 SI (Continued) *Superheated R-134a*

Temp. C	*v* m³/kg	*h* kJ/kg	*s* kJ/kg K	*v* m³/kg	*h* kJ/kg	*s* kJ/kg K	*v* m³/kg	*h* kJ/kg	*s* kJ/kg K
	3000 kPa (86.20)			3500 kPa (93.72)			4000 kPa (100.33)		
Sat.	0.00528	427.67	1.6759	0.00396	422.43	1.6552	0.00252	404.94	1.6036
90	0.00575	436.19	1.6995	—	—	—	—	—	—
100	0.00665	453.73	1.7472	0.00484	440.43	1.7039	—	—	—
110	0.00734	468.50	1.7862	0.00567	459.21	1.7535	0.00428	446.84	1.7148
120	0.00792	482.04	1.8211	0.00629	474.70	1.7935	0.00500	465.99	1.7642
130	0.00845	494.91	1.8535	0.00681	488.77	1.8288	0.00556	481.87	1.8040
140	0.00893	507.39	1.8840	0.00728	502.08	1.8614	0.00603	496.29	1.8394
150	0.00937	519.62	1.9133	0.00771	514.93	1.8922	0.00644	509.92	1.8720
160	0.00980	531.70	1.9415	0.00810	527.50	1.9215	0.00683	523.07	1.9027
170	0.01021	543.71	1.9689	0.00848	539.89	1.9498	0.00718	535.92	1.9320
180	0.01060	555.69	1.9956	0.00884	552.18	1.9772	0.00752	548.57	1.9603
	5000 kPa			6000 kPa			7000 kPa		
90	0.001089	336.61	1.4163	0.001059	334.70	1.4081	0.001037	333.29	1.4013
100	0.001216	357.68	1.4735	0.001150	353.61	1.4595	0.001110	351.10	1.4497
110	0.001659	392.10	1.5644	0.001307	375.90	1.5184	0.001215	370.68	1.5015
120	0.002969	440.47	1.6892	0.001698	406.78	1.5979	0.001393	393.45	1.5601
130	0.003705	464.63	1.7499	0.002396	441.18	1.6843	0.001720	420.73	1.6286
140	0.004226	482.86	1.7946	0.002985	466.25	1.7458	0.002169	448.28	1.6961
150	0.004652	498.77	1.8327	0.003439	485.82	1.7926	0.002599	471.55	1.7518
160	0.005023	513.48	1.8670	0.003814	502.77	1.8322	0.002968	491.16	1.7976
170	0.005357	527.47	1.8990	0.004141	518.30	1.8676	0.003287	508.52	1.8373
180	0.005665	541.00	1.9292	0.004435	532.96	1.9004	0.003569	524.51	1.8729
	8000 kPa			10000 kPa			20000 kPa		
90	0.001019	332.20	1.3955	0.000991	330.62	1.3856	0.000912	327.89	1.3520
100	0.001081	349.30	1.4420	0.001040	346.85	1.4297	0.000939	342.49	1.3917
110	0.001163	367.57	1.4903	0.001100	363.73	1.4744	0.000969	357.33	1.4309
120	0.001282	387.56	1.5417	0.001175	381.44	1.5200	0.001002	372.38	1.4697
130	0.001465	409.98	1.5981	0.001272	400.16	1.5670	0.001037	387.65	1.5081
140	0.001736	434.40	1.6579	0.001400	419.98	1.6155	0.001076	403.13	1.5460
150	0.002061	458.21	1.7148	0.001564	440.63	1.6649	0.001118	418.83	1.5836
160	0.002384	479.59	1.7648	0.001758	461.34	1.7133	0.001164	434.72	1.6207
170	0.002680	498.61	1.8082	0.001965	481.30	1.7589	0.001214	450.80	1.6574
180	0.002946	515.93	1.8469	0.002172	500.12	1.8009	0.001268	467.03	1.6936

TABLE B.22 SI *Thermodynamic Properties of R-152a*
TABLE B.22.1 SI *Saturated R-152a*

Temp.	Press.	Specific Volume, m³/kg			Internal Energy, kJ/kg		
C	kPa	Sat. Liquid	Evap.	Sat. Vapor	Sat. Liquid	Evap.	Sat. Vapor
T	P	v_f	v_{fg}	v_g	u_f	u_{fg}	u_g
-50	27.4	0.000941	1.00596	1.00690	118.05	324.13	442.18
-45	36.2	0.000949	0.77512	0.77607	125.94	319.60	445.55
-40	47.2	0.000958	0.60490	0.60586	133.89	315.01	448.90
-35	60.7	0.000968	0.47763	0.47859	141.90	310.35	452.25
-30	77.2	0.000977	0.38122	0.38219	149.96	305.63	455.58
-25	97.0	0.000987	0.30730	0.30829	158.08	300.82	458.90
-24.0	101.3	0.000989	0.29488	0.29587	159.69	299.86	459.55
-20	120.7	0.000998	0.24999	0.25098	166.27	295.93	462.20
-15	148.7	0.001008	0.20508	0.20609	174.53	290.95	465.48
-10	181.6	0.001019	0.16954	0.17056	182.86	285.88	468.73
-5	219.9	0.001031	0.14115	0.14218	191.25	280.71	471.96
0	264.1	0.001043	0.11828	0.11932	199.72	275.43	475.15
5	315.0	0.001055	0.09970	0.10076	208.28	270.04	478.31
10	373.0	0.001069	0.08450	0.08557	216.91	264.52	481.43
15	438.9	0.001082	0.07196	0.07304	225.63	258.87	484.50
20	513.3	0.001097	0.06154	0.06264	234.44	253.07	487.51
25	596.9	0.001112	0.05284	0.05395	243.35	247.11	490.47
30	690.4	0.001128	0.04552	0.04665	252.37	240.98	493.35
35	794.6	0.001145	0.03933	0.04048	261.49	234.66	496.15
40	910.1	0.001163	0.03406	0.03522	270.74	228.12	498.85
45	1037.8	0.001183	0.02955	0.03073	280.11	221.34	501.45
50	1178.4	0.001204	0.02567	0.02687	289.62	214.30	503.92
55	1332.9	0.001226	0.02231	0.02354	299.29	206.95	506.24
60	1502.0	0.001251	0.01939	0.02064	309.13	199.26	508.39
65	1686.7	0.001278	0.01684	0.01812	319.17	191.16	510.33
70	1888.0	0.001308	0.01459	0.01590	329.42	182.60	512.02
75	2106.8	0.001341	0.01260	0.01394	339.94	173.48	513.42
80	2344.3	0.001379	0.01083	0.01220	350.75	163.69	514.44
85	2601.8	0.001422	0.00923	0.01065	361.92	153.06	514.98
90	2880.5	0.001473	0.00778	0.00925	373.56	141.34	514.91
95	3182.1	0.001536	0.00644	0.00797	385.82	128.16	513.98
100	3508.4	0.001616	0.00517	0.00679	398.96	112.82	511.78
105	3861.9	0.001728	0.00391	0.00564	413.58	93.86	507.44
110	4246.2	0.001926	0.00251	0.00443	431.70	66.46	498.17
113.3	4516.8	0.002717	0	0.00272	465.35	0	465.35

(Continued)

TABLE B.22.1 SI (Continued) *Saturated R-152a*

Temp.	Press.	Enthalpy, kJ/kg			Entropy, kJ/kg K		
C	kPa	Sat. Liquid	Evap.	Sat. Vapor	Sat. Liquid	Evap.	Sat. Vapor
T	P	h_f	h_{fg}	h_g	s_f	s_{fg}	s_g
-50	27.4	118.08	351.70	469.77	0.6701	1.5761	2.2462
-45	36.2	125.98	347.67	473.65	0.7051	1.5239	2.2290
-40	47.2	133.94	343.55	477.49	0.7396	1.4735	2.2131
-35	60.7	141.95	339.34	481.30	0.7735	1.4249	2.1985
-30	77.2	150.03	335.04	485.07	0.8071	1.3779	2.1850
-25	97.0	158.18	330.62	488.80	0.8401	1.3324	2.1725
-24.0	101.3	159.79	329.74	489.53	0.8466	1.3236	2.1702
-20	120.7	166.39	326.09	492.49	0.8728	1.2881	2.1610
-15	148.7	174.68	321.44	496.12	0.9051	1.2452	2.1503
-10	181.6	183.04	316.66	499.70	0.9371	1.2034	2.1404
-5	219.9	191.48	311.74	503.22	0.9687	1.1626	2.1312
0	264.1	200.00	306.67	506.67	1.0000	1.1227	2.1227
5	315.0	208.61	301.44	510.05	1.0310	1.0837	2.1148
10	373.0	217.31	296.04	513.35	1.0618	1.0455	2.1073
15	438.9	226.10	290.45	516.56	1.0924	1.0080	2.1004
20	513.3	235.01	284.66	519.67	1.1227	0.9711	2.0938
25	596.9	244.02	278.66	522.67	1.1529	0.9346	2.0875
30	690.4	253.15	272.41	525.56	1.1829	0.8986	2.0815
35	794.6	262.40	265.91	528.31	1.2128	0.8629	2.0757
40	910.1	271.79	259.12	530.91	1.2426	0.8275	2.0700
45	1037.8	281.34	252.01	533.35	1.2723	0.7921	2.0645
50	1178.4	291.04	244.55	535.59	1.3021	0.7568	2.0589
55	1332.9	300.93	236.69	537.62	1.3319	0.7213	2.0532
60	1502.0	311.01	228.38	539.40	1.3617	0.6855	2.0473
65	1686.7	321.33	219.56	540.89	1.3918	0.6493	2.0411
70	1888.0	331.89	210.14	542.04	1.4220	0.6124	2.0344
75	2106.8	342.76	200.03	542.79	1.4526	0.5745	2.0272
80	2344.3	353.98	189.07	543.05	1.4837	0.5354	2.0191
85	2601.8	365.62	177.07	542.69	1.5154	0.4944	2.0098
90	2880.5	377.81	163.75	541.56	1.5481	0.4509	1.9990
95	3182.1	390.70	148.65	539.35	1.5821	0.4038	1.9859
100	3508.4	404.63	130.96	535.59	1.6183	0.3510	1.9693
105	3861.9	420.25	108.97	529.23	1.6583	0.2882	1.9465
110	4246.2	439.88	77.11	516.98	1.7080	0.2012	1.9093
113.3	4516.8	477.62	0.00	477.62	1.8045	0.0000	1.8045

TABLE B.22.2 SI *Superheated R-152a*

Temp. C	v m³/kg	h kJ/kg	s kJ/kg K	v m³/kg	h kJ/kg	s kJ/kg K	v m³/kg	h kJ/kg	s kJ/kg K
	100 kPa (-24.31)			200 kPa (-7.50)			300 kPa (3.60)		
Sat.	0.29956	489.31	2.1709	0.15559	501.47	2.1357	0.10560	509.11	2.1169
-10	0.31999	503.94	2.2280	—	—	—	—	—	—
0	0.33377	514.10	2.2659	0.16143	509.86	2.1669	—	—	—
10	0.34734	524.37	2.3028	0.16887	520.76	2.2061	0.10914	516.72	2.1441
20	0.36076	534.79	2.3390	0.17608	531.61	2.2438	0.11437	528.21	2.1840
30	0.37406	545.38	2.3745	0.18314	542.53	2.2804	0.11940	539.55	2.2221
40	0.38728	556.15	2.4095	0.19009	553.58	2.3162	0.12429	550.91	2.2589
50	0.40042	567.12	2.4440	0.19696	564.77	2.3514	0.12908	562.36	2.2949
60	0.41350	578.28	2.4780	0.20376	576.13	2.3860	0.13380	573.93	2.3302
70	0.42652	589.64	2.5116	0.21050	587.67	2.4202	0.13846	585.65	2.3648
80	0.43950	601.21	2.5448	0.21719	599.39	2.4538	0.14306	597.53	2.3990
90	0.45243	612.98	2.5777	0.22384	611.29	2.4871	0.14762	609.58	2.4326
100	0.46534	624.96	2.6102	0.23046	623.40	2.5199	0.15215	621.81	2.4658
110	0.47821	637.16	2.6425	0.23704	635.70	2.5525	0.15664	634.22	2.4986
120	0.49106	649.56	2.6744	0.24360	648.20	2.5847	0.16110	646.82	2.5311
130	0.50388	662.18	2.7061	0.25013	660.90	2.6166	0.16554	659.61	2.5632
140	0.51669	675.01	2.7375	0.25664	673.81	2.6482	0.16996	672.59	2.5950
	400 kPa (12.12)			500 kPa (19.15)			600 kPa (25.17)		
Sat.	0.07996	514.72	2.1043	0.06429	519.15	2.0949	0.05368	522.78	2.0873
20	0.08337	524.47	2.1381	0.06461	520.28	2.0987	—	—	—
30	0.08743	536.38	2.1780	0.06817	532.98	2.1413	0.05522	529.27	2.1089
40	0.09133	548.13	2.2161	0.07149	545.20	2.1810	0.05820	542.11	2.1506
50	0.09510	559.87	2.2530	0.07467	557.28	2.2190	0.06101	554.59	2.1898
60	0.09879	571.67	2.2890	0.07775	569.35	2.2558	0.06370	566.95	2.2275
70	0.10241	583.59	2.3242	0.08076	581.48	2.2916	0.06630	579.31	2.2640
80	0.10598	595.64	2.3589	0.08371	593.71	2.3268	0.06885	591.74	2.2997
90	0.10950	607.84	2.3929	0.08661	606.06	2.3613	0.07134	604.25	2.3347
100	0.11298	620.19	2.4265	0.08947	618.55	2.3952	0.07378	616.89	2.3690
110	0.11643	632.72	2.4596	0.09229	631.20	2.4286	0.07619	629.66	2.4027
120	0.11984	645.42	2.4923	0.09508	644.01	2.4616	0.07857	642.57	2.4360
130	0.12324	658.30	2.5247	0.09785	656.98	2.4942	0.08092	655.65	2.4689
140	0.12661	671.37	2.5567	0.10060	670.13	2.5264	0.08325	668.88	2.5013
150	0.12996	684.62	2.5884	0.10332	683.46	2.5583	0.08556	682.29	2.5334
160	0.13330	698.07	2.6198	0.10603	696.97	2.5899	0.08785	695.87	2.5651

(Continued)

TABLE B.22.2 SI (Continued) ***Superheated R-152a***

Temp. C	v m³/kg	h kJ/kg	s kJ/kg K	v m³/kg	h kJ/kg	s kJ/kg K	v m³/kg	h kJ/kg	s kJ/kg K
	700 kPa (30.48)			800 kPa (35.25)			900 kPa (39.58)		
Sat.	0.04601	525.83	2.0809	0.04020	528.44	2.0754	0.03563	530.70	2.0705
40	0.04865	538.81	2.1230	0.04143	535.24	2.0973	0.03573	531.33	2.0725
50	0.05121	551.77	2.1638	0.04382	548.80	2.1399	0.03804	545.66	2.1176
60	0.05363	564.47	2.2025	0.04605	561.89	2.1798	0.04014	559.20	2.1589
70	0.05596	577.09	2.2398	0.04818	574.80	2.2180	0.04211	572.43	2.1980
80	0.05821	589.72	2.2761	0.05023	587.65	2.2549	0.04400	585.53	2.2356
90	0.06041	602.41	2.3115	0.05221	600.53	2.2909	0.04582	598.61	2.2721
100	0.06257	615.19	2.3462	0.05415	613.47	2.3260	0.04760	611.72	2.3077
110	0.06469	628.09	2.3804	0.05605	626.51	2.3605	0.04933	624.89	2.3426
120	0.06677	641.12	2.4139	0.05791	639.65	2.3944	0.05102	638.16	2.3768
130	0.06883	654.30	2.4470	0.05975	652.93	2.4277	0.05269	651.55	2.4104
140	0.07086	667.62	2.4797	0.06156	666.35	2.4606	0.05433	665.06	2.4435
150	0.07287	681.11	2.5119	0.06335	679.92	2.4931	0.05594	678.72	2.4762
160	0.07486	694.76	2.5438	0.06512	693.65	2.5251	0.05754	692.52	2.5084
	1000 kPa (43.57)			1200 kPa (50.73)			1400 kPa (57.04)		
Sat.	0.03195	532.67	2.0661	0.02636	535.90	2.0580	0.02231	538.37	2.0508
50	0.03336	542.30	2.0961	—	—	—	—	—	—
60	0.03537	556.39	2.1391	0.02815	550.31	2.1019	0.02286	543.38	2.0659
70	0.03724	569.98	2.1793	0.02988	564.80	2.1448	0.02454	559.15	2.1125
80	0.03901	583.36	2.2177	0.03148	578.80	2.1850	0.02605	573.95	2.1550
90	0.04070	596.65	2.2548	0.03300	592.58	2.2235	0.02746	588.29	2.1951
100	0.04235	609.93	2.2909	0.03445	606.25	2.2606	0.02878	602.41	2.2334
110	0.04394	623.26	2.3262	0.03585	619.90	2.2967	0.03006	616.42	2.2705
120	0.04550	636.65	2.3607	0.03722	633.57	2.3319	0.03128	630.39	2.3065
130	0.04703	650.15	2.3946	0.03854	647.30	2.3664	0.03247	644.38	2.3416
140	0.04854	663.76	2.4279	0.03984	661.12	2.4003	0.03363	658.42	2.3760
150	0.05002	677.50	2.4608	0.04112	675.04	2.4336	0.03476	672.53	2.4098
160	0.05148	691.38	2.4932	0.04238	689.08	2.4664	0.03587	686.74	2.4430
	1600 kPa (62.71)			2000 kPa (72.61)			4000 kPa (106.85)		
Sat.	0.01923	540.24	2.0440	0.01485	542.49	2.0307	0.00521	525.75	1.9354
70	0.02046	552.85	2.0811	—	—	—	—	—	—
80	0.02193	568.71	2.1267	0.01598	556.59	2.0711	—	—	—
90	0.02327	583.76	2.1687	0.01729	573.70	2.1189	—	—	—
100	0.02451	598.39	2.2084	0.01845	589.70	2.1624	—	—	—
110	0.02569	612.81	2.2466	0.01953	605.12	2.2031	0.00598	542.43	1.9791
120	0.02682	627.11	2.2834	0.02053	620.21	2.2420	0.00732	572.89	2.0576
130	0.02790	641.38	2.3193	0.02148	635.11	2.2794	0.00825	595.53	2.1145
140	0.02896	655.65	2.3542	0.02240	649.92	2.3157	0.00902	615.48	2.1634
150	0.02998	669.97	2.3885	0.02328	664.69	2.3511	0.00970	634.05	2.2078
160	0.03099	684.36	2.4221	0.02414	679.47	2.3856	0.01032	651.82	2.2493

TABLE B.23 SI *Thermodynamic Properties of R-C318*
TABLE B.23.1 SI *Saturated R-C318*

Temp.	Press.	Specific Volume, m^3/kg			Internal Energy, kJ/kg		
C T	kPa P	Sat. Liquid v_f	Evap. v_{fg}	Sat. Vapor v_g	Sat. Liquid u_f	Evap. u_{fg}	Sat. Vapor u_g
-20	54.2	0.000602	0.18727	0.18787	18.26	110.81	129.07
-15	68.2	0.000608	0.15068	0.15129	23.06	109.17	132.23
-10	85.0	0.000615	0.12231	0.12292	27.94	107.46	135.41
-5.9	101.3	0.000620	0.10350	0.10412	32.07	105.99	138.06
-5	105.0	0.000621	0.10008	0.10070	32.93	105.68	138.60
0	128.4	0.000628	0.08252	0.08314	38.00	103.81	141.81
5	155.9	0.000635	0.06850	0.06914	43.16	101.87	145.03
10	187.6	0.000642	0.05724	0.05788	48.42	99.84	148.26
15	224.2	0.000650	0.04810	0.04875	53.76	97.73	151.49
20	266.1	0.000658	0.04064	0.04130	59.19	95.54	154.73
25	313.7	0.000667	0.03450	0.03517	64.71	93.26	157.97
30	367.4	0.000676	0.02942	0.03009	70.31	90.89	161.21
35	427.9	0.000685	0.02518	0.02587	76.00	88.44	164.44
40	495.4	0.000696	0.02162	0.02232	81.77	85.89	167.66
45	570.7	0.000707	0.01862	0.01933	87.62	83.24	170.86
50	654.0	0.000719	0.01607	0.01679	93.55	80.50	174.05
55	746.1	0.000731	0.01389	0.01463	99.56	77.65	177.21
60	847.2	0.000745	0.01202	0.01277	105.65	74.68	180.34
65	958.1	0.000761	0.01041	0.01117	111.83	71.60	183.42
70	1079.2	0.000777	0.00900	0.00977	118.08	68.37	186.45
75	1211.2	0.000796	0.00776	0.00856	124.43	64.98	189.41
80	1354.6	0.000817	0.00668	0.00749	130.87	61.40	192.28
85	1510.2	0.000841	0.00571	0.00655	137.43	57.59	195.02
90	1678.9	0.000869	0.00483	0.00570	144.12	53.47	197.59
95	1861.5	0.000903	0.00403	0.00494	151.00	48.93	199.93
100	2059.3	0.000945	0.00329	0.00423	158.13	43.77	201.90
105	2274.0	0.001002	0.00256	0.00357	165.71	37.57	203.28
110	2507.9	0.001091	0.00180	0.00289	174.22	29.18	203.40
115.3	2783.5	0.001613	0	0.00161	192.47	0	192.47

(Continued)

TABLE B.23.1 SI (Continued) *Saturated R-C318*							
Temp.	Press.	Enthalpy, kJ/kg			Entropy, kJ/kg K		
C	kPa	Sat. Liquid	Evap.	Sat. Vapor	Sat. Liquid	Evap.	Sat. Vapor
T	P	h_f	h_{fg}	h_g	s_f	s_{fg}	s_g
-20	54.2	18.29	120.95	139.25	0.0751	0.4778	0.5529
-15	68.2	23.10	119.45	142.55	0.0939	0.4627	0.5566
-10	85.0	28.00	117.86	145.86	0.1127	0.4479	0.5605
-5.9	101.3	32.13	116.48	148.61	0.1282	0.4357	0.5640
-5	105.0	32.99	116.18	149.17	0.1314	0.4333	0.5647
0	128.4	38.08	114.41	152.49	0.1502	0.4188	0.5690
5	155.9	43.26	112.55	155.81	0.1689	0.4046	0.5735
10	187.6	48.54	110.58	159.12	0.1876	0.3905	0.5781
15	224.2	53.90	108.52	162.42	0.2063	0.3766	0.5829
20	266.1	59.36	106.36	165.72	0.2250	0.3628	0.5878
25	313.7	64.92	104.08	169.00	0.2437	0.3491	0.5928
30	367.4	70.56	101.70	172.26	0.2623	0.3355	0.5978
35	427.9	76.29	99.21	175.50	0.2810	0.3219	0.6029
40	495.4	82.11	96.60	178.71	0.2995	0.3085	0.6080
45	570.7	88.02	93.87	181.89	0.3181	0.2950	0.6131
50	654.0	94.02	91.01	185.03	0.3366	0.2816	0.6182
55	746.1	100.11	88.01	188.12	0.3551	0.2682	0.6233
60	847.2	106.29	84.87	191.16	0.3736	0.2547	0.6283
65	958.1	112.56	81.56	194.12	0.3920	0.2412	0.6332
70	1079.2	118.92	78.08	197.00	0.4104	0.2275	0.6379
75	1211.2	125.39	74.38	199.78	0.4288	0.2136	0.6425
80	1354.6	131.98	70.44	202.42	0.4473	0.1995	0.6468
85	1510.2	138.70	66.20	204.90	0.4658	0.1848	0.6507
90	1678.9	145.58	61.58	207.16	0.4845	0.1696	0.6541
95	1861.5	152.68	56.44	209.12	0.5035	0.1533	0.6568
100	2059.3	160.08	50.54	210.62	0.5229	0.1354	0.6584
105	2274.0	167.99	43.40	211.38	0.5434	0.1148	0.6582
110	2507.9	176.96	33.68	210.64	0.5664	0.0879	0.6543
115.3	2783.5	196.96	0	196.96	0.6172	0	0.6172

TABLE B.23.2 SI *Superheated R-C318*

Temp. C	v m³/kg	h kJ/kg	s kJ/kg K	v m³/kg	h kJ/kg	s kJ/kg K	v m³/kg	h kJ/kg	s kJ/kg K
	50 kPa (-21.70)			100 kPa (-6.17)			200 kPa (11.77)		
Sat.	0.20262	138.13	0.5517	0.10543	148.40	0.5637	0.05444	160.29	0.5798
20	0.23953	169.39	0.6666	0.11761	168.60	0.6358	0.05654	166.91	0.6027
30	0.24821	177.29	0.6931	0.12215	176.57	0.6626	0.05903	175.05	0.6300
40	0.25684	185.33	0.7192	0.12664	184.67	0.6889	0.06147	183.31	0.6568
50	0.26544	193.52	0.7449	0.13109	192.92	0.7148	0.06386	191.68	0.6832
60	0.27400	201.85	0.7703	0.13550	201.30	0.7403	0.06621	200.17	0.7090
70	0.28254	210.32	0.7954	0.13989	209.82	0.7655	0.06853	208.79	0.7345
80	0.29105	218.94	0.8201	0.14425	218.47	0.7904	0.07082	217.52	0.7596
90	0.29954	227.68	0.8446	0.14859	227.26	0.8149	0.07309	226.38	0.7843
100	0.30801	236.57	0.8687	0.15291	236.17	0.8391	0.07533	235.36	0.8087
110	0.31647	245.58	0.8925	0.15721	245.21	0.8630	0.07756	244.46	0.8328
120	0.32492	254.72	0.9161	0.16150	254.38	0.8866	0.07978	253.67	0.8565
130	0.33335	263.99	0.9393	0.16578	263.66	0.9100	0.08198	263.01	0.8800
140	0.34178	273.38	0.9624	0.17005	273.07	0.9330	0.08418	272.46	0.9031
150	0.35019	282.88	0.9851	0.17431	282.60	0.9558	0.08636	282.02	0.9260
160	0.35860	292.51	1.0076	0.17856	292.24	0.9783	0.08854	291.69	0.9486
170	0.36700	302.24	1.0298	0.18280	301.99	1.0006	0.09070	301.47	0.9709
180	0.37539	312.09	1.0518	0.18704	311.85	1.0226	0.09286	311.36	0.9930
	300 kPa (23.63)			400 kPa (32.77)			500 kPa (40.32)		
Sat.	0.03674	168.10	0.5914	0.02766	174.06	0.6006	0.02211	178.92	0.6083
40	0.03968	181.86	0.6365	0.02871	180.31	0.6208	—	—	—
50	0.04140	190.38	0.6633	0.03011	189.00	0.6481	0.02330	187.53	0.6354
60	0.04307	199.00	0.6896	0.03146	197.76	0.6748	0.02446	196.45	0.6626
70	0.04471	207.71	0.7153	0.03277	206.60	0.7010	0.02558	205.42	0.6891
80	0.04632	216.54	0.7407	0.03404	215.52	0.7266	0.02666	214.47	0.7151
90	0.04790	225.48	0.7657	0.03529	224.55	0.7518	0.02770	223.58	0.7406
100	0.04946	234.53	0.7902	0.03651	233.67	0.7766	0.02872	232.79	0.7656
110	0.05100	243.69	0.8145	0.03771	242.89	0.8010	0.02972	242.08	0.7901
120	0.05253	252.96	0.8383	0.03889	252.22	0.8250	0.03071	251.47	0.8143
130	0.05404	262.34	0.8619	0.04007	261.65	0.8487	0.03167	260.96	0.8382
140	0.05555	271.83	0.8852	0.04123	271.19	0.8721	0.03263	270.54	0.8616
150	0.05704	281.43	0.9081	0.04237	280.83	0.8951	0.03357	280.22	0.8848
160	0.05852	291.13	0.9308	0.04351	290.57	0.9179	0.03450	290.00	0.9076
170	0.06000	300.95	0.9532	0.04464	300.42	0.9403	0.03543	299.88	0.9302
180	0.06147	310.86	0.9753	0.04577	310.36	0.9625	0.03635	309.85	0.9524
190	0.06293	320.88	0.9972	0.04689	320.40	0.9844	0.03726	319.92	0.9744
200	0.06439	330.99	1.0188	0.04800	330.54	1.0061	0.03816	330.08	0.9961
210	0.06584	341.20	1.0401	0.04911	340.77	1.0275	0.03907	340.34	1.0176

(Continued)

TABLE B.23.2 SI (Continued) ***Superheated R-C318***

Temp. C	v m³/kg	h kJ/kg	s kJ/kg K	v m³/kg	h kJ/kg	s kJ/kg K	v m³/kg	h kJ/kg	s kJ/kg K
	600 kPa (46.82)			800 kPa (57.72)			1000 kPa (66.78)		
Sat.	0.01836	183.04	0.6150	0.01358	189.78	0.6260	0.01065	195.16	0.6349
60	0.01976	195.06	0.6518	0.01379	191.97	0.6326	—	—	—
80	0.02172	213.36	0.7052	0.01549	210.99	0.6881	0.01168	208.33	0.6729
100	0.02352	231.88	0.7562	0.01700	229.96	0.7403	0.01305	227.89	0.7268
120	0.02524	250.70	0.8053	0.01839	249.11	0.7903	0.01426	247.42	0.7778
140	0.02689	269.88	0.8529	0.01971	268.52	0.8385	0.01539	267.10	0.8266
160	0.02850	289.42	0.8991	0.02098	288.24	0.8851	0.01646	287.02	0.8737
180	0.03006	309.34	0.9440	0.02221	308.30	0.9303	0.01749	307.23	0.9193
200	0.03161	329.63	0.9878	0.02341	328.69	0.9744	0.01849	327.74	0.9636
220	0.03313	350.27	1.0305	0.02459	349.43	1.0173	0.01946	348.58	1.0067
240	0.03463	371.26	1.0723	0.02575	370.50	1.0592	0.02042	369.73	1.0488
260	0.03613	392.59	1.1130	0.02689	391.89	1.1001	0.02136	391.18	1.0898
280	0.03761	414.23	1.1529	0.02803	413.59	1.1400	0.02228	412.94	1.1298
300	0.03908	436.18	1.1918	0.02916	435.58	1.1791	0.02320	434.98	1.1690
320	0.04054	458.40	1.2300	0.03027	457.84	1.2172	0.02411	457.28	1.2072
340	0.04200	480.88	1.2672	0.03139	480.36	1.2546	0.02502	479.84	1.2446
360	0.04345	503.62	1.3037	0.03249	503.13	1.2911	0.02592	502.64	1.2812
380	0.04490	526.57	1.3394	0.03359	526.12	1.3269	0.02681	525.66	1.3170
400	0.04634	549.74	1.3744	0.03469	549.31	1.3618	0.02770	548.88	1.3520
	1500 kPa (84.68)			2000 kPa (98.54)			3000 kPa		
Sat.	0.00660	204.75	0.6504	0.00443	210.24	0.6580	—	—	—
100	0.00763	221.74	0.6969	0.00456	212.35	0.6637	—	—	—
120	0.00869	242.70	0.7516	0.00578	236.86	0.7277	0.00190	207.80	0.6440
140	0.00960	263.25	0.8026	0.00665	258.86	0.7823	0.00354	247.20	0.7422
160	0.01042	283.78	0.8512	0.00738	280.23	0.8328	0.00428	271.86	0.8006
180	0.01119	304.43	0.8978	0.00803	301.45	0.8807	0.00485	294.80	0.8523
200	0.01192	325.29	0.9428	0.00864	322.71	0.9267	0.00535	317.17	0.9006
220	0.01263	346.39	0.9865	0.00921	344.13	0.9710	0.00580	339.36	0.9466
240	0.01331	367.76	1.0289	0.00976	365.74	1.0139	0.00622	361.55	0.9907
260	0.01398	389.39	1.0703	0.01029	387.57	1.0557	0.00662	383.84	1.0333
280	0.01463	411.30	1.1106	0.01081	409.64	1.0963	0.00700	406.27	1.0746
300	0.01527	433.47	1.1500	0.01131	431.94	1.1359	0.00737	428.88	1.1147
320	0.01591	455.88	1.1884	0.01181	454.48	1.1746	0.00773	451.67	1.1538
340	0.01654	478.54	1.2260	0.01230	477.24	1.2123	0.00808	474.65	1.1919
360	0.01716	501.43	1.2627	0.01278	500.21	1.2492	0.00843	497.81	1.2291
380	0.01777	524.52	1.2986	0.01326	523.39	1.2852	0.00876	521.15	1.2654
400	0.01839	547.81	1.3338	0.01373	546.75	1.3204	0.00910	544.66	1.3008

TABLE B.24 SI *Thermodynamic Properties of R-500*
TABLE B.24.1 SI *Saturated R-500*

Temp.	Press.	Specific Volume, m³/kg			Internal Energy, kJ/kg		
C	kPa	Sat. Liquid	Evap.	Sat. Vapor	Sat. Liquid	Evap.	Sat. Vapor
T	P	v_f	v_{fg}	v_g	u_f	u_{fg}	u_g
-50	46.4	0.000726	0.39175	0.39248	-9.35	189.46	180.12
-45	59.5	0.000733	0.31026	0.31100	-4.74	187.21	182.47
-40	75.6	0.000741	0.24832	0.24906	-0.06	184.87	184.82
-35	94.9	0.000748	0.20067	0.20142	4.72	182.45	187.16
-30	117.9	0.000756	0.16363	0.16438	9.58	179.92	189.50
-25	145.1	0.000764	0.13453	0.13529	14.53	177.31	191.83
-20	177.0	0.000772	0.11145	0.11222	19.56	174.59	194.15
-15	214.1	0.000781	0.09298	0.09376	24.68	171.77	196.46
-10	257.1	0.000790	0.07807	0.07886	29.89	168.85	198.75
-5	306.3	0.000799	0.06594	0.06674	35.19	165.83	201.02
0	362.5	0.000809	0.05600	0.05680	40.58	162.69	203.27
5	426.1	0.000819	0.04779	0.04861	46.06	159.44	205.50
10	497.9	0.000830	0.04096	0.04179	51.62	156.08	207.70
15	578.5	0.000841	0.03525	0.03609	57.28	152.59	209.87
20	668.4	0.000853	0.03045	0.03130	63.03	148.98	212.01
25	768.4	0.000865	0.02638	0.02724	68.87	145.23	214.10
30	879.1	0.000878	0.02291	0.02379	74.81	141.34	216.15
35	1001.2	0.000893	0.01994	0.02083	80.85	137.30	218.15
40	1135.3	0.000908	0.01738	0.01829	86.99	133.10	220.09
45	1282.2	0.000924	0.01516	0.01609	93.24	128.71	221.96
50	1442.6	0.000942	0.01324	0.01418	99.61	124.13	223.74
55	1617.3	0.000961	0.01155	0.01251	106.10	119.32	225.42
60	1807.1	0.000983	0.01006	0.01104	112.72	114.26	226.98
65	2012.9	0.001007	0.00874	0.00974	119.50	108.89	228.39
70	2235.5	0.001033	0.00756	0.00859	126.46	103.16	229.62
75	2476.0	0.001064	0.00650	0.00756	133.64	96.98	230.62
80	2735.7	0.001100	0.00553	0.00663	141.09	90.22	231.30
85	3015.8	0.001144	0.00464	0.00578	148.90	82.67	231.57
90	3318.0	0.001198	0.00380	0.00499	157.23	73.98	231.21
95	3644.6	0.001272	0.00297	0.00424	166.42	63.43	229.86
100	3998.6	0.001387	0.00209	0.00347	177.37	49.14	226.51

(Continued)

TABLE B.24.1 SI (Continued) *Saturated R-500*

Temp.	Press.	Enthalpy, kJ/kg			Entropy, kJ/kg K		
C	kPa	Sat. Liquid	Evap.	Sat. Vapor	Sat. Liquid	Evap.	Sat. Vapor
T	P	h_f	h_{fg}	h_g	s_f	s_{fg}	s_g
-50	46.4	-9.31	207.63	198.32	-0.0407	0.9304	0.8897
-45	59.5	-4.70	205.69	200.99	-0.0203	0.9015	0.8812
-40	75.6	0	203.64	203.64	0	0.8734	0.8734
-35	94.9	4.79	201.48	206.27	0.0203	0.8460	0.8663
-30	117.9	9.67	199.21	208.88	0.0405	0.8193	0.8598
-25	145.1	14.64	196.83	211.46	0.0606	0.7932	0.8538
-20	177.0	19.70	194.32	214.02	0.0807	0.7676	0.8483
-15	214.1	24.85	191.69	216.54	0.1007	0.7425	0.8433
-10	257.1	30.10	188.92	219.02	0.1207	0.7179	0.8387
-5	306.3	35.44	186.03	221.46	0.1407	0.6937	0.8344
0	362.5	40.87	182.99	223.86	0.1606	0.6699	0.8305
5	426.1	46.41	179.81	226.21	0.1805	0.6464	0.8269
10	497.9	52.04	176.47	228.51	0.2003	0.6233	0.8236
15	578.5	57.77	172.99	230.75	0.2202	0.6003	0.8205
20	668.4	63.60	169.33	232.93	0.2400	0.5776	0.8176
25	768.4	69.54	165.50	235.03	0.2598	0.5551	0.8149
30	879.1	75.58	161.48	237.06	0.2796	0.5327	0.8122
35	1001.2	81.74	157.26	239.01	0.2994	0.5103	0.8097
40	1135.3	88.02	152.83	240.85	0.3192	0.4880	0.8072
45	1282.2	94.43	148.16	242.58	0.3390	0.4657	0.8047
50	1442.6	100.96	143.23	244.19	0.3590	0.4432	0.8022
55	1617.3	107.65	138.00	245.65	0.3790	0.4205	0.7995
60	1807.1	114.50	132.44	246.93	0.3991	0.3975	0.7967
65	2012.9	121.53	126.48	248.01	0.4195	0.3740	0.7935
70	2235.5	128.77	120.06	248.83	0.4401	0.3499	0.7900
75	2476.0	136.27	113.07	249.34	0.4611	0.3248	0.7858
80	2735.7	144.10	105.35	249.45	0.4826	0.2983	0.7809
85	3015.8	152.34	96.66	249.01	0.5049	0.2699	0.7748
90	3318.0	161.20	86.58	247.78	0.5284	0.2384	0.7669
95	3644.6	171.06	74.25	245.31	0.5543	0.2017	0.7560
100	3998.6	182.92	57.49	240.40	0.5850	0.1541	0.7391

TABLE B.24.2 SI *Superheated R-500*

Temp. C	v m³/kg	h kJ/kg	s kJ/kg K	v m³/kg	h kJ/kg	s kJ/kg K	v m³/kg	h kJ/kg	s kJ/kg K
	50 kPa (-48.52)			100 kPa (-33.81)			200 kPa (-16.82)		
Sat.	0.36589	199.11	0.8871	0.19172	206.90	0.8647	0.10001	215.62	0.8451
-10	0.43414	224.47	0.9912	0.21376	223.22	0.9297	0.10342	220.59	0.8642
0	0.45152	231.35	1.0169	0.22279	230.23	0.9558	0.10832	227.92	0.8915
10	0.46882	238.34	1.0420	0.23173	237.35	0.9814	0.11310	235.29	0.9180
20	0.48603	245.47	1.0668	0.24059	244.57	1.0065	0.11780	242.74	0.9439
30	0.50319	252.71	1.0911	0.24938	251.91	1.0311	0.12242	250.25	0.9691
40	0.52029	260.08	1.1150	0.25812	259.35	1.0553	0.12699	257.86	0.9938
50	0.53735	267.58	1.1385	0.26681	266.91	1.0790	0.13151	265.55	1.0179
60	0.55437	275.19	1.1617	0.27546	274.58	1.1024	0.13598	273.34	1.0417
70	0.57136	282.92	1.1846	0.28408	282.36	1.1254	0.14042	281.22	1.0650
80	0.58832	290.77	1.2071	0.29268	290.25	1.1481	0.14484	289.20	1.0879
90	0.60526	298.73	1.2294	0.30125	298.25	1.1704	0.14923	297.29	1.1105
100	0.62218	306.81	1.2513	0.30980	306.36	1.1924	0.15359	305.47	1.1327
110	0.63908	315.00	1.2730	0.31833	314.58	1.2142	0.15794	313.75	1.1546
120	0.65597	323.30	1.2943	0.32685	322.91	1.2356	0.16228	322.13	1.1762
130	0.67285	331.70	1.3155	0.33535	331.34	1.2568	0.16660	330.61	1.1975
140	0.68971	340.22	1.3363	0.34385	339.87	1.2777	0.17091	339.18	1.2185
160	0.72342	357.55	1.3773	0.36081	357.24	1.3187	0.17950	356.62	1.2597
	300 kPa (-5.60)			400 kPa (3.02)			500 kPa (10.14)		
Sat.	0.06807	221.17	0.8349	0.05167	225.29	0.8283	0.04162	228.57	0.8235
0	0.07005	225.47	0.8508	—	—	—	—	—	—
10	0.07348	233.14	0.8784	0.05359	230.87	0.8483	—	—	—
20	0.07681	240.82	0.9051	0.05626	238.82	0.8759	0.04388	236.73	0.8518
30	0.08006	248.54	0.9310	0.05883	246.77	0.9025	0.04606	244.92	0.8793
40	0.08324	256.32	0.9562	0.06134	254.73	0.9284	0.04817	253.09	0.9058
50	0.08638	264.16	0.9809	0.06379	262.73	0.9535	0.05021	261.25	0.9315
60	0.08947	272.07	1.0050	0.06619	270.77	0.9780	0.05220	269.44	0.9564
70	0.09252	280.06	1.0286	0.06855	278.88	1.0020	0.05416	277.67	0.9808
80	0.09554	288.14	1.0518	0.07088	287.06	1.0255	0.05608	285.95	1.0045
90	0.09854	296.30	1.0746	0.07319	295.30	1.0485	0.05797	294.29	1.0278
100	0.10152	304.56	1.0970	0.07547	303.63	1.0712	0.05983	302.69	1.0507
110	0.10447	312.90	1.1191	0.07773	312.04	1.0934	0.06168	311.17	1.0731
120	0.10741	321.34	1.1408	0.07998	320.53	1.1153	0.06351	319.72	1.0951
130	0.11034	329.86	1.1622	0.08221	329.11	1.1368	0.06533	328.36	1.1168
140	0.11326	338.48	1.1833	0.08443	337.78	1.1581	0.06713	337.07	1.1381
150	0.11617	347.19	1.2042	0.08664	346.53	1.1790	0.06892	345.86	1.1592
160	0.11906	356.00	1.2247	0.08884	355.37	1.1996	0.07071	354.74	1.1799
170	0.12195	364.89	1.2450	0.09103	364.29	1.2200	0.07248	363.69	1.2003

(Continued)

TABLE B.24.2 SI (Continued) ***Superheated R-500***

Temp. C	v m³/kg	h kJ/kg	s kJ/kg K	v m³/kg	h kJ/kg	s kJ/kg K	v m³/kg	h kJ/kg	s kJ/kg K
	600 kPa (16.25)			800 kPa (26.48)			1000 kPa (34.95)		
Sat.	0.03482	231.30	0.8198	0.02616	235.64	0.8141	0.02086	238.99	0.8097
20	0.03558	234.52	0.8308	—	—	—	—	—	—
40	0.03936	251.38	0.8865	0.02829	247.77	0.8537	0.02155	243.80	0.8252
60	0.04287	268.08	0.9382	0.03116	265.23	0.9077	0.02409	262.20	0.8822
80	0.04620	284.82	0.9870	0.03382	282.50	0.9581	0.02637	280.06	0.9343
100	0.04940	301.74	1.0336	0.03635	299.79	1.0057	0.02851	297.77	0.9831
120	0.05253	318.90	1.0784	0.03879	317.23	1.0512	0.03054	315.52	1.0294
140	0.05559	336.35	1.1216	0.04117	334.89	1.0950	0.03251	333.40	1.0737
160	0.05861	354.10	1.1636	0.04350	352.81	1.1374	0.03442	351.50	1.1165
180	0.06160	372.16	1.2043	0.04579	371.00	1.1784	0.03630	369.83	1.1579
200	0.06456	390.53	1.2440	0.04805	389.49	1.2183	0.03815	388.43	1.1980
220	0.06750	409.21	1.2827	0.05030	408.26	1.2572	0.03998	407.29	1.2371
240	0.07042	428.19	1.3204	0.05253	427.31	1.2951	0.04179	426.42	1.2751
260	0.07333	447.45	1.3572	0.05474	446.64	1.3320	0.04358	445.82	1.3122
280	0.07623	466.98	1.3932	0.05694	466.22	1.3681	0.04537	465.46	1.3484
300	0.07912	486.77	1.4283	0.05913	486.06	1.4033	0.04714	485.34	1.3837
320	0.08200	506.79	1.4627	0.06132	506.12	1.4377	0.04890	505.45	1.4181
340	0.08488	527.02	1.4962	0.06349	526.39	1.4713	0.05066	525.77	1.4518
360	0.08775	547.46	1.5290	0.06566	546.86	1.5042	0.05241	546.27	1.4847
	1500 kPa (51.69)			2000 kPa (64.70)			3000 kPa (84.73)		
Sat.	0.01359	244.70	0.8013	0.00982	247.95	0.7937	0.00583	249.05	0.7751
60	0.01448	253.51	0.8281	—	—	—	—	—	—
80	0.01635	273.41	0.8861	0.01119	265.57	0.8448	—	—	—
100	0.01799	292.39	0.9384	0.01266	286.40	0.9021	0.00710	271.33	0.8362
120	0.01951	311.02	0.9870	0.01395	306.16	0.9537	0.00829	294.94	0.8978
140	0.02094	329.54	1.0330	0.01513	325.46	1.0016	0.00927	316.43	0.9512
160	0.02231	348.12	1.0769	0.01624	344.60	1.0469	0.01014	337.02	0.9998
180	0.02364	366.84	1.1191	0.01730	363.74	1.0900	0.01095	357.20	1.0454
200	0.02494	385.73	1.1600	0.01833	382.97	1.1316	0.01171	377.21	1.0886
220	0.02621	404.84	1.1995	0.01933	402.35	1.1717	0.01244	397.19	1.1300
240	0.02747	424.18	1.2379	0.02031	421.90	1.2106	0.01315	417.24	1.1698
260	0.02871	443.75	1.2753	0.02127	441.65	1.2483	0.01384	437.39	1.2083
280	0.02993	463.54	1.3118	0.02222	461.60	1.2851	0.01451	457.68	1.2457
300	0.03115	483.55	1.3473	0.02316	481.75	1.3208	0.01517	478.12	1.2820
320	0.03236	503.77	1.3820	0.02409	502.09	1.3557	0.01583	498.71	1.3173
340	0.03356	524.19	1.4159	0.02501	522.61	1.3897	0.01647	519.46	1.3517
360	0.03475	544.79	1.4489	0.02593	543.31	1.4230	0.01711	540.35	1.3852

TABLE B.25 SI *Thermodynamic Properties of R-502*
TABLE B.25.1 SI *Saturated R-502*

Temp.	Press.	Specific Volume, m³/kg			Internal Energy, kJ/kg		
C	kPa	Sat. Liquid	Evap.	Sat. Vapor	Sat. Liquid	Evap.	Sat. Vapor
T	P	v_f	v_{fg}	v_g	u_f	u_{fg}	u_g
-50	81.4	0.000668	0.19660	0.19726	-9.37	158.39	149.02
-45	103.3	0.000676	0.15724	0.15791	-4.79	156.06	151.27
-40	129.6	0.000683	0.12700	0.12769	-0.09	153.60	153.51
-35	161.0	0.000691	0.10351	0.10420	4.73	151.02	155.75
-30	197.9	0.000699	0.08507	0.08577	9.66	148.31	157.97
-25	241.0	0.000707	0.07045	0.07116	14.70	145.48	160.18
-20	291.0	0.000716	0.05875	0.05946	19.86	142.52	162.38
-15	348.6	0.000725	0.04930	0.05002	25.11	139.43	164.54
-10	414.3	0.000735	0.04161	0.04234	30.47	136.21	166.69
-5	488.9	0.000745	0.03530	0.03604	35.93	132.87	168.80
0	573.1	0.000756	0.03008	0.03084	41.48	129.39	170.88
5	667.6	0.000768	0.02574	0.02651	47.12	125.79	172.91
10	773.0	0.000780	0.02210	0.02288	52.84	122.06	174.90
15	890.2	0.000793	0.01903	0.01983	58.64	118.20	176.84
20	1019.7	0.000807	0.01643	0.01723	64.52	114.20	178.71
25	1162.3	0.000822	0.01420	0.01502	70.47	110.05	180.52
30	1318.9	0.000838	0.01228	0.01312	76.49	105.75	182.24
35	1490.1	0.000857	0.01062	0.01148	82.58	101.29	183.86
40	1677.0	0.000877	0.00918	0.01005	88.74	96.63	185.37
45	1880.3	0.000899	0.00790	0.00880	94.99	91.74	186.73
50	2101.3	0.000925	0.00678	0.00770	101.33	86.58	187.91
55	2341.1	0.000954	0.00577	0.00672	107.81	81.05	188.86
60	2601.4	0.000990	0.00485	0.00584	114.47	75.03	189.50
65	2884.0	0.001033	0.00401	0.00504	121.43	68.27	189.69
70	3191.8	0.001091	0.00320	0.00429	128.90	60.29	189.19
75	3528.5	0.001175	0.00237	0.00355	137.47	49.94	187.41
80	3900.4	0.001342	0.00136	0.00271	149.50	32.53	182.03

(Continued)

TABLE B.25.1 SI (Continued) ***Saturated R-502***

Temp.	Press.	Enthalpy, kJ/kg			Entropy, kJ/kg K		
C	kPa	Sat. Liquid	Evap.	Sat. Vapor	Sat. Liquid	Evap.	Sat. Vapor
T	P	h_f	h_{fg}	h_g	s_f	s_{fg}	s_g
-50	81.4	-9.32	174.40	165.08	-0.0407	0.7815	0.7408
-45	103.3	-4.72	172.30	167.58	-0.0204	0.7552	0.7348
-40	129.6	0	170.06	170.06	0	0.7294	0.7294
-35	161.0	4.84	167.68	172.52	0.0204	0.7041	0.7245
-30	197.9	9.80	165.15	174.94	0.0409	0.6792	0.7201
-25	241.0	14.87	162.46	177.33	0.0615	0.6547	0.7161
-20	291.0	20.06	159.61	179.68	0.0820	0.6305	0.7126
-15	348.6	25.37	156.61	181.98	0.1026	0.6067	0.7093
-10	414.3	30.78	153.45	184.23	0.1232	0.5831	0.7063
-5	488.9	36.30	150.13	186.42	0.1438	0.5599	0.7036
0	573.1	41.92	146.64	188.55	0.1643	0.5368	0.7011
5	667.6	47.63	142.98	190.61	0.1848	0.5140	0.6988
10	773.0	53.44	139.15	192.59	0.2052	0.4914	0.6966
15	890.2	59.35	135.14	194.49	0.2255	0.4690	0.6945
20	1019.7	65.34	130.95	196.29	0.2458	0.4467	0.6925
25	1162.3	71.42	126.56	197.98	0.2660	0.4245	0.6904
30	1318.9	77.59	121.95	199.55	0.2861	0.4023	0.6883
35	1490.1	83.85	117.12	200.97	0.3061	0.3801	0.6861
40	1677.0	90.21	112.02	202.23	0.3260	0.3577	0.6837
45	1880.3	96.68	106.61	203.28	0.3459	0.3351	0.6810
50	2101.3	103.28	100.82	204.10	0.3659	0.3120	0.6779
55	2341.1	110.04	94.56	204.60	0.3860	0.2882	0.6741
60	2601.4	117.05	87.65	204.70	0.4064	0.2631	0.6695
65	2884.0	124.41	79.82	204.22	0.4275	0.2360	0.6635
70	3191.8	132.38	70.49	202.87	0.4499	0.2054	0.6553
75	3528.5	141.61	58.31	199.93	0.4755	0.1675	0.6430
80	3900.4	154.74	37.85	192.59	0.5116	0.1072	0.6188

TABLE B.25.2 SI *Superheated R-502*

Temp. C	v m³/kg	h kJ/kg	s kJ/kg K	v m³/kg	h kJ/kg	s kJ/kg K	v m³/kg	h kJ/kg	s kJ/kg K
	50 kPa (-59.52)			100 kPa (-45.70)			200 kPa (-29.73)		
Sat.	0.31069	160.27	0.7541	0.16281	167.23	0.7356	0.08490	175.07	0.7199
-20	0.37206	184.04	0.8561	0.18347	183.19	0.8020	0.08908	181.40	0.7454
-10	0.38740	190.38	0.8806	0.19137	189.60	0.8269	0.09328	187.99	0.7709
0	0.40268	196.84	0.9047	0.19922	196.13	0.8512	0.09743	194.67	0.7958
10	0.41792	203.43	0.9284	0.20701	202.78	0.8751	0.10151	201.44	0.8202
20	0.43311	210.14	0.9517	0.21477	209.54	0.8986	0.10556	208.32	0.8440
30	0.44828	216.97	0.9746	0.22249	216.42	0.9217	0.10956	215.30	0.8674
40	0.46341	223.92	0.9971	0.23018	223.41	0.9444	0 .11354	222.38	0.8904
50	0.47852	230.99	1.0194	0.23784	230.52	0.9667	0.11749	229.56	0.9130
60	0.49361	238.17	1.0413	0.24549	237.73	0.9887	0.12141	236.84	0.9352
70	0.50868	245.47	1.0628	0.25311	245.06	1.0104	0.12532	244.23	0.9570
80	0.52373	252.87	1.0841	0.26072	252.49	1.0317	0.12920	251.72	0.9786
90	0.53877	260.39	1.1051	0.26831	260.03	1.0528	0 .13308	259.30	0.9997
100	0.55380	268.01	1.1258	0.27589	267.67	1.0735	0.13694	266.99	1.0206
110	0.56882	275.73	1.1462	0.28346	275.41	1.0940	0.14078	274.77	1.0412
120	0.58382	283.56	1.1664	0.29102	283.26	1.1142	0.14462	282.65	1.0615
130	0.59882	291.48	1.1863	0.29858	291.20	1.1341	0.14845	290.62	1.0815
140	0.61381	299.50	1.2059	0.30612	299.23	1.1538	0.15227	298.69	1.1013
150	0.62880	307.61	1.2253	0.31366	307.36	1.1733	0.15608	306.84	1.1208
	300 kPa (-19.17)			400 kPa (-11.03)			500 kPa (-4.31)		
Sat.	0.05776	180.06	0.7120	0.04381	183.77	0.7069	0.03526	186.72	0.7033
0	0.06343	193.14	0.7616	0.04639	191.54	0.7359	0.03611	189.85	0.7148
10	0.06630	200.06	0.7865	0.04866	198.62	0.7614	0.03803	197.11	0.7409
20	0.06912	207.06	0.8108	0.05087	205.75	0.7862	0.03989	204.40	0.7662
30	0.07189	214.14	0.8345	0.05304	212.95	0.8103	0.04170	211.72	0.7908
40	0.07464	221.31	0.8578	0.05517	220.22	0.8339	0.04347	219.11	0.8147
50	0.07735	228.58	0.8807	0.05727	227.58	0.8570	0.04520	226.55	0.8381
60	0.08004	235.94	0.9031	0.05934	235.01	0.8797	0.04691	234.07	0.8610
70	0.08271	243.39	0.9251	0.06139	242.53	0.9019	0.04860	241.65	0.8835
80	0.08536	250.93	0.9468	0.06342	250.13	0.9238	0.05026	249.32	0 .9055
90	0.08799	258.57	0.9681	0.06544	257.82	0.9452	0.05191	257.07	0.9271
100	0.09061	266.30	0.9891	0.06744	265.60	0.9664	0.05354	264.90	0.9484
110	0.09322	274.13	1.0098	0.06943	273.47	0.9872	0.05516	272.81	0.9693
120	0.09582	282.04	1.0302	0.07141	281.42	1.0077	0.05676	280.80	0.9899
130	0.09840	290.05	1.0503	0.07338	289.47	1.0279	0.05836	288.88	1.0102
140	0.10098	298.14	1.0701	0.07534	297.59	1.0478	0.05995	297.04	1.0302
150	0.10356	306.32	1.0897	0.07729	305.80	1.0674	0.06153	305.28	1.0499
160	0.10613	314.59	1.1090	0.07924	314.10	1.0868	0.06311	313.60	1.0693
170	0.10869	322.95	1.1281	0.08118	322.47	1.1059	0.06468	322.00	1.0885

(Continued)

TABLE B.252 SI (Continued) *Superheated R-502*

Temp. C	v m³/kg	h kJ/kg	s kJ/kg K	v m³/kg	h kJ/kg	s kJ/kg K	v m³/kg	h kJ/kg	s kJ/kg K
	600 kPa (1.48)			800 kPa (11.20)			1000 kPa (19.27)		
Sat.	0.02948	189.17	0.7004	0.02210	193.05	0.6961	0.01759	196.03	0.6928
20	0.03255	202.99	0.7491	0.02331	199.98	0.7201	0.01767	196.64	0.6948
30	0.03413	210.46	0.7742	0.02461	207.77	0.7462	0.01884	204.86	0.7224
40	0.03566	217.95	0.7985	0.02586	215.54	0.7714	0.01993	212.96	0.7487
50	0.03715	225.50	0.8222	0.02706	223.31	0.7959	0.02097	221.00	0.7740
60	0.03862	233.10	0.8454	0.02823	231.11	0.8196	0 .02197	229.01	0.7984
70	0.04006	240.77	0.8680	0.02937	238.93	0.8428	0.02294	237.02	0.8221
80	0.04148	248.50	0.8903	0.03049	246.81	0.8654	0.02388	245.06	0.8452
90	0.04288	256.30	0.9120	0.03158	254.74	0.8875	0.02480	253.12	0.8677
100	0.04426	264.18	0.9335	0.03266	262.72	0.9092	0.02570	261.23	0.8897
110	0.04564	272.14	0.9545	0.03373	270.78	0.9305	0.02658	269.38	0.9113
120	0.04700	280.17	0.9752	0.03479	278.90	0.9514	0.02745	277.59	0.9324
130	0.04835	288.29	0.9956	0.03583	287.09	0.9720	0.02831	285.86	0.9532
140	0.04969	296.48	1.0156	0.03686	295.35	0.9923	0.02916	294.20	0.9736
150	0.05103	304.75	1.0354	0.03789	303.68	1.0122	0.03001	302.60	0.9937
160	0.05235	313.10	1.0549	0.03891	312.08	1.0318	0.03084	311.06	1.0135
170	0.05368	321.52	1.0742	0.03992	320.56	1.0512	0.03167	319.59	1.0329
180	0.05499	330.03	1.0931	0.04093	329.11	1.0702	0.03249	328.19	1.0521
190	0.05630	338.60	1.1119	0.04193	337.73	1.0891	0.03331	336.85	1.0710
	1500 kPa (35.27)			2000 kPa (47.76)			3000 kPa (66.94)		
Sat.	0.01140	201.04	0.6860	0.00818	203.77	0.6793	0.00474	203.83	0.6607
40	0.01185	205.45	0.7002	—	—	—	—	—	—
60	0.01353	223.23	0.7552	0.00917	216.30	0.7177	—	—	—
80	0.01501	240.36	0.8052	0.01052	235.09	0.7725	0.00578	221.47	0.7116
100	0.01638	257.28	0.8518	0.01168	253.01	0.8218	0.00688	243.06	0.7711
120	0.01766	274.21	0.8960	0.01274	270.62	0.8678	0.00778	262.66	0.8223
140	0.01889	291.24	0.9382	0.01374	288.14	0.9113	0.00857	281.50	0.8690
160	0.02008	308.44	0.9789	0.01469	305.72	0.9528	0.00929	300.01	0.9128
180	0.02124	325.84	1.0181	0.01561	323.43	0.9928	0.00998	318.42	0.9543
200	0.02237	343.46	1.0562	0.01650	341.29	1.0314	0.01063	336.84	0.9941
220	0.02349	361.31	1.0931	0.01737	359.34	1.0687	0.01126	355.33	1.0324
240	0.02460	379.39	1.1291	0.01823	377.59	1.1050	0.01187	373.95	1.0694
260	0.02569	397.70	1.1641	0.01908	396.05	1.1403	0.01247	392.71	1.1053
280	0.02677	416.23	1.1982	0.01991	414.70	1.1746	0.01306	411.63	1.1401
300	0.02785	434.98	1.2315	0.02074	433.56	1.2081	0.01364	430.72	1.1740
320	0.02891	453.94	1.2640	0.02156	452.61	1.2408	0.01421	449.97	1.2070
340	0.02997	473.09	1.2958	0.02237	471.85	1.2727	0.01478	469.38	1.2392

SECTION C

IDEAL GAS TABLES

A short explanation for the ideal gas tables is given here to help in the use and prevent misunderstanding of the table entries. The air tables are done on a mass basis, properties per kg, whereas the additional tables are done on a mole basis, properties per kmol, so division with molecular weight will yield the property per kg. Because of the common appearance of air both internal energy, u, and the enthalpy, h, are listed in the air table and only enthalpy for the remaining tables. We do have ideal gas so

$$h = u + Pv = u + RT.$$

The entropy is listed as the absolute standard entropy (entropy at a pressure of 100 kPa) as a function of temperature. The change in entropy between two different states is

$$s_2 - s_1 = s_2^o - s_1^o - R\ln(P_2 / P_1)$$

so the mathematical explicit pressure dependence should be included besides the table entries.

For the air tables the two dimensionless functions, P_r and v_r are special functions relating to an isentropic process.

$$s_2 = s_1 \Rightarrow \frac{P_2}{P_1} = \frac{P_{r2}}{P_{r1}} \quad \text{and} \quad \frac{v_2}{v_1} = \frac{v_{r2}}{v_{r1}}$$

For constant heat capacity (i.e. not a function of temperature) these relations simplify to the power functions.

TABLE C.1 *Ideal-Gas Properties of Air, Standard Entropy at 0.1 MPa (1 bar) Pressure*

T K	u kJ/kg	h kJ/kg	s^o kJ/kg K	P_r	v_r
200	142.768	200.174	6.46260	0.27027	493.466
220	157.071	220.218	6.55812	0.37700	389.150
240	171.379	240.267	6.64535	0.51088	313.274
260	185.695	260.323	6.72562	0.67573	256.584
280	200.022	280.390	6.79998	0.87556	213.257
290	207.191	290.430	6.83521	0.98990	195.361
298.15	213.036	298.615	6.86305	1.09071	182.288
300	214.364	300.473	6.86926	1.11458	179.491
320	228.726	320.576	6.93413	1.39722	152.728
340	243.113	340.704	6.99515	1.72814	131.200
360	257.532	360.863	7.05276	2.11226	113.654
380	271.988	381.060	7.10735	2.55479	99.1882
400	286.487	401.299	7.15926	3.06119	87.1367
420	301.035	421.589	7.20875	3.63727	77.0025
440	315.640	441.934	7.25607	4.28916	68.4088
460	330.306	462.340	7.30142	5.02333	61.0658
480	345.039	482.814	7.34499	5.84663	54.7479
500	359.844	503.360	7.38692	6.76629	49.2777
520	374.726	523.982	7.42736	7.78997	44.5143
540	389.689	544.686	7.46642	8.92569	40.3444
560	404.736	565.474	7.50422	10.18197	36.6765
580	419.871	586.350	7.54084	11.56771	33.4358
600	435.097	607.316	7.57638	13.09232	30.5609
620	450.415	628.375	7.61090	14.76564	28.0008
640	465.828	649.528	7.64448	16.59801	25.7132
660	481.335	670.776	7.67717	18.60025	23.6623
680	496.939	692.120	7.70903	20.78367	21.8182
700	512.639	713.561	7.74010	23.16010	20.1553
720	528.435	735.098	7.77044	25.74188	18.6519
740	544.328	756.731	7.80008	28.54188	17.2894
760	560.316	778.460	7.82905	31.57347	16.0518
780	576.400	800.284	7.85740	34.85061	14.9250
800	592.577	822.202	7.88514	38.38777	13.8972
850	633.422	877.397	7.95207	48.46828	11.6948
900	674.824	933.152	8.01581	60.51977	9.91692
950	716.756	989.436	8.07667	74.81519	8.46770
1000	759.189	1046.221	8.13493	91.65077	7.27604
1050	802.095	1103.478	8.19081	111.3467	6.28845
1100	845.445	1161.180	8.24449	134.2478	5.46408

(Continued)

TABLE C.1 (Continued) ***Ideal-Gas Properties of Air, Standard Entropy at 0.1 MPa (1 bar) Pressure***

T K	u kJ/kg	h kJ/kg	s^o kJ/kg K	P_r	v_r
1100	845.445	1161.180	8.24449	134.2478	5.46408
1150	889.211	1219.298	8.29616	160.7245	4.77141
1200	933.367	1277.805	8.34596	191.1736	4.18586
1250	977.888	1336.677	8.39402	226.0192	3.68804
1300	1022.751	1395.892	8.44046	265.7145	3.26257
1350	1067.936	1455.429	8.48539	310.7426	2.89711
1400	1113.426	1515.270	8.52891	361.6192	2.58171
1450	1159.202	1575.398	8.57111	418.8942	2.30831
1500	1205.253	1635.800	8.61208	483.1554	2.07031
1550	1251.547	1696.446	8.65185	554.9577	1.86253
1600	1298.079	1757.329	8.69051	634.9670	1.68035
1650	1344.834	1818.436	8.72811	723.8560	1.52007
1700	1391.801	1879.755	8.76472	822.3320	1.37858
1750	1438.970	1941.275	8.80039	931.1376	1.25330
1800	1486.331	2002.987	8.83516	1051.051	1.14204
1850	1533.873	2064.882	8.86908	1182.888	1.04294
1900	1581.591	2126.951	8.90219	1327.498	0.95445
1950	1629.474	2189.186	8.93452	1485.772	0.87521
2000	1677.518	2251.581	8.96611	1658.635	0.80410
2050	1725.714	2314.128	8.99699	1847.077	0.74012
2100	1774.057	2376.823	9.02721	2052.109	0.68242
2150	1822.541	2439.659	9.05678	2274.789	0.63027
2200	1871.161	2502.630	9.08573	2516.217	0.58305
2250	1919.912	2565.733	9.11409	2777.537	0.54020
2300	1968.790	2628.962	9.14189	3059.939	0.50124
2350	2017.789	2692.313	9.16913	3364.658	0.46576
2400	2066.907	2755.782	9.19586	3692.974	0.43338
2450	2116.138	2819.366	9.22208	4046.215	0.40378
2500	2165.480	2883.059	9.24781	4425.759	0.37669
2550	2214.929	2946.859	9.27308	4833.031	0.35185
2600	2264.481	3010.763	9.29790	5269.505	0.32903
2650	2314.133	3074.767	9.32228	5736.707	0.30805
2700	2363.883	3138.868	9.34625	6236.215	0.28872
2750	2413.727	3203.064	9.36980	6769.657	0.27089
2800	2463.663	3267.351	9.39297	7338.715	0.25443
2850	2513.687	3331.726	9.41576	7945.124	0.23921
2900	2563.797	3396.188	9.43818	8590.676	0.22511
2950	2613.990	3460.733	9.46025	9277.216	0.21205
3000	2664.265	3525.359	9.48198	10006.645	0.19992

TABLE C.2 ***Ideal-Gas Properties of Various Substances, Entropies at 0.1-MPa (1-bar) Pressure***

	Nitrogen, Diatomic (N_2) $\bar{h}^o_{f,298}$ = 0 kJ/kmol M = 28.013		Nitrogen, Monatomic (N) $\bar{h}^o_{f,298}$ = 472 680 kJ/kmol M = 14.007	
T K	$(\bar{h}-\bar{h}^o_{298})$ kJ/kmol	$\bar{s}^o$ kJ/kmol K	$(\bar{h}-\bar{h}^o_{298})$ kJ/kmol	$\bar{s}^o$ kJ/kmol K
0	−8670	0	−6197	0
100	−5768	159.812	−4119	130.593
200	−2857	179.985	−2040	145.001
298	0	191.609	0	153.300
300	54	191.789	38	153.429
400	2971	200.181	2117	159.409
500	5911	206.740	4196	164.047
600	8894	212.177	6274	167.837
700	11937	216.865	8353	171.041
800	15046	221.016	10431	173.816
900	18223	224.757	12510	176.265
1000	21463	228.171	14589	178.455
1100	24760	231.314	16667	180.436
1200	28109	234.227	18746	182.244
1300	31503	236.943	20825	183.908
1400	34936	239.487	22903	185.448
1500	38405	241.881	24982	186.883
1600	41904	244.139	27060	188.224
1700	45430	246.276	29139	189.484
1800	48979	248.304	31218	190.672
1900	52549	250.234	33296	191.796
2000	56137	252.075	35375	192.863
2200	63362	255.518	39534	194.845
2400	70640	258.684	43695	196.655
2600	77963	261.615	47860	198.322
2800	85323	264.342	52033	199.868
3000	92715	266.892	56218	201.311
3200	100134	269.286	60420	202.667
3400	107577	271.542	64646	203.948
3600	115042	273.675	68902	205.164
3800	122526	275.698	73194	206.325
4000	130027	277.622	77532	207.437
4400	145078	281.209	86367	209.542
4800	160188	284.495	95457	211.519
5200	175352	287.530	104843	213.397
5600	190572	290.349	114550	215.195
6000	205848	292.984	124590	216.926

(Continued)

TABLE C.2 (Continued) ***Ideal-Gas Properties of Various Substances, Entropies at 0.1-MPa (1-bar) Pressure***

	Oxygen, Diatomic (O_2) $\bar{h}^o_{f,298} = 0$ kJ/kmol $M = 31.999$		Oxygen, Monatomic (O) $\bar{h}^o_{f,298} = 249\ 170$ kJ/kmol $M = 16.00$	
T K	$(\bar{h}-\bar{h}^o_{298})$ kJ/kmol	$\bar{s}^o$ kJ/kmol K	$(\bar{h}-\bar{h}^o_{298})$ kJ/kmol	$\bar{s}^o$ kJ/kmol K
0	−8683	0	−6725	0
100	−5777	173.308	−4518	135.947
200	−2868	193.483	−2186	152.153
298	0	205.148	0	161.059
300	54	205.329	41	161.194
400	3027	213.873	2207	167.431
500	6086	220.693	4343	172.198
600	9245	226.450	6462	176.060
700	12499	231.465	8570	179.310
800	15836	235.920	10671	182.116
900	19241	239.931	12767	184.585
1000	22703	243.579	14860	186.790
1100	26212	246.923	16950	188.783
1200	29761	250.011	19039	190.600
1300	33345	252.878	21126	192.270
1400	36958	255.556	23212	193.816
1500	40600	258.068	25296	195.254
1600	44267	260.434	27381	196.599
1700	47959	262.673	29464	197.862
1800	51674	264.797	31547	199.053
1900	55414	266.819	33630	200.179
2000	59176	268.748	35713	201.247
2200	66770	272.366	39878	203.232
2400	74453	275.708	44045	205.045
2600	82225	278.818	48216	206.714
2800	90080	281.729	52391	208.262
3000	98013	284.466	56574	209.705
3200	106022	287.050	60767	211.058
3400	114101	289.499	64971	212.332
3600	122245	291.826	69190	213.538
3800	130447	294.043	73424	214.682
4000	138705	296.161	77675	215.773
4400	155374	300.133	86234	217.812
4800	172240	303.801	94873	219.691
5200	189312	307.217	103592	221.435
5600	206618	310.423	112391	223.066
6000	224210	313.457	121264	224.597

(Continued)

TABLE C.2 (Continued) ***Ideal-Gas Properties of Various Substances, Entropies at 0.1-MPa (1-bar) Pressure***

	Carbon Dioxide (CO_2) $\bar{h}^o_{f,298} = -393\ 522$ kJ/kmol $M = 44.01$		Carbon Monoxide (CO) $\bar{h}^o_{f,298} = -110\ 527$ kJ/kmol $M = 28.01$	
T K	**$(\bar{h}-\bar{h}^o_{298})$ kJ/kmol**	**$\bar{s}^o$ kJ/kmol K**	**$(\bar{h}-\bar{h}^o_{298})$ kJ/kmol**	**$\bar{s}^o$ kJ/kmol K**
0	−9364	0	−8671	0
100	−6457	179.010	−5772	165.852
200	−3413	199.976	−2860	186.024
298	0	213.794	0	197.651
300	69	214.024	54	197.831
400	4003	225.314	2977	206.240
500	8305	234.902	5932	212.833
600	12906	243.284	8942	218.321
700	17754	250.752	12021	223.067
800	22806	257.496	15174	227.277
900	28030	263.646	18397	231.074
1000	33397	269.299	21686	234.538
1100	38885	274.528	25031	237.726
1200	44473	279.390	28427	240.679
1300	50148	283.931	31867	243.431
1400	55895	288.190	35343	246.006
1500	61705	292.199	38852	248.426
1600	67569	295.984	42388	250.707
1700	73480	299.567	45948	252.866
1800	79432	302.969	49529	254.913
1900	85420	306.207	53128	256.860
2000	91439	309.294	56743	258.716
2200	103562	315.070	64012	262.182
2400	115779	320.384	71326	265.361
2600	128074	325.307	78679	268.302
2800	140435	329.887	86070	271.044
3000	152853	334.170	93504	273.607
3200	165321	338.194	100962	276.012
3400	177836	341.988	108440	278.279
3600	190394	345.576	115938	280.422
3800	202990	348.981	123454	282.454
4000	215624	352.221	130989	284.387
4400	240992	358.266	146108	287.989
4800	266488	363.812	161285	291.290
5200	292112	368.939	176510	294.337
5600	317870	373.711	191782	297.167
6000	343782	378.180	207105	299.809

(Continued)

TABLE C.2 (Continued) ***Ideal-Gas Properties of Various Substances, Entropies at 0.1-MPa (1-bar) Pressure***

	Water (H_2O) $\bar{h}^o_{f,298} = -241\ 826$ kJ/kmol $M = 18.015$		Hydroxyl (OH) $\bar{h}^o_{f,298} = 38\ 987$ kJ/kmol $M = 17.007$	
T K	$(\bar{h}-\bar{h}^o_{298})$ kJ/kmol	$\bar{s}^o$ kJ/kmol K	$(\bar{h}-\bar{h}^o_{298})$ kJ/kmol	$\bar{s}^o$ kJ/kmol K
0	−9904	0	−9172	0
100	−6617	152.386	−6140	149.591
200	−3282	175.488	−2975	171.592
298	0	188.835	0	183.709
300	62	189.043	55	183.894
400	3450	198.787	3034	192.466
500	6922	206.532	5991	199.066
600	10499	213.051	8943	204.448
700	14190	218.739	11902	209.008
800	18002	223.826	14881	212.984
900	21937	228.460	17889	216.526
1000	26000	232.739	20935	219.735
1100	30190	236.732	24024	222.680
1200	34506	240.485	27159	225.408
1300	38941	244.035	30340	227.955
1400	43491	247.406	33567	230.347
1500	48149	250.620	36838	232.604
1600	52907	253.690	40151	234.741
1700	57757	256.631	43502	236.772
1800	62693	259.452	46890	238.707
1900	67706	262.162	50311	240.556
2000	72788	264.769	53763	242.328
2200	83153	269.706	60751	245.659
2400	93741	274.312	67840	248.743
2600	104520	278.625	75018	251.614
2800	115463	282.680	82268	254.301
3000	126548	286.504	89585	256.825
3200	137756	290.120	96960	259.205
3400	149073	293.550	104388	261.456
3600	160484	296.812	111864	263.592
3800	171981	299.919	119382	265.625
4000	183552	302.887	126940	267.563
4400	206892	308.448	142165	271.191
4800	230456	313.573	157522	274.531
5200	254216	318.328	173002	277.629
5600	278161	322.764	188598	280.518
6000	302295	326.926	204309	283.227

(Continued)

TABLE C.2 (Continued) ***Ideal-Gas Properties of Various Substances, Entropies at 0.1-MPa (1-bar) Pressure***

	Hydrogen (H_2) $\bar{h}^{o}_{f,298} = 0$ kJ/kmol $M = 2.016$		Hydrogen, Monatomic (H) $\bar{h}^{o}_{f,298} = 217\ 999$ kJ/kmol $M = 1.008$	
T K	**$(\bar{h}-\bar{h}^{o}_{298})$ kJ/kmol**	**$\bar{s}^{o}$ kJ/kmol K**	**$(\bar{h}-\bar{h}^{o}_{298})$ kJ/kmol**	**$\bar{s}^{o}$ kJ/kmol K**
0	−8467	0	−6197	0
100	−5467	100.727	−4119	92.009
200	−2774	119.410	−2040	106.417
298	0	130.678	0	114.716
300	53	130.856	38	114.845
400	2961	139.219	2117	120.825
500	5883	145.738	4196	125.463
600	8799	151.078	6274	129.253
700	11730	155.609	8353	132.457
800	14681	159.554	10431	135.233
900	17657	163.060	12510	137.681
1000	20663	166.225	14589	139.871
1100	23704	169.121	16667	141.852
1200	26785	171.798	18746	143.661
1300	29907	174.294	20825	145.324
1400	33073	176.637	22903	146.865
1500	36281	178.849	24982	148.299
1600	39533	180.946	27060	149.640
1700	42826	182.941	29139	150.900
1800	46160	184.846	31218	152.089
1900	49532	186.670	33296	153.212
2000	52942	188.419	35375	154.279
2200	59865	191.719	39532	156.260
2400	66915	194.789	43689	158.069
2600	74082	197.659	47847	159.732
2800	81355	200.355	52004	161.273
3000	88725	202.898	56161	162.707
3200	96187	205.306	60318	164.048
3400	103736	207.593	64475	165.308
3600	111367	209.773	68633	166.497
3800	119077	211.856	72790	167.620
4000	126864	213.851	76947	168.687
4400	142658	217.612	85261	170.668
4800	158730	221.109	93576	172.476
5200	175057	224.379	101890	174.140
5600	191607	227.447	110205	175.681
6000	208332	230.322	118519	177.114

(Continued)

TABLE C.2 (Continued) ***Ideal-Gas Properties of Various Substances, Entropies at 0.1-MPa (1-bar) Pressure***

	Nitric Oxide (NO) $\bar{h}^o_{f,298}$ = 90 291 kJ/kmol M = 30.006		Nitrogen Dioxide (NO_2) $\bar{h}^o_{f,298}$ = 33 100 kJ/kmol M = 46.005	
T K	$(\bar{h}-\bar{h}^o_{298})$ kJ/kmol	$\bar{s}^o$ kJ/kmol K	$(\bar{h}-\bar{h}^o_{298})$ kJ/kmol	$\bar{s}^o$ kJ/kmol K
0	−9192	0	−10186	0
100	−6073	177.031	−6861	202.563
200	−2951	198.747	−3495	225.852
298	0	210.759	0	240.034
300	55	210.943	68	240.263
400	3040	219.529	3927	251.342
500	6059	226.263	8099	260.638
600	9144	231.886	12555	268.755
700	12308	236.762	17250	275.988
800	15548	241.088	22138	282.513
900	18858	244.985	27180	288.450
1000	22229	248.536	32344	293.889
1100	25653	251.799	37606	298.904
1200	29120	254.816	42946	303.551
1300	32626	257.621	48351	307.876
1400	36164	260.243	53808	311.920
1500	39729	262.703	59309	315.715
1600	43319	265.019	64846	319.289
1700	46929	267.208	70414	322.664
1800	50557	269.282	76008	325.861
1900	54201	271.252	81624	328.898
2000	57859	273.128	87259	331.788
2200	65212	276.632	98578	337.182
2400	72606	279.849	109948	342.128
2600	80034	282.822	121358	346.695
2800	87491	285.585	132800	350.934
3000	94973	288.165	144267	354.890
3200	102477	290.587	155756	358.597
3400	110000	292.867	167262	362.085
3600	117541	295.022	178783	365.378
3800	125099	297.065	190316	368.495
4000	132671	299.007	201860	371.456
4400	147857	302.626	224973	376.963
4800	163094	305.940	248114	381.997
5200	178377	308.998	271276	386.632
5600	193703	311.838	294455	390.926
6000	209070	314.488	317648	394.926

TABLE C.3. ***Enthalpy of Combustion of Some Hydrocarbons at 25°C, H_{RP} [kJ/kg]***

		Liquid H_2O in Products		Gas H_2O in Products	
Hydrocarbon	**Formula**	**Liq. HC**	**Gas HC**	**Liq. HC**	**Gas HC**
Paraffins	C_nH_{2n+2}				
Methane	CH_4		–55 496		–50 010
Ethane	C_2H_6		–51 875		–47 484
Propane	C_3H_8	–49 973	–50 343	–45 982	–46 352
n-Butane	C_4H_{10}	–49 130	–49 500	–45 344	–45 714
n-Pentane	C_5H_{12}	–48 643	–49 011	–44 983	–45 351
n-Hexane	C_6H_{14}	–48 308	–48 676	–44 733	–45 101
n-Heptane	C_7H_{16}	–48 071	–48 436	–44 557	–44 922
n-Octane	C_8H_{18}	–47 893	–48 256	–44 425	–44 788
n-Decane	$C_{10}H_{22}$	–47 641	–48 000	–44 239	–44 598
n-Dodecane	$C_{12}H_{26}$	–47 470	–47 828	–44 109	–44 467
n-Cetane	$C_{16}H_{34}$	–47 300	–47 658	–44 000	–44 358
Olefins	C_nH_{2n}				
Ethene	C_2H_4		–50 296		–47 158
Propene	C_3H_6		–48 917		–45 780
Butene	C_4H_8		–48 453		–45 316
Pentene	C_5H_{10}		–48 134		–44 996
Hexene	C_6H_{12}		–47 937		–44 800
Heptene	C_7H_{14}		–47 800		–44 662
Octene	C_8H_{16}		–47 693		–44 556
Nonene	C_9H_{18}		–47 612		–44 475
Decene	$C_{10}H_{20}$		–47 547		–44 410
Alkylbenzenes	$C_{6+n}H_{6+2n}$				
Benzene	C_6H_6	–41 831	–42 266	–40 141	–40 576
Methylbenzene	C_7H_8	–42 437	–42 847	–40 527	–40 937
Ethylbenzene	C_8H_{10}	–42 997	–43 395	–40 924	–41 322
Propylbenzene	C_9H_{12}	–43 416	–43 800	–41 219	–41 603
Butylbenzene	$C_{10}H_{14}$	–43 748	–44 123	–41 453	–41 828
Other fuels					
Gasoline	C_7H_{17}	–48 201	–48 582	–44 506	–44 886
Diesel	$C_{14.4}H_{24.9}$	–45 700	–46 074	–42 934	–43 308
Methanol	CH_3OH	–22 657	–23 840	–19 910	–21 093
Ethanol	C_2H_5OH	–29 676	–30 596	–26 811	–27 731
Nitromethane	CH_3NO_2	–11 618	–12 247	–10 537	–11 165
Phenol	C_6H_5OH	–32 520	–33 176	–31 117	–31 774
Hydrogen	H_2		–60 955		–51 750

TABLE C.4 ***Enthalpy of Formation, Gibbs Function of Formation, and Absolute Entropy of Various Substances at 25°C, 100 kPa Pressure.***

Substance	Formula	M	State	$\bar{h}_f^0$ kJ/kmol	$\bar{g}_f^0$ kJ/kmol	$\bar{s}_f^0$ kJ/kmol K
Water	H_2O	18.015	gas	−241 826	−228 582	188.834
Water	H_2O	18.015	liq	−285 830	−237 141	69.950
Hydrogen peroxide	H_2O_2	34.015	gas	−136 106	−105 445	232.991
Ozone	O_3	47.998	gas	+142 674	+163 184	238.932
Carbon (graphite)	C	12.011	solid	0	0	5.740
Carbon monoxide	CO	28.011	gas	−110 527	−137 163	197.653
Carbon dioxide	CO_2	44.010	gas	−393 522	−394 389	213.795
Methane	CH_4	16.043	gas	−74 873	−50 768	186.251
Acetylene	C_2H_2	26.038	gas	+226 731	+209 200	200.958
Ethene	C_2H_4	28.054	gas	+52 467	+68 421	219.330
Ethane	C_2H_6	30.070	gas	−84 740	−32 885	229.597
Propene	C_3H_6	42.081	gas	+20 430	+62 825	267.066
Propane	C_3H_8	44.094	gas	−103 900	−23 393	269.917
Butane	C_4H_{10}	58.124	gas	−126 200	−15 970	306.647
Pentane	C_5H_{12}	72.151	gas	−146 500	−8 208	348.945
Benzene	C_6H_6	78.114	gas	+82 980	+129 765	269.562
Hexane	C_6H_{14}	86.178	gas	−167 300	+28	387.979
Heptane	C_7H_{16}	100.205	gas	−187 900	+8 227	427.805
n-Octane	C_8H_{18}	114.232	gas	−208 600	+16 660	466.514
n-Octane	C_8H_{18}	114.232	liq	−250 105	+6 741	360.575
Methanol	CH_3OH	32.042	gas	−201 300	−162 551	239.709
Ethanol	C_2H_5OH	46.069	gas	−235 000	−168 319	282.444
Ammonia	NH_3	17.031	gas	−45 720	−16 128	192.572
T-T-Diesel	$C_{14.4}H_{24.9}$	198.06	liq	−174 000	+178 919	525.90
Sulfur	S	32.06	solid	0	0	32.056
Sulfur dioxide	SO_2	64.059	gas	−296 842	−300 125	248.212
Sulfur trioxide	SO_3	80.058	gas	−395 765	−371 016	256.769
Nitrogen oxide	N_2O	44.013	gas	+82 050	+104 179	219.957
Nitromethane	CH_3NO_2	61.04	liq	−113 100	−14 439	171.80

TABLE C.5 ***Logarithms to the Base e of the Equilibrium Constant K***

For the reaction $v_A A + v_b B \rightleftharpoons v_C C + v_D D$, the equilibrium constant K is defined as

$$K = \frac{y_C^{v_C} y_D^{v_D}}{y_A^{v_A} y_B^{v_B}} \left(\frac{P}{P^{\circ}} \right)^{v_C + v_D - v_A - v_B}, P^{\circ} = 0.1\text{MPa}$$

Temp K	$H_2 \rightleftharpoons 2H$	$O_2 \rightleftharpoons 2O$	$N_2 \rightleftharpoons 2N$	$2H_2O \rightleftharpoons 2H2 + O_2$	$2H_2O \rightleftharpoons H_2 + 2OH$	$2CO_2 \rightleftharpoons 2CO + O_2$	$N_2 + O_2 \rightleftharpoons 2NO$	$N_2 + 2O_2 \rightleftharpoons 2NO_2$
298	−164.003	−186.963	−367.528	−184.420	−212.075	−207.529	−69.868	−41.355
500	−92.830	−105.623	−213.405	−105.385	−120.331	−115.234	−40.449	−30.725
1000	−39.810	−45.146	−99.146	−46.321	−51.951	−47.052	−18.709	−23.039
1200	−30.878	−35.003	−80.025	−36.363	−40.467	−35.736	−15.082	−21.752
1400	−24.467	−27.741	−66.345	−29.222	−32.244	−27.679	−12.491	−20.826
1600	−19.638	−22.282	−56.069	−23.849	−26.067	−21.656	−10.547	−20.126
1800	−15.868	−18.028	−48.066	−19.658	−21.258	−16.987	−9.035	−19.577
2000	−12.841	−14.619	−41.655	−16.299	−17.406	−13.266	−7.825	−19.136
2200	−10.356	−11.826	−36.404	−13.546	−14.253	−10.232	−6.836	−18.773
2400	−8.280	−9.495	−32.023	−11.249	−11.625	−7.715	−6.012	−18.470
2600	−6.519	−7.520	−28.313	−9.303	−9.402	−5.594	−5.316	−18.214
2800	−5.005	−5.826	−25.129	−7.633	−7.496	−3.781	−4.720	−17.994
3000	−3.690	−4.356	−22.367	−6.184	−5.845	−2.217	−4.205	−17.805
3200	−2.538	−3.069	−19.947	−4.916	−4.401	−0.853	−3.755	−17.640
3400	−1.519	−1.932	−17.810	−3.795	−3.128	0.346	−3.359	−17.496
3600	−0.611	−0.922	−15.909	−2.799	−1.996	1.408	−3.008	−17.369
3800	0.201	−0.017	−14.205	−1.906	−0.984	2.355	−2.694	−17.257
4000	0.934	0.798	−12.671	−1.101	−0.074	3.204	−2.413	−17.157
4500	2.483	2.520	−9.423	0.602	1.847	4.985	−1.824	−16.953
5000	3.724	3.898	−6.816	1.972	3.383	6.397	−1.358	−16.797
5500	4.739	5.027	−4.672	3.098	4.639	7.542	−0.980	−16.678
6000	5.587	5.969	−2.876	4.040	5.684	8.488	−0.671	−16.588

Source: Consistent with thermodynamic data in *JANAF Thermochemical Tables*, third edition, Thermal Group, Dow Chemical U.S.A., Midland, MI 1985.

TABLE C.6 *One-Dimensional Isentropic Compressible-Flow Functions for an Ideal Gas with Constant Specific Heat and Molecular Weight and k = 1.4*

M	M^*	A/A^*	P/P_o	ρ/ρ_o	T/T_o
0.0	0.00000	∞	1.00000	1.00000	1.00000
0.1	0.10944	5.82183	0.99303	0.99502	0.99800
0.2	0.21822	2.96352	0.97250	0.98028	0.99206
0.3	0.32572	2.03506	0.93947	0.95638	0.98232
0.4	0.43133	1.59014	0.89561	0.92427	0.96899
0.5	0.53452	1.33984	0.84302	0.88517	0.95238
0.6	0.63481	1.18820	0.78400	0.84045	0.93284
0.7	0.73179	1.09437	0.72093	0.79158	0.91075
0.8	0.82514	1.03823	0.65602	0.73999	0.88652
0.9	0.91460	1.00886	0.59126	0.68704	0.86059
1.0	1.0000	1.00000	0.52828	0.63394	0.83333
1.1	1.0812	1.00793	0.46835	0.58170	0.80515
1.2	1.1583	1.03044	0.41238	0.53114	0.77640
1.3	1.2311	1.06630	0.36091	0.48290	0.74738
1.4	1.2999	1.11493	0.31424	0.43742	0.71839
1.5	1.3646	1.17617	0.27240	0.39498	0.68966
1.6	1.4254	1.25023	0.23527	0.35573	0.66138
1.7	1.4825	1.33761	0.20259	0.31969	0.63371
1.8	1.5360	1.43898	0.17404	0.28682	0.60680
1.9	1.5861	1.55526	0.14924	0.25699	0.58072
2.0	1.6330	1.68750	0.12780	0.23005	0.55556
2.1	1.6769	1.83694	0.10935	0.20580	0.53135
2.2	1.7179	2.00497	0.93522E-01	0.18405	0.50813
2.3	1.7563	2.19313	0.79973E-01	0.16458	0.48591
2.4	1.7922	2.40310	0.68399E-01	0.14720	0.46468
2.5	1.8257	2.63672	0.58528E-01	0.13169	0.44444
2.6	1.8571	2.89598	0.50115E-01	0.11787	0.42517
2.7	1.8865	3.18301	0.42950E-01	0.10557	0.40683
2.8	1.9140	3.50012	0.36848E-01	0.94626E-01	0.38941
2.9	1.9398	3.84977	0.31651E-01	0.84889E-01	0.37286
3.0	1.9640	4.23457	0.27224E-01	0.76226E-01	0.35714
3.5	2.0642	6.78962	0.13111E-01	0.45233E-01	0.28986
4.0	2.1381	10.7188	0.65861E-02	0.27662E-01	0.23810
4.5	2.1936	16.5622	0.34553E-02	0.17449E-01	0.19802
5.0	2.2361	25.0000	0.18900E-02	0.11340E-01	0.16667
6.0	2.2953	53.1798	0.63336E-03	0.51936E-02	0.12195
7.0	2.3333	104.143	0.24156E-03	0.26088E-02	0.09259
8.0	2.3591	190.109	0.10243E-03	0.14135E-02	0.07246
9.0	2.3772	327.189	0.47386E-04	0.81504E-03	0.05814
10.0	2.3905	535.938	0.23563E-04	0.49482E-03	0.04762
∞	2.4495	∞	0.0	0.0	0.0

TABLE C.7 ***One-Dimensional Normal Shock Functions for an Ideal Gas with Constant Specific Heat and Molecular Weight and k = 1.4***

M_x	M_y	P_y/P_x	ρ_y/ρ_x	T_y/T_x	P_{0y}/P_{0x}	P_{0y}/P_x
1.00	1.00000	1.0000	1.0000	1.0000	1.00000	1.8929
1.10	0.91177	1.2450	1.1691	1.0649	0.99893	2.1328
1.20	0.84217	1.5133	1.3416	1.1280	0.99280	2.4075
1.30	0.78596	1.8050	1.5157	1.1909	0.97937	2.7136
1.40	0.73971	2.1200	1.6897	1.2547	0.95819	3.0492
1.50	0.70109	2.4583	1.8621	1.3202	0.92979	3.4133
1.60	0.66844	2.8200	2.0317	1.3880	0.89520	3.8050
1.70	0.64054	3.2050	2.1977	1.4583	0.85572	4.2238
1.80	0.61650	3.6133	2.3592	1.5316	0.81268	4.6695
1.90	0.59562	4.0450	2.5157	1.6079	0.76736	5.1418
2.00	0.57735	4.5000	2.6667	1.6875	0.72087	5.6404
2.10	0.56128	4.9783	2.8119	1.7705	0.67420	6.1654
2.20	0.54706	5.4800	2.9512	1.8569	0.62814	6.7165
2.30	0.53441	6.0050	3.0845	1.9468	0.58329	7.2937
2.40	0.52312	6.5533	3.2119	2.0403	0.54014	7.8969
2.50	0.51299	7.1250	3.3333	2.1375	0.49901	8.5261
2.60	0.50387	7.7200	3.4490	2.2383	0.46012	9.1813
2.70	0.49563	8.3383	3.5590	2.3429	0.42359	9.8624
2.80	0.48817	8.9800	3.6636	2.4512	0.38946	10.569
2.90	0.48138	9.6450	3.7629	2.5632	0.35773	11.302
3.00	0.47519	10.333	3.8571	2.6790	0.32834	12.061
4.00	0.43496	18.500	4.5714	4.0469	0.13876	21.068
5.00	0.41523	29.000	5.0000	5.8000	0.06172	32.653
10.00	0.38758	116.50	5.7143	20.387	0.00304	129.22
∞	0.37796	∞	6.0000	∞	0.0	∞

SECTION D

MISCELLANEOUS TABLES AND FIGURES

THE PSYCHROMETRIC CHART

Properties of moist-air are determined from three independent properties (two for the thermodynamic state and one relating to the mixture composition). For a fixed total pressure (here a pressure of 100 kPa) the states can be mapped out as a function of temperature and humidity ratio (w) as shown in figure D.1. With the mass of water vapor in the moist air as, m_v, then the humidity ratio and the relative humidity, Φ , are

$$w = \frac{m_v}{m_{dryair}} = \frac{R_{air} P_v}{R_v P_{air}} \cong 0.622 \frac{P_v}{P_{air}} \quad ; \quad \Phi = \frac{P_v}{P_g}$$

The saturated water vapor pressure, P_g, is from the steam tables at the given temperature. A value for the enthalpy per kg dry air is shown in the chart with an offset with the dry air enthalpy at -20C so it becomes

$$\tilde{h} \equiv h_{air} - h_{air\ at -20C} + w h_v$$

using the value of h_v from the steam tables. Any term with liquid water uses h_f from the saturated liquid water table at the temperature of the liquid.

EQUATIONS OF STATE

Some of the most used pressure explicit equations of state can be shown in a form with two parameters. This form is known as a cubic equation of state and contains as a special case the ideal gas law.

$$P = \frac{RT}{v - b} - \frac{a}{v^2 + cbv + db^2}$$

where (a,b) are parameters and (c,d) defines the model as shown in the following table with the acentric factor (ω) and

$$b = b_o R T_c / P_c \qquad \text{and} \qquad a = a_o R^2 T_c^2 / P_c$$

TABLE D.1 Equations Of State

Model	c	d	b_o	a_o
Ideal Gas	0	0	0	0
van der Waal	0	0	1/8	27/64
Redlich-Kwong	1	0	0.08664	$0.42748\ T_r^{-1/2}$
Soave	1	0	0.08664	$0.42748[1+f(1-T_r^{1/2})]^2$
Peng-Robinson	2	-1	0.0778	$0.45724[1+f(1-T_r^{1/2})]^2$

$f = 0.48 + 1.574\ \omega - 0.176\ \omega^2$ for Soave

$f = 0.37464 + 1.54226\omega - 0.26992\ \omega^2$ for Peng-Robinson

More accurate equations of state are developed with parameters obtained from extensive curvefits to real substance behavior. One example is the Benedict-Webb-Rubin equation of state (BWR) containing eight empirical constants with pressure given as:

$$P = \frac{RT}{v} + \frac{RTB_o - A_o - C_o/T^2}{v^2} + \frac{RTb - a}{v^3} + \frac{a\alpha}{v^6} + \frac{c}{v^3 T^2}(1 + \frac{\gamma}{v^2})e^{-\gamma/v^2}$$

The constants for several substances are listed in Table D.2, with the units used in the form shown in the heading of the table. An extended form of this equation is the Lee-Kesler equation of state with twelve constants used for a simple fluid shown in table D.4 which is used with the corresponding states principle (i.e. same reduced properties). For further accuracy and for extension to mixtures a three parameter model gives any property as the simple fluid value, $Z^{(o)}$, plus a correction term, $Z^{(1)}$, weighted with the accentric factor, ω, as

$$Z = Z^{(o)} + \omega\, Z^{(1)}$$

$$\frac{h^* - h}{RT_c} = \left(\frac{h^* - h}{RT_c}\right)^{(0)} + \omega\left(\frac{h^* - h}{RT_c}\right)^{(1)} ; \frac{s_P^* - s_P}{R} = \left(\frac{s_P^* - s_P}{R}\right)^{(0)} + \omega\left(\frac{s_P^* - s_P}{R}\right)^{(1)}$$

$$\ln\frac{f}{P} = \ln\left(\frac{f}{P}\right)^{(0)} + \omega \ln\left(\frac{f}{P}\right)^{(1)}$$

When used for a mixture Lee-Kesler's model finds the mixture critical point properties as:

$$\bar{v}_{c,mix} = \sum_j \sum_k y_j y_k \bar{v}_{c,jk} \quad ; \quad \bar{v}_{c,jk} = \frac{1}{8}(\bar{v}_{c,j}^{1/3} + \bar{v}_{c,k}^{1/3})^3$$

$$(\bar{v}_c T_c)_{mix} = \sum_j \sum_k y_j y_k \bar{v}_{c,jk} T_{c,jk} \quad ; \quad T_{c,jk} = \sqrt{T_{c,j} T_{c,k}}$$

Once evaluated, the pseudocritical pressure is found from:

$$\omega_{mix} = \sum y_j \omega_j \quad ; \quad Z_{c,mix} = 0.2905 - 0.085\,\omega_{mix}$$

$$P_{c,mix} = Z_{c,mix} \bar{R} T_{c,mix} / \bar{v}_{c,mix}$$

Given the pseudocritical properties the mixture properties can be found from table D.4 and D.5 as for a pure substance.

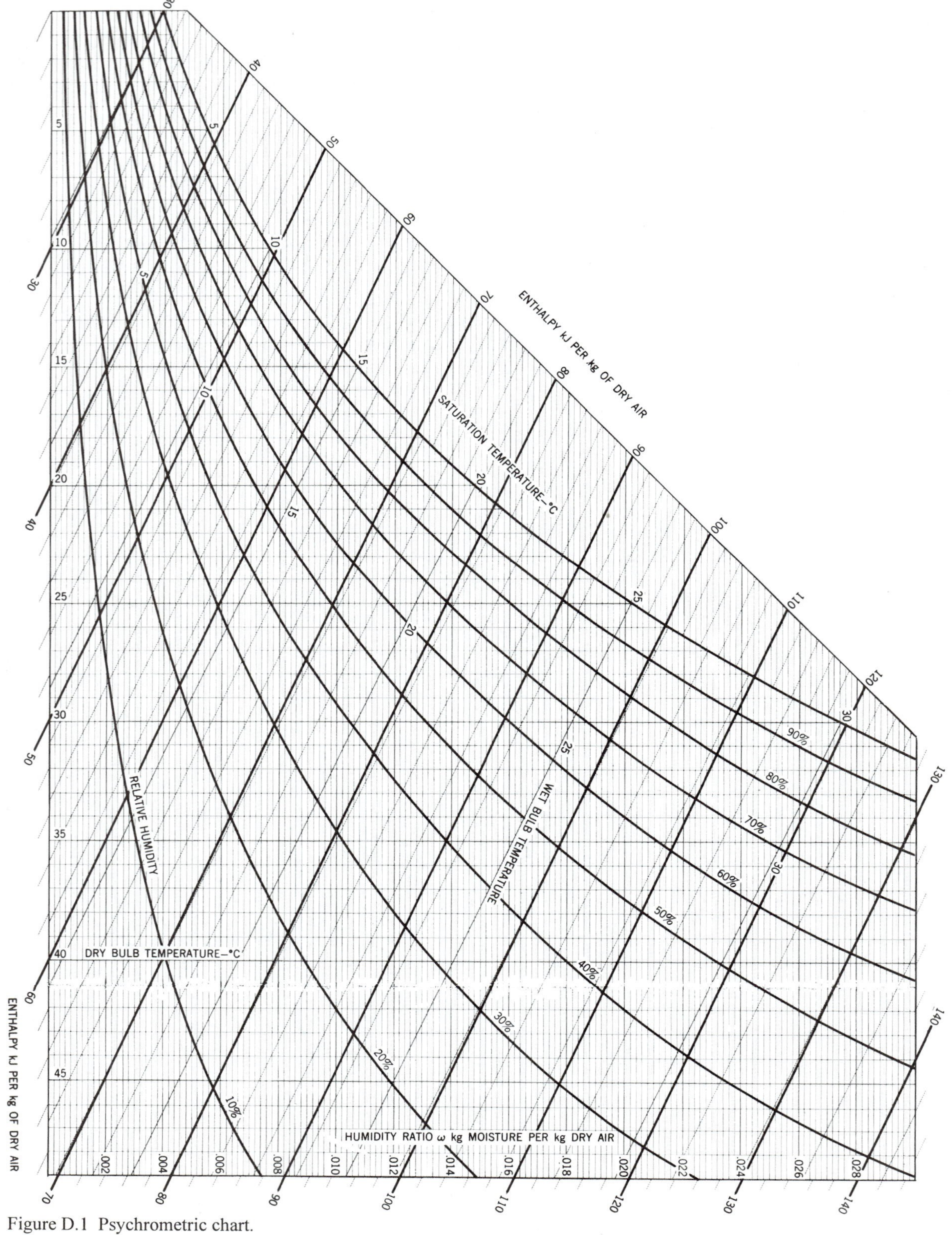

Figure D.1 Psychrometric chart.

TABLE D.2 Empirical Constants for Benedict-Webb-Rubin Equation
Units: Atmospheres, Liters, Moles, K, Gas constant R = 0.08206.

Gas	Formula	A_0	B_0	$C_0 \times 10^{-6}$	a
Methane	CH_4	1.85500	0.042600	0.022570	0.49400
Ethylene	C_2H_4	3.33958	0.0556833	0.131140	0.25900
Ethane	C_2H_6	4.15556	0.0627724	0.179592	0.34516
Propylene	C_3H_6	6.11220	0.0850647	0.439182	0.774056
Propane	C_3H_8	6.872	0.097313	0.508256	0.94770
n-Butane	C_4H_{10}	10.0847	0.124361	0.992830	1.88231
n-Pentane	C_5H_{12}	12.1794	0.156751	2.12121	4.07480
n-Hexane	C_6H_{14}	14.4373	0.177813	3.31935	7.11671
n-Heptane	C_7H_{16}	17.5206	0.199005	4.74574	10.36475
Nitrogen	N_2	1.19250	0.04580	0.0058891	0.01490
Oxugen	O_2	1.49880	0.046524	0.0038617	-0.040507
Ammonia	NH_4	3.78928	0.0516461	0.178567	0.10354
Carbon dioxide	CO_2	2.67340	0.045628	0.11333	0.051689

TABLE D.2 (Continued)

Gas	Formula	b	$c \times 10^{-6}$	$\alpha \times 10^3$	$\gamma \times 10^2$
Methane	CH_4	0.00338004	0.002545	0.124359	0.600
Ethylene	C_2H_4	0.008600	0.021120	0.17800	0.923
Ethane	C_2H_6	0.011122	0.032767	0.243389	1.180
Propylene	C_3H_6	0.0187059	0.102611	0.455696	1.829
Propane	C_3H_8	0.022500	0.12900	0.607175	2.200
n-Butane	C_4H_{10}	0.0399983	0.316400	1.10132	3.400
n-Pentane	C_5H_{12}	0.066812	0.82417	1.81000	4.750
n-Hexane	C_6H_{14}	0.109131	1.51276	2.81086	6.66849
n-Heptane	C_7H_{16}	0.151954	2.47000	4.35611	9.000
Nitrogen	N_2	0.00198154	0.000548064	0.291545	0.750
Oxugen	O_2	-0.000027963	-0.00020376	0.008641	0.359
Ammonia	NH_4	0.000719561	0.000157536	0.00465189	1.980
Carbon dioxide	CO_2	0.0030819	0.0070672	0.11271	0.494

TABLE D.3 ***The Reduced Second and Third Virial Coefficients and Force Constants for the Lennard–Jones (6-12) Potential***

TABLE D.3.1
The Reduced Second and Third Coefficients and their Derivatives

T^*	B^*	$B_1^* = T^* \frac{dB^*}{dT^*}$	C^*	$C_1^* = T^* \frac{dC^*}{dT^*}$
0.3	−27.88061	76.60701		
0.4	−13.79885	30.26698		
0.5	−8.72022	16.92367		
0.6	−6.19798	11.24883		
0.7	−4.71004	8.25711	−3.44223	29.02471
0.8	−3.73423	6.45414	−0.87753	11.80911
0.9	−3.04712	5.26492	0.06579	5.05023
1.0	−2.53809	4.42826	0.42600	2.12100
1.1	−2.14638	3.81063	0.55670	0.76761
1.2	−1.83595	3.33749	0.59235	0.12051
1.3	−1.58411	2.96421	0.58821	−0.18965
1.4	−1.37585	2.66262	0.56823	−0.33189
1.5	−1.20089	2.41414	0.54307	−0.38813
1.6	−1.05191	2.20602	0.51748	−0.39994
1.7	−0.92362	2.02926	0.49348	−0.38906
1.8	−0.81203	1.87733	0.47183	−0.36719
1.9	−0.71415	1.74537	0.45267	−0.34065
2.0	−0.62763	1.62972	0.43590	−0.31290
2.2	−0.48171	1.43663	0.40861	−0.26013
2.4	−0.36358	1.28190	0.38797	−0.21492
2.6	−0.26613	1.15517	0.37228	−0.17792
2.8	−0.18451	1.04948	0.36022	−0.14821
3.0	−0.11523	0.96000	0.35084	−0.12454
3.2	−0.05579	0.88328	0.34342	−0.10574
3.4	−0.00428	0.81676	0.33748	−0.09081
3.6	0.04072	0.75854	0.33264	−0.07895
3.8	0.08033	0.70716	0.32863	−0.06955
4.0	0.11542	0.66148	0.32526	−0.06209
4.2	0.14668	0.62060	0.32238	−0.05619
4.4	0.17469	0.58381	0.31988	−0.05154
4.6	0.19990	0.55051	0.31767	−0.04789
4.8	0.22268	0.52024	0.31569	−0.04506
5.0	0.24334	0.49260	0.31390	−0.04288
6.0	0.32290	0.38397	0.30661	−0.03831
7.0	0.37609	0.30826	0.30069	−0.03899
8.0	0.41343	0.25248	0.29533	−0.04152
9.0	0.44060	0.20970	0.29027	−0.04456
10.0	0.46088	0.17587	0.28541	−0.04758
20.0	0.52538	0.02866	0.24609	−0.06402
30.0	0.52693	-0.01749	0.21930	−0.06728

TABLE D.3.2
Force Constants from Experimental Virial Coefficient Data

Substance	ε/k, K	b_0, m^3/kmol
Ne	35.8	0.0262
Ar	119.0	0.0502
Kr	173.0	0.0583
Xe	225.3	0.0854
N_2	95.05	0.0635
O_2	117.5	0.0578
CO	100.2	0.0675
NO	131.0	0.0402
CO_2	186.0	0.118
N_2O	193.0	0.118
CH_4	148.1	0.0698
CF_4	152.0	0.131

TABLE D.3.3
Formula for the derivatives.

$$B^* = B / \mathrm{b_o} \qquad C^* = C / \mathrm{b_o^2}$$

$$T^* = T / (\varepsilon / k)$$

$$Z = P\bar{v} / \bar{R}T$$

$$= 1 + \frac{B(T)}{\bar{v}} + \frac{C(T)}{\bar{v}^2} + \cdots$$

$$\left(\frac{\partial P}{\partial T}\right)_v = \frac{Z\bar{R}}{\bar{v}} + \frac{\bar{R}T}{\bar{v}}\left(\frac{\partial Z}{\partial T}\right)_v$$

$$T\left(\frac{\partial Z}{\partial T}\right)_v = \frac{T}{\bar{v}}\frac{dB}{dT} + \frac{T}{\bar{v}^2}\frac{dC}{dT} + \cdots$$

$$T\frac{dB}{dT} = \mathrm{b_o} T^* \frac{dB^*}{dT^*} = \mathrm{b_o} B_1^*$$

$$T\frac{dC}{dT} = \mathrm{b_o^2} T^* \frac{dC^*}{dT^*} = \mathrm{b_o^2} C_1^*$$

TABLE D.4 ***Generalized Compressibility Factor Tables.***
TABLE D.4.1 ***Generalized Saturation Pressure, Compressibility Factor, Enthalpy Departure, Entropy Departure, Fugacity Coefficient Table of a Simple Fluid***

T_r	$\ln(P_r)$	Z_f	Z_g	$\left(\frac{h^*-h}{RT_c}\right)_f$	$\left(\frac{h^*-h}{RT_c}\right)_g$	$\left(\frac{s_p^*-s_p}{R}\right)_f$	$\left(\frac{s_p^*-s_p}{R}\right)_g$	$\ln\left(\frac{f}{P}\right)$
0.30	−13.14053	0.00000	0.99998	6.04616	0.00002	20.09953	0.00005	−0.00002
0.32	−11.89025	0.00000	0.99993	5.99061	0.00007	18.06562	0.00014	−0.00007
0.34	−10.79655	0.00001	0.99983	5.93515	0.00018	16.51869	0.00035	−0.00017
0.36	−9.83281	0.00002	0.99963	5.87895	0.00040	16.01837	0.00076	−0.00037
0.38	−8.97801	0.00003	0.99927	5.82177	0.00085	15.31191	0.00150	−0.00073
0.40	−8.21540	0.00006	0.99865	5.76367	0.00163	14.41129	0.00272	−0.00134
0.42	−7.53140	0.00012	0.99770	5.70481	0.00291	13.58378	0.00463	−0.00230
0.44	−6.91492	0.00022	0.99628	5.64539	0.00489	12.82092	0.00741	−0.00371
0.46	−6.35683	0.00038	0.99430	5.58558	0.00781	12.13784	0.01129	−0.00568
0.48	−5.84950	0.00061	0.99165	5.52553	0.01191	11.50344	0.01648	−0.00832
0.50	−5.38653	0.00095	0.98820	5.46534	0.01745	10.91801	0.02317	−0.01173
0.52	−4.96253	0.00141	0.98389	5.40509	0.02472	10.37636	0.03155	−0.01600
0.54	−4.57289	0.00204	0.97862	5.34481	0.03399	9.87361	0.04176	−0.02118
0.56	−4.21367	0.00286	0.97233	5.28447	0.04552	9.40554	0.05395	−0.02734
0.58	−3.88146	0.00390	0.96498	5.22403	0.05958	8.96810	0.06823	−0.03449
0.60	−3.57331	0.00521	0.95652	5.16343	0.07641	8.55883	0.08470	−0.04265
0.62	−3.28666	0.00682	0.94695	5.10255	0.09626	8.17415	0.10344	−0.05182
0.64	−3.01926	0.00877	0.93623	5.04126	0.11938	7.81156	0.12455	−0.06198
0.66	−2.76913	0.01110	0.92436	4.97940	0.14601	7.46873	0.14811	−0.07312
0.68	−2.53452	0.01384	0.91133	4.91678	0.17640	7.14358	0.17423	−0.08519
0.70	−2.31388	0.01705	0.89711	4.85318	0.21084	6.83415	0.20302	−0.09817
0.72	−2.10584	0.02077	0.88170	4.78832	0.24961	6.53867	0.23465	−0.11203
0.74	−1.90915	0.02504	0.86506	4.72190	0.29309	6.25549	0.26934	−0.12674
0.76	−1.72272	0.02994	0.84712	4.65355	0.34170	5.98304	0.30734	−0.14227
0.78	−1.54554	0.03552	0.82782	4.58283	0.39595	5.71981	0.34902	−0.15861
0.80	−1.37672	0.04186	0.80704	4.50922	0.45650	5.46432	0.39486	−0.17576
0.82	−1.21545	0.04906	0.78463	4.43203	0.52418	5.21508	0.44552	−0.19373
0.84	−1.06097	0.05725	0.76036	4.35043	0.60010	4.97049	0.50187	−0.21254
0.86	−0.91263	0.06658	0.73394	4.26329	0.68574	4.72880	0.56514	−0.23224
0.88	−0.76980	0.07727	0.70491	4.16906	0.78319	4.48793	0.63709	−0.25290
0.90	−0.63192	0.08961	0.67262	4.06548	0.89550	4.24517	0.72040	−0.27461
0.92	−0.49847	0.10407	0.63605	3.94904	1.02744	3.99668	0.81928	−0.29750
0.94	−0.36898	0.12143	0.59347	3.81358	1.18719	3.73613	0.94120	−0.32176
0.96	−0.24301	0.14328	0.54146	3.64643	1.39116	3.45088	1.10146	−0.34766
0.98	−0.12014	0.17412	0.47112	3.41136	1.68360	3.10521	1.34235	−0.37560
1.00	0.00000	0.29010	0.29010	2.58438	2.58438	2.17799	2.17799	−0.40639

TABLE D.4.2 *Compressibility Factor of a Simple Fluid, Z*

$T_r \backslash P_r$	0.10	0.20	0.40	0.60	0.80	1.00	1.20	1.40	1.70	2.00	2.50	3.00	5.00	7.00	10.00
0.30	0.0290	0.0579	0.1158	0.1737	0.2315	0.2892	0.3470	0.4047	0.4911	0.5775	0.7213	0.8648	1.4366	2.0048	2.8507
0.40	0.0239	0.0477	0.0953	0.1429	0.1904	0.2379	0.2853	0.3327	0.4036	0.4744	0.5921	0.7095	1.1758	1.6373	2.3211
0.50	0.0207	0.0413	0.0825	0.1236	0.1647	0.2056	0.2465	0.2873	0.3483	0.4092	0.5103	0.6110	1.0094	1.4017	1.9801
0.60	0.0186	0.0371	0.0741	0.1109	0.1476	0.1842	0.2207	0.2571	0.3115	0.3657	0.4554	0.5446	0.8959	1.2398	1.7440
0.70	0.0172	0.0344	0.0687	0.1027	0.1366	0.1703	0.2038	0.2372	0.2869	0.3364	0.4181	0.4991	0.8161	1.1241	1.5729
0.75	0.9165	0.0336	0.0670	0.1001	0.1330	0.1656	0.1981	0.2303	0.2784	0.3260	0.4046	0.4823	0.7854	1.0787	1.5047
0.80	0.9319	0.8539	0.0661	0.0985	0.1307	0.1626	0.1942	0.2255	0.2721	0.3182	0.3942	0.4690	0.7598	1.0400	1.4456
0.85	0.9436	0.8810	0.0661	0.0983	0.1301	0.1614	0.1924	0.2230	0.2684	0.3132	0.3868	0.4591	0.7388	1.0071	1.3943
0.90	0.9528	0.9015	0.7800	0.1006	0.1321	0.1630	0.1935	0.2235	0.2678	0.3114	0.3828	0.4527	0.7220	0.9793	1.3496
0.95	0.9600	0.9174	0.8206	0.6967	0.1410	0.1705	0.1998	0.2288	0.2717	0.3138	0.3827	0.4501	0.7092	0.9561	1.3108
1.00	0.9659	0.9300	0.8509	0.7574	0.6353	0.2901	0.2237	0.2459	0.2839	0.3229	0.3880	0.4522	0.7004	0.9372	1.2772
1.05	0.9707	0.9401	0.8743	0.8002	0.7130	0.6026	0.4437	0.3246	0.3182	0.3452	0.4014	0.4604	0.6956	0.9222	1.2481
1.10	0.9747	0.9485	0.8930	0.8323	0.7649	0.6880	0.5984	0.5003	0.4086	0.3953	0.4277	0.4770	0.6950	0.9110	1.2232
1.15	0.9780	0.9554	0.9081	0.8576	0.8032	0.7443	0.6803	0.6129	0.5227	0.4760	0.4718	0.5042	0.6987	0.9033	1.2021
1.20	0.9808	0.9611	0.9205	0.8779	0.8330	0.7858	0.7363	0.6856	0.6135	0.5605	0.5295	0.5425	0.7069	0.8990	1.1844
1.30	0.9852	0.9702	0.9396	0.9083	0.8764	0.8438	0.8111	0.7784	0.7316	0.6908	0.6467	0.6344	0.7358	0.8998	1.1580
1.40	0.9884	0.9768	0.9534	0.9298	0.9062	0.8827	0.8595	0.8367	0.8043	0.7753	0.7387	0.7202	0.7761	0.9112	1.1419
1.50	0.9909	0.9818	0.9636	0.9456	0.9278	0.9103	0.8933	0.8768	0.8536	0.8328	0.8052	0.7887	0.8200	0.9297	1.1339
1.60	0.9928	0.9856	0.9714	0.9575	0.9439	0.9308	0.9180	0.9059	0.8889	0.8738	0.8537	0.8410	0.8617	0.9518	1.1320
1.80	0.9955	0.9910	0.9823	0.9739	0.9659	0.9583	0.9511	0.9444	0.9353	0.9275	0.9176	0.9118	0.9297	0.9961	1.1391
2.00	0.9972	0.9944	0.9892	0.9842	0.9796	0.9754	0.9715	0.9680	0.9635	0.9599	0.9561	0.9550	0.9772	1.0328	1.1516
2.50	0.9994	0.9989	0.9981	0.9975	0.9971	0.9969	0.9970	0.9973	0.9982	0.9996	1.0031	1.0080	1.0395	1.0866	1.1763
3.00	1.0004	1.0008	1.0018	1.0030	1.0043	1.0057	1.0074	1.0091	1.0121	1.0153	1.0215	1.0284	1.0635	1.1075	1.1848
3.50	1.0008	1.0017	1.0035	1.0055	1.0075	1.0097	1.0120	1.0143	1.0181	1.0221	1.0292	1.0368	1.0723	1.1138	1.1834
4.00	1.0010	1.0021	1.0043	1.0066	1.0090	1.0115	1.0140	1.0166	1.0207	1.0249	1.0323	1.0401	1.0747	1.1136	1.1773
5.00	1.0012	1.0024	1.0048	1.0073	1.0098	1.0124	1.0150	1.0176	1.0217	1.0259	1.0331	1.0405	1.0722	1.1064	1.1611

TABLE D.4.3 *Enthalpy Departure of a Simple Fluid, $(h^* - h)/RT_c$*

$T_r \backslash P_r$	0.10	0.20	0.40	0.60	0.80	1.00	1.20	1.40	1.70	2.00	2.50	3.00	5.00	7.00	10.00
0.30	6.040	6.034	6.022	6.011	5.999	5.987	5.975	5.963	5.945	5.927	5.898	5.868	5.748	5.628	5.446
0.40	5.757	5.751	5.738	5.726	5.713	5.700	5.687	5.675	5.655	5.636	5.604	5.572	5.442	5.311	5.113
0.50	5.459	5.453	5.440	5.427	5.414	5.401	5.388	5.375	5.355	5.336	5.303	5.270	5.135	4.999	4.791
0.60	5.159	5.153	5.141	5.129	5.116	5.104	5.091	5.079	5.060	5.041	5.008	4.976	4.842	4.704	4.492
0.70	4.853	4.848	4.839	4.828	4.818	4.808	4.797	4.786	4.769	4.752	4.723	4.693	4.566	4.432	4.221
0.75	0.183	4.687	4.679	4.672	4.664	4.655	4.646	4.637	4.622	4.607	4.581	4.554	4.434	4.303	4.095
0.80	0.160	0.345	4.507	4.504	4.499	4.494	4.488	4.481	4.470	4.459	4.437	4.413	4.303	4.178	3.974
0.85	0.141	0.300	4.308	4.313	4.316	4.316	4.316	4.314	4.309	4.302	4.287	4.269	4.173	4.056	3.857
0.90	0.126	0.264	0.596	4.074	4.094	4.108	4.118	4.125	4.130	4.132	4.129	4.119	4.043	3.935	3.744
0.95	0.113	0.235	0.516	0.885	3.763	3.825	3.865	3.893	3.922	3.939	3.955	3.958	3.910	3.815	3.634
1.00	0.103	0.212	0.455	0.750	1.151	2.584	3.441	3.560	3.653	3.706	3.757	3.782	3.774	3.695	3.526
1.05	0.094	0.192	0.407	0.654	0.955	1.359	2.034	2.831	3.243	3.398	3.521	3.583	3.632	3.575	3.420
1.10	0.086	0.175	0.367	0.581	0.827	1.120	1.487	1.955	2.609	2.965	3.231	3.353	3.484	3.453	3.315
1.15	0.079	0.160	0.334	0.523	0.732	0.968	1.239	1.550	2.059	2.479	2.888	3.091	3.329	3.329	3.211
1.20	0.073	0.148	0.305	0.474	0.657	0.857	1.076	1.315	1.704	2.079	2.537	2.807	3.166	3.202	3.107
1.30	0.063	0.127	0.259	0.399	0.545	0.698	0.860	1.029	1.293	1.560	1.964	2.274	2.825	2.942	2.899
1.40	0.055	0.110	0.224	0.341	0.463	0.588	0.716	0.848	1.050	1.253	1.576	1.857	2.486	2.679	2.692
1.50	0.048	0.097	0.196	0.297	0.400	0.505	0.611	0.719	0.883	1.046	1.309	1.549	2.175	2.421	2.486
1.60	0.043	0.086	0.173	0.261	0.350	0.440	0.531	0.622	0.759	0.894	1.114	1.318	1.904	2.177	2.285
1.80	0.034	0.068	0.137	0.206	0.275	0.344	0.413	0.481	0.583	0.683	0.844	0.996	1.476	1.751	1.908
2.00	0.028	0.056	0.111	0.167	0.222	0.276	0.330	0.384	0.463	0.541	0.665	0.782	1.167	1.411	1.577
2.50	0.018	0.035	0.070	0.104	0.137	0.170	0.203	0.234	0.281	0.326	0.398	0.465	0.687	0.838	0.954
3.00	0.011	0.023	0.045	0.067	0.088	0.109	0.129	0.149	0.177	0.205	0.248	0.288	0.415	0.495	0.545
3.50	0.007	0.015	0.029	0.043	0.056	0.069	0.081	0.093	0.111	0.127	0.152	0.174	0.239	0.270	0.264
4.00	0.005	0.009	0.017	0.026	0.033	0.041	0.048	0.054	0.064	0.072	0.085	0.095	0.116	0.110	0.061
5.00	0.001	0.001	0.002	0.003	0.004	0.004	0.004	0.004	0.003	0.001	−0.003	−0.009	−0.045	−0.101	−0.213

TABLE D.4.4 *Entropy Departure of a Simple Fluid*, $(s_P^* - s_P)/R$

$T_r \backslash P_r$	0.10	0.20	0.40	0.60	0.80	1.00	1.20	1.40	1.70	2.00	2.50	3.00	5.00	7.00	10.00
0.30	9.319	8.635	7.961	7.574	7.304	7.099	6.935	6.799	6.633	6.497	6.319	6.182	5.847	5.683	5.578
0.40	8.506	7.821	7.144	6.755	6.483	6.275	6.109	5.970	5.799	5.660	5.475	5.330	4.967	4.772	4.619
0.50	7.842	7.156	6.479	6.089	5.816	5.608	5.441	5.302	5.130	4.989	4.802	4.656	4.282	4.074	3.899
0.60	7.294	6.610	5.933	5.544	5.273	5.066	4.900	4.762	4.591	4.451	4.266	4.120	3.747	3.537	3.353
0.70	6.823	6.140	5.467	5.082	4.814	4.610	4.446	4.310	4.143	4.007	3.826	3.684	3.322	3.117	2.935
0.75	0.164	5.917	5.248	4.866	4.600	4.399	4.238	4.104	3.940	3.807	3.630	3.491	3.138	2.939	2.761
0.80	0.134	0.294	5.026	4.649	4.388	4.191	4.034	3.904	3.744	3.615	3.444	3.310	2.970	2.777	2.605
0.85	0.111	0.239	4.785	4.418	4.166	3.976	3.825	3.701	3.548	3.425	3.262	3.135	2.812	2.629	2.463
0.90	0.094	0.199	0.463	4.145	3.912	3.738	3.599	3.484	3.344	3.231	3.081	2.963	2.663	2.491	2.334
0.95	0.080	0.168	0.377	0.671	3.556	3.433	3.326	3.235	3.119	3.023	2.893	2.790	2.520	2.361	2.215
1.00	0.069	0.144	0.315	0.532	0.847	2.178	2.893	2.893	2.843	2.784	2.690	2.609	2.380	2.239	2.105
1.05	0.060	0.124	0.267	0.439	0.656	0.965	1.523	2.185	2.444	2.483	2.461	2.415	2.242	2.121	2.001
1.10	0.053	0.108	0.230	0.371	0.537	0.742	1.012	1.368	1.855	2.081	2.191	2.202	2.104	2.007	1.903
1.15	0.047	0.096	0.201	0.319	0.452	0.607	0.790	1.007	1.366	1.649	1.885	1.968	1.966	1.897	1.810
1.20	0.042	0.085	0.177	0.277	0.389	0.512	0.651	0.807	1.063	1.308	1.587	1.727	1.827	1.789	1.722
1.30	0.033	0.068	0.140	0.217	0.298	0.385	0.478	0.576	0.732	0.891	1.127	1.299	1.554	1.581	1.556
1.40	0.027	0.056	0.114	0.174	0.237	0.303	0.372	0.442	0.552	0.663	0.839	0.990	1.303	1.386	1.402
1.50	0.023	0.046	0.094	0.143	0.194	0.246	0.299	0.353	0.436	0.520	0.654	0.777	1.088	1.208	1.260
1.60	0.019	0.039	0.079	0.120	0.162	0.204	0.247	0.290	0.356	0.421	0.528	0.628	0.913	1.050	1.130
1.80	0.014	0.029	0.058	0.088	0.117	0.147	0.177	0.207	0.252	0.296	0.369	0.438	0.661	0.799	0.908
2.00	0.011	0.022	0.044	0.067	0.089	0.111	0.134	0.156	0.189	0.221	0.274	0.325	0.497	0.620	0.733
2.50	0.006	0.013	0.026	0.038	0.051	0.064	0.076	0.088	0.106	0.124	0.153	0.181	0.281	0.361	0.453
3.00	0.004	0.008	0.017	0.025	0.033	0.041	0.049	0.057	0.068	0.080	0.098	0.116	0.181	0.236	0.303
3.50	0.003	0.006	0.012	0.017	0.023	0.029	0.034	0.040	0.048	0.056	0.068	0.081	0.126	0.166	0.216
4.00	0.002	0.004	0.009	0.013	0.017	0.021	0.025	0.029	0.035	0.041	0.050	0.059	0.093	0.123	0.162
5.00	0.001	0.003	0.005	0.008	0.010	0.013	0.015	0.018	0.021	0.025	0.031	0.036	0.057	0.075	0.100

TABLE D.4.5 *Fugacity Coefficent of a Simple Fluid, ln(f/P)*

$T_r \backslash P_r$	0.10	0.20	0.40	0.60	0.80	1.00	1.20	1.40	1.70	2.00	2.50	3.00	5.00	7.00	10.00
0.30	−10.815	−11.479	−12.114	−12.462	−12.692	−12.857	−12.981	−13.078	−13.185	−13.261	−13.340	−13.378	−13.313	−13.076	−12.576
0.40	−5.887	−6.556	−7.202	−7.559	−7.800	−7.975	−8.110	−8.216	−8.339	−8.431	−8.535	−8.599	−8.638	−8.506	−8.164
0.50	−3.077	−3.749	−4.401	−4.766	−5.012	−5.194	−5.335	−5.448	−5.581	−5.682	−5.803	−5.883	−5.989	−5.923	−5.682
0.60	−1.304	−1.979	−2.635	−3.003	−3.254	−3.440	−3.586	−3.703	−3.842	−3.950	−4.082	−4.173	−4.323	−4.304	−4.133
0.70	−0.110	−0.786	−1.445	−1.816	−2.069	−2.258	−2.407	−2.527	−2.670	−2.782	−2.922	−3.021	−3.202	−3.215	−3.095
0.75	−0.080	−0.332	−0.991	−1.363	−1.618	−1.808	−1.957	−2.078	−2.223	−2.336	−2.478	−2.580	−2.773	−2.799	−2.699
0.80	−0.066	0.137	−0.608	−0.981	−1.236	−1.426	−1.576	−1.698	−1.844	−1.959	−2.103	−2.206	−2.409	−2.445	−2.362
0.85	−0.055	−0.113	−0.284	−0.656	−0.911	−1.102	−1.252	−1.374	−1.521	−1.636	−1.782	−1.887	−2.097	−2.142	−2.074
0.90	−0.046	−0.095	−0.198	−0.382	−0.636	−0.826	−0.976	−1.098	−1.245	−1.360	−1.506	−1.613	−1.829	−1.881	−1.826
0.95	−0.039	−0.080	−0.166	−0.261	−0.405	−0.594	−0.742	−0.864	−1.009	−1.124	−1.270	−1.376	−1.596	−1.655	−1.610
1.00	−0.034	−0.068	−0.140	−0.218	−0.303	−0.406	−0.548	−0.667	−0.809	−0.923	−1.067	−1.173	−1.394	−1.457	−1.422
1.05	−0.029	−0.059	−0.120	−0.185	−0.254	−0.329	−0.414	−0.511	−0.644	−0.753	−0.893	−0.997	−1.217	−1.284	−1.256
1.10	−0.025	−0.051	−0.103	−0.158	−0.215	−0.276	−0.340	−0.410	−0.517	−0.615	−0.747	−0.847	−1.063	−1.131	−1.110
1.15	−0.022	−0.044	−0.089	−0.136	−0.184	−0.235	−0.287	−0.341	−0.425	−0.507	−0.626	−0.719	−0.929	−0.997	−0.981
1.20	−0.019	−0.038	−0.078	−0.118	−0.159	−0.202	−0.245	−0.289	−0.357	−0 .425	−0.527	−0.612	−0.811	−0.879	−0.867
1.30	−0.015	−0.030	−0.060	−0.090	−0.121	−0.152	−0.183	−0.215	−0.262	−0.309	−0.384	−0.450	−0.619	−0.682	−0.674
1.40	−0.012	−0.023	−0.046	−0.070	−0.093	−0.117	−0.140	−0.163	−0.198	−0.232	−0.287	−0.336	−0.472	−0.527	−0.521
1.50	−0.009	−0.018	−0.036	−0.055	−0.073	−0.091	−0.109	−0.126	−0.152	−0.178	−0.218	−0.255	−0.361	−0.406	−0.397
1.60	−0.007	−0.014	−0.029	−0.043	−0.057	−0.071	−0.085	−0.098	−0.118	−0.137	−0.168	−0.196	−0.277	−0.310	−0.298
1.80	−0.005	−0.009	−0.018	−0.027	−0.035	−0.044	−0.052	−0.060	−0.072	−0.083	−0.100	−0.116	−0.159	−0.174	−0.152
2.00	−0.003	−0.006	−0.011	−0.016	−0.022	−0.027	−0.032	−0.036	−0.043	−0.049	−0.058	−0.067	−0.086	−0.086	−0.055
2.50	−0.001	−0.001	−0.002	−0.003	−0.004	−0.005	−0.005	−0.006	−0.006	−0.006	−0.006	−0.005	0.006	0.026	0.072
3.00	0.000	0.001	0.002	0.003	0.004	0.005	0.006	0.007	0.009	0.011	0.016	0.020	0.042	0.070	0.121
3.50	0.001	0.002	0.003	0.005	0.007	0.009	0.011	0.013	0.016	0.019	0.025	0.031	0.058	0.089	0.141
4.00	0.001	0.002	0.004	0.006	0.009	0.011	0.013	0.016	0.019	0.023	0.029	0.036	0.064	0.095	0.146
5.00	0.001	0.002	0.005	0.007	0.010	0.012	0.015	0.017	0.021	0.025	0.031	0.038	0.066	0.096	0.143

TABLE D.4.6 ***The Lee-Kesler Equation of State, The Lee-Kesler Generalized Equation of State is***

$$Z = \frac{P_r v_r'}{T_r} = 1 + \frac{B}{v_r'} + \frac{C}{v_r'^2} + \frac{D}{v_r'^5} + \frac{c_4}{T_r^3 v_r'^2}\left(\beta + \frac{\gamma}{v_r'^2}\right)\exp\left(-\frac{\gamma}{v_r'^2}\right)$$

$$B = b_1 - \frac{b_2}{T_r} - \frac{b_3}{T_r^2} - \frac{b_4}{T_r^3}$$

$$C = c_1 - \frac{c_2}{T_r} + \frac{c_3}{T_r^3}$$

$$D = d_1 + \frac{d_2}{T_r}$$

in which

$$T_r = \frac{T}{T_c}, \qquad P_r = \frac{P}{P_c}, \qquad v_r' = \frac{v}{RT_c / P_c}$$

The two sets of constants are as follows:

Constant	Simple Fluids	Reference Fluid
b_1	0.118 119 3	0.202 657 9
b_2	0.265 728	0.331 511
b_3	0.154 790	0.027 655
b_4	0.030 323	0.203 488
c_1	0.023 674 4	0.031 338 5
c_2	0.018 698 4	0.050 361 8
c_3	0.0	0.016 901
c_4	0.042 724	0.041 577
$d_1 \times 10^4$	0.155 488	0.487 36
$d_2 \times 10^4$	0.623 689	0.074 033 6
β	0.653 92	1.226
γ	0.060 167	0.037 54

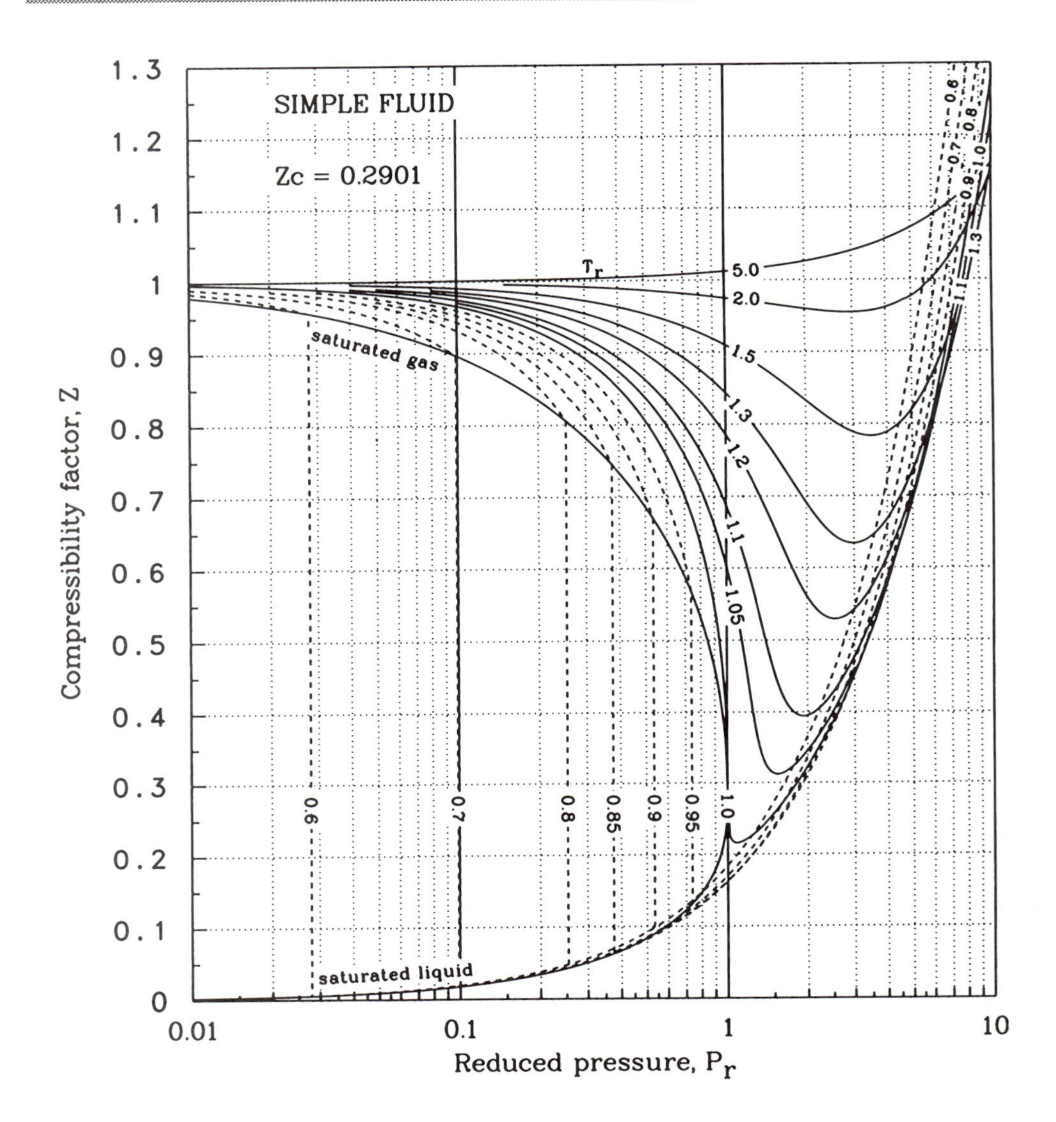

Figure D.2 Lee-Kesler Simple Fluid Compressibility Factor, Graph of Table D.4.2.

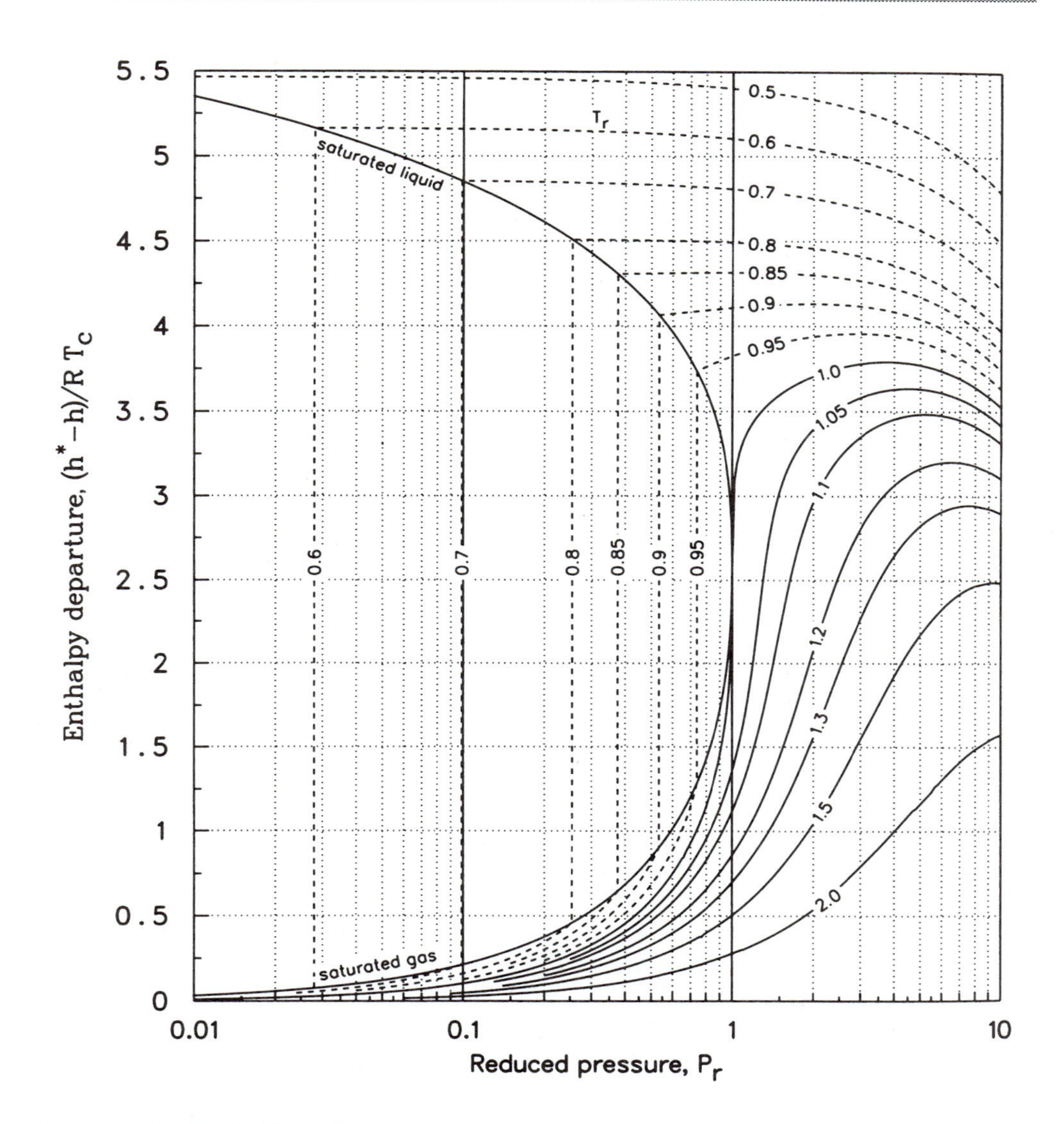

Figure D.3 Lee-Kesler Simple Fluid Entropy Departure, Graph of Table D.4.3.

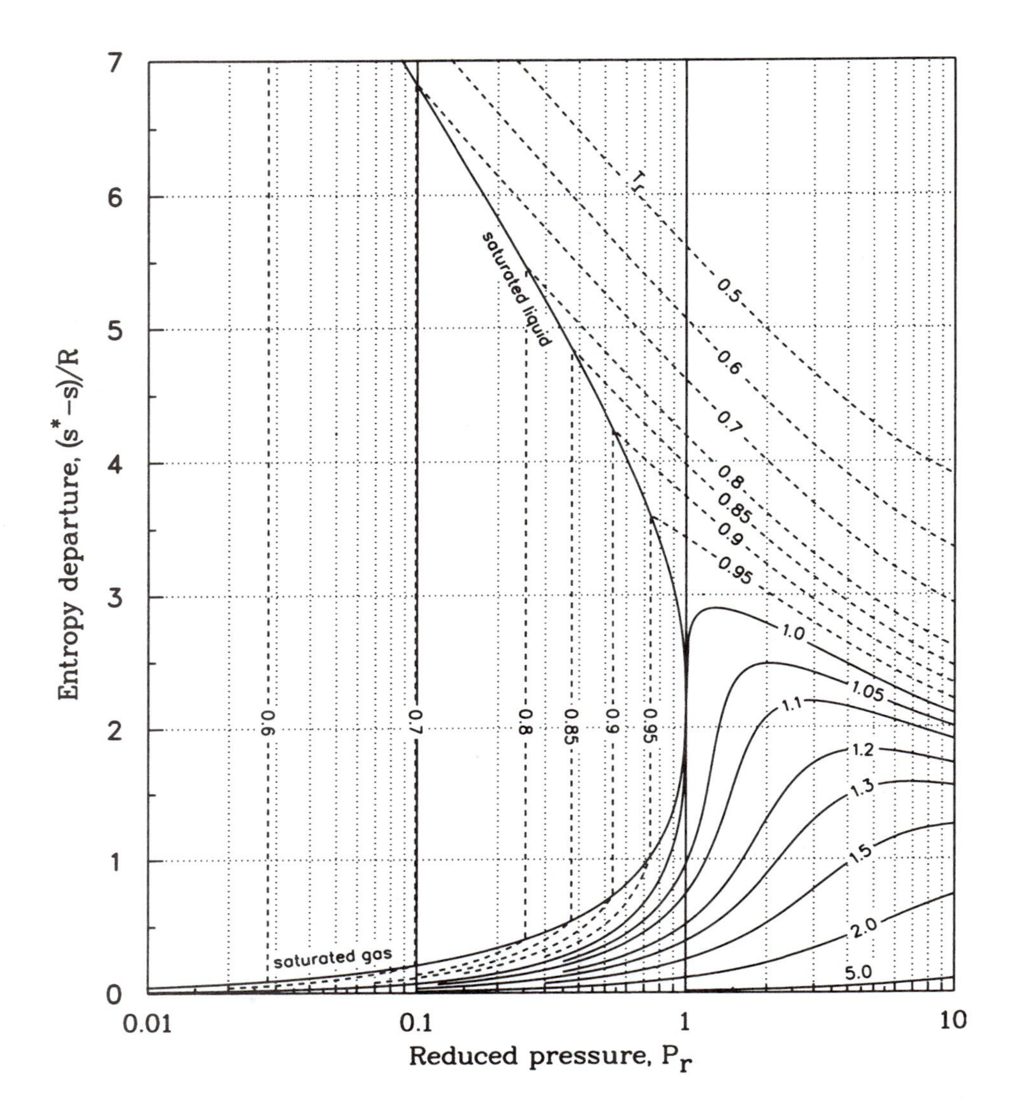

Figure D.4 Lee-Kesler Simple Fluid Entropy Departure, Graph from Table D.4.4.

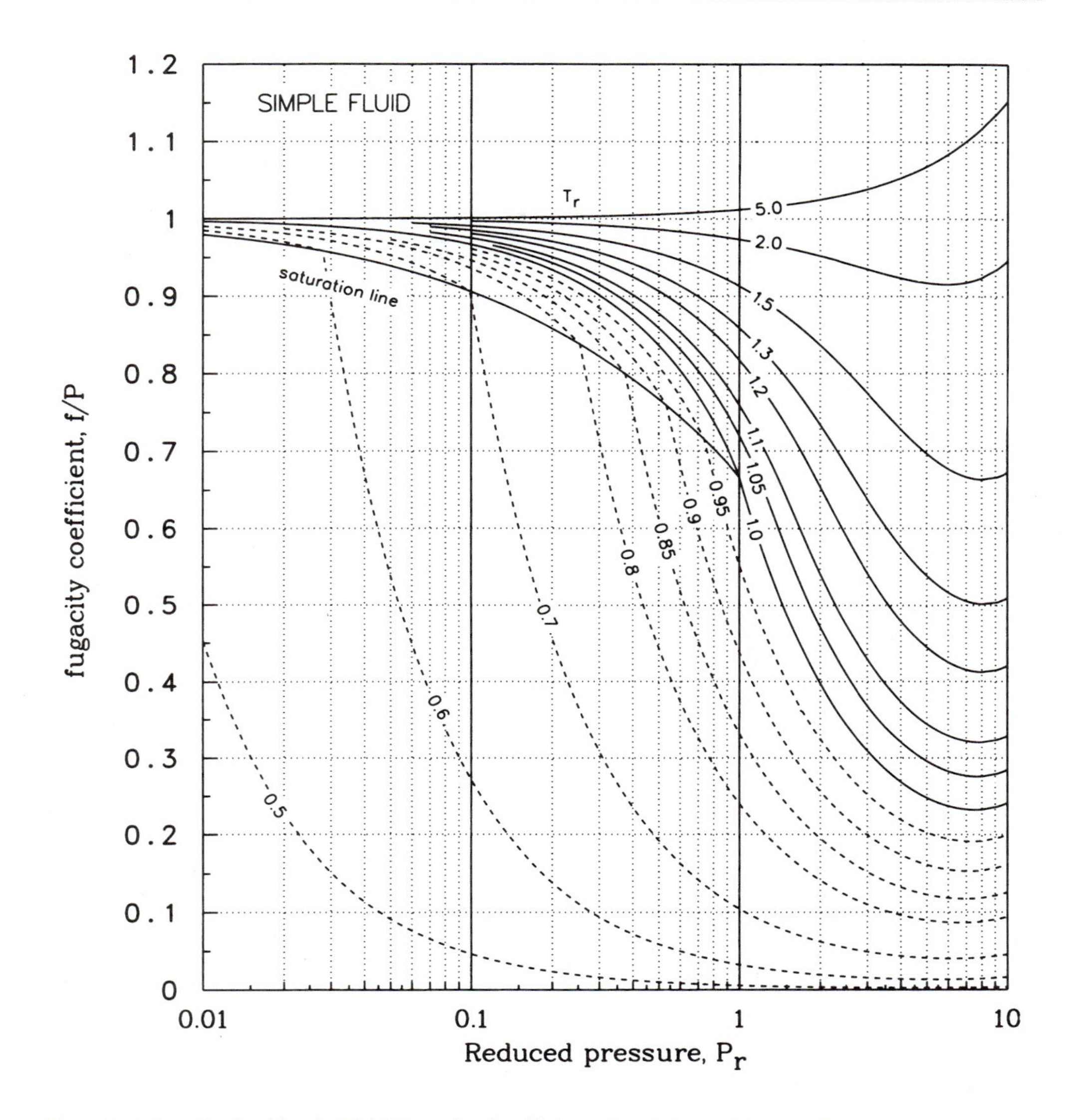

FigureD.5 Lee-Kesler Simple Fluid Fugacity Coefficient, Graph from Table D.4.5.

TABLE D.5.1 *Generalized Three-Parameter (T_r, P_r, ω) Compressibility Factor Tables Compressibility Factor Correction, $Z^{(1)}$*

$T_r \backslash P_r$	0.10	0.20	0.40	0.60	0.80	1.00	1.20	1.40	1.70	2.00	2.50	3.00	5.00	7.00	10.00
0.30	–.0081	–.0162	–.0323	–.0484	–.0645	–.0806	–.0967	–.1127	–.1368	–.1608	–.2008	–.2407	–.3996	–.5573	–.7916
0.40	–.0095	–.0190	–.0380	–.0569	–.0758	–.0946	–.1134	–.1321	–.1601	–.1879	–.2340	–.2799	–.4603	–.6365	–.8936
0.50	–.0090	–.0181	–.0360	–.0539	–.0716	–.0893	–.1069	–.1243	–.1504	–.1762	–.2189	–.2611	–.4253	–.5831	–.8099
0.60	–.0082	–.0164	–.0326	–.0487	–.0646	–.0803	–.0960	–.1115	–.1345	–.1572	–.1945	–.2312	–.3718	–.5047	–.6929
0.70	–.0075	–.0148	–.0294	–.0438	–.0579	–.0718	–.0855	–.0990	–.1189	–.1385	–.1703	–.2013	–.3184	–.4270	–.5785
0.75	–.0744	–.0143	–.0282	–.0417	–.0550	–.0681	–.0808	–.0934	–.1118	–.1298	–.1590	–.1872	–.2929	–.3901	–.5250
0.80	–.0487	–.1160	–.0272	–.0401	–.0526	–.0648	–.0767	–.0883	–.1052	–.1217	–.1481	–.1736	–.2682	–.3545	–.4740
0.85	–.0319	–.0715	–.0268	–.0391	–.0509	–.0622	–.0731	–.0837	–.0990	–.1138	–.1375	–.1602	–.2439	–.3201	–.4254
0.90	–.0205	–.0442	–.1118	–.0396	–.0503	–.0604	–.0701	–.0795	–.0929	–.1059	–.1265	–.1463	–.2195	–.2862	–.3788
0.95	–.0126	–.0262	–.0589	–.1110	–.0540	–.0607	–.0678	–.0751	–.0860	–.0967	–.1141	–.1310	–.1943	–.2526	–.3339
1.00	–.0069	–.0140	–.0285	–.0435	–.0588	–.0879	–.0609	–.0652	–.0735	–.0824	–.0972	–.1118	–.1672	–.2185	–.2902
1.05	–.0029	–.0054	–.0092	–.0097	–.0032	0.0220	0.1059	0.0951	–.0117	–.0432	–.0671	–.0838	–.1370	–.1835	–.2476
1.10	0.0001	0.0007	0.0038	0.0106	0.0236	0.0476	0.0897	0.1468	0.1418	0.0698	–.0033	–.0373	–.1021	–.1469	–.2056
1.15	0.0023	0.0052	0.0127	0.0237	0.0396	0.0625	0.0943	0.1345	0.1815	0.1667	0.0906	0.0332	–.0611	–.1084	–.1642
1.20	0.0040	0.0084	0.0190	0.0326	0.0499	0.0719	0.0991	0.1310	0.1780	0.1990	0.1651	0.1095	–.0141	–.0678	–.1231
1.30	0.0061	0.0125	0.0267	0.0429	0.0612	0.0819	0.1048	0.1294	0.1669	0.1991	0.2223	0.2079	0.0875	0.0176	–.0423
1.40	0.0072	0.0147	0.0306	0.0477	0.0661	0.0857	0.1063	0.1276	0.1596	0.1894	0.2259	0.2397	0.1737	0.1008	0.0350
1.50	0.0078	0.0158	0.0323	0.0497	0.0677	0.0864	0.1055	0.1248	0.1535	0.1806	0.2186	0.2433	0.2309	0.1717	0.1058
1.60	0.0080	0.0162	0.0330	0.0501	0.0677	0.0855	0.1035	0.1214	0.1478	0.1729	0.2098	0.2381	0.2631	0.2255	0.1673
1.80	0.0081	0.0162	0.0325	0.0488	0.0652	0.0816	0.0978	0.1137	0.1370	0.1593	0.1932	0.2224	0.2846	0.2871	0.2576
2.00	0.0078	0.0155	0.0310	0.0464	0.0617	0.0767	0.0916	0.1061	0.1273	0.1476	0.1789	0.2069	0.2820	0.3097	0.3096
2.50	0.0068	0.0135	0.0268	0.0399	0.0528	0.0654	0.0778	0.0899	0.1075	0.1245	0.1511	0.1757	0.2542	0.3052	0.3475
3.00	0.0059	0.0117	0.0232	0.0345	0.0456	0.0565	0.0672	0.0776	0.0929	0.1076	0.1310	0.1529	0.2268	0.2817	0.3385
3.50	0.0052	0.0103	0.0204	0.0303	0.0401	0.0497	0.0591	0.0683	0.0818	0.0949	0.1158	0.1356	0.2042	0.2584	0.3194
4.00	0.0046	0.0091	0.0182	0.0270	0.0357	0.0443	0.0527	0.0610	0.0731	0.0849	0.1038	0.1219	0.1857	0.2378	0.2994
5.00	0.0038	0.0075	0.0149	0.0222	0.0294	0.0365	0.0434	0.0503	0.0604	0.0703	0.0863	0.1016	0.1573	0.2047	0.2637

TABLE D.5.2 *Generalized Three-Parameter (T_r, P_r, ω) Compressibility Factor Tables Enthalpy Departure Correction, $((h^*-h)/RT_c)^{(1)}$*

$T_r \backslash P_r$	0.10	0.20	0.40	0.60	0.80	1.00	1.20	1.40	1.70	2.00	2.50	3.00	5.00	7.00	10.00
0.30	11.098	11.095	11.088	11.081	11.074	11.067	11.061	11.054	11.044	11.034	11.017	11.001	10.936	10.873	10.782
0.40	10.120	10.120	10.120	10.120	10.120	10.120	10.120	10.120	10.120	10.121	10.121	10.122	10.127	10.135	10.150
0.50	8.869	8.871	8.875	8.879	8.883	8.887	8.891	8.895	8.902	8.908	8.919	8.931	8.979	9.030	9.111
0.60	7.570	7.573	7.579	7.585	7.592	7.598	7.605	7.611	7.622	7.632	7.650	7.668	7.744	7.825	7.950
0.70	6.356	6.360	6.366	6.373	6.381	6.388	6.396	6.404	6.417	6.430	6.452	6.475	6.573	6.677	6.837
0.75	0.306	5.796	5.803	5.809	5.816	5.824	5.832	5.841	5.854	5.868	5.892	5.918	6.027	6.141	6.317
0.80	0.234	0.542	5.266	5.271	5.277	5.285	5.292	5.301	5.315	5.330	5.356	5.384	5.506	5.632	5.824
0.85	0.182	0.401	4.753	4.754	4.758	4.763	4.771	4.779	4.794	4.810	4.840	4.871	5.008	5.149	5.358
0.90	0.144	0.308	0.751	4.254	4.248	4.249	4.255	4.263	4.279	4.298	4.333	4.371	4.530	4.688	4.916
0.95	0.115	0.241	0.542	0.994	3.737	3.713	3.713	3.723	3.746	3.773	3.822	3.873	4.068	4.248	4.498
1.00	0.093	0.191	0.410	0.675	1.034	2.471	2.952	3.033	3.119	3.186	3.279	3.358	3.615	3.825	4.100
1.05	0.075	0.153	0.318	0.498	0.691	0.877	0.878	1.113	2.027	2.381	2.645	2.800	3.167	3.418	3.722
1.10	0.061	0.123	0.251	0.381	0.507	0.617	0.673	0.631	0.780	1.261	1.853	2.167	2.720	3.023	3.362
1.15	0.050	0.099	0.199	0.296	0.385	0.459	0.503	0.502	0.468	0.604	1.083	1.497	2.275	2.641	3.019
1.20	0.040	0.080	0.158	0.232	0.297	0.349	0.381	0.388	0.358	0.361	0.591	0.934	1.840	2.273	2.692
1.30	0.026	0.052	0.100	0.142	0.177	0.203	0.218	0.221	0.205	0.178	0.182	0.300	1.066	1.592	2.086
1.40	0.016	0.032	0.060	0.083	0.100	0.111	0.115	0.112	0.096	0.070	0.034	0.044	0.504	1.012	1.547
1.50	0.009	0.018	0.032	0.042	0.048	0.049	0.046	0.038	0.018	−0.008	−0.052	−0.078	0.142	0.556	1.080
1.60	0.004	0.007	0.012	0.013	0.011	0.005	−0.004	−0.016	−0.038	−0.065	−0.113	−0.151	−0.082	0.217	0.689
1.80	−0.003	−0.006	−0.015	−0.025	−0.037	−0.051	−0.067	−0.084	−0.113	−0.143	−0.194	−0.241	−0.317	−0.203	0.112
2.00	−0.007	−0.015	−0.030	−0.047	−0.065	−0.085	−0.105	−0.125	−0.157	−0.190	−0.244	−0.295	−0.428	−0.424	−0.255
2.50	−0.012	−0.025	−0.049	−0.075	−0.100	−0.125	−0.150	−0.176	−0.213	−0.250	−0.310	−0.367	−0.552	−0.661	−0.704
3.00	−0.014	−0.029	−0.058	−0.086	−0.114	−0.142	−0.170	−0.198	−0.238	−0.278	−0.342	−0.403	−0.611	−0.763	−0.899
3.50	−0.016	−0.031	−0.062	−0.092	−0.122	−0.152	−0.181	−0.210	−0.253	−0.294	−0.361	−0.425	−0.650	−0.827	−1.015
4.00	−0.016	−0.032	−0.064	−0.096	−0.127	−0.158	−0.188	−0.218	−0.262	−0.306	−0.375	−0.442	−0.680	−0.874	−1.097
5.00	−0.017	−0.034	−0.068	−0.101	−0.133	−0.166	−0.198	−0.229	−0.276	−0.321	−0.395	−0.466	−0.726	−0.947	−1.219

TABLE D.5.3 *Generalized Three-Parameter (T_r, P_r, ω) Compressibility Factor Tables Entropy Departure Correction, $((s_P^* - s_P)/R)^{(1)}$*

$T_r \backslash P_r$	0.10	0.20	0.40	0.60	0.80	1.00	1.20	1.40	1.70	2.00	2.50	3.00	5.00	7.00	10.00
0.30	16.773	16.754	16.714	16.675	16.637	16.598	16.559	16.520	16.463	16.405	16.310	16.214	15.838	15.469	14.927
0.40	13.980	13.970	13.951	13.932	13.914	13.895	13.876	13.858	13.830	13.803	13.757	13.713	13.540	13.376	13.144
0.50	11.196	11.191	11.181	11.171	11.161	11.151	11.142	11.132	11.118	11.105	11.083	11.063	10.986	10.920	10.836
0.60	8.827	8.823	8.817	8.811	8.806	8.800	8.795	8.790	8.784	8.777	8.768	8.759	8.735	8.723	8.720
0.70	6.954	6.951	6.946	6.941	6.937	6.933	6.930	6.928	6.924	6.922	6.919	6.919	6.928	6.952	7.002
0.75	0.340	6.173	6.167	6.162	6.158	6.155	6.152	6.150	6.148	6.146	6.147	6.149	6.174	6.213	6.285
0.80	0.246	0.578	5.474	5.467	5.462	5.458	5.455	5.453	5.452	5.452	5.455	5.461	5.501	5.555	5.648
0.85	0.183	0.408	4.853	4.841	4.832	4.826	4.822	4.820	4.820	4.822	4.828	4.839	4.898	4.969	5.083
0.90	0.140	0.301	0.744	4.269	4.250	4.238	4.232	4.230	4.231	4.236	4.250	4.267	4.351	4.442	4.578
0.95	0.109	0.228	0.517	0.961	3.697	3.658	3.647	3.646	3.655	3.669	3.697	3.728	3.851	3.966	4.125
1.00	0.086	0.177	0.382	0.632	0.977	2.399	2.868	2.940	3.012	3.067	3.140	3.200	3.387	3.532	3.717
1.05	0.069	0.140	0.292	0.460	0.642	0.820	0.831	1.073	1.951	2.283	2.522	2.655	2.949	3.134	3.348
1.10	0.055	0.112	0.229	0.350	0.470	0.577	0.640	0.620	0.786	1.241	1.786	2.067	2.534	2.767	3.013
1.15	0.045	0.091	0.183	0.275	0.361	0.437	0.489	0.506	0.507	0.654	1.100	1.471	2.138	2.428	2.708
1.20	0.037	0.075	0.149	0.220	0.286	0.343	0.385	0.408	0.414	0.447	0.680	0.991	1.767	2.115	2.430
1.30	0.026	0.052	0.102	0.148	0.190	0.226	0.254	0.275	0.291	0.300	0.351	0.481	1.147	1.569	1.944
1.40	0.019	0.037	0.072	0.104	0.133	0.158	0.178	0.194	0.210	0.220	0.240	0.290	0.730	1.138	1.544
1.50	0.014	0.027	0.053	0.076	0.097	0.115	0.130	0.142	0.156	0.166	0.181	0.206	0.479	0.823	1.222
1.60	0.011	0.021	0.040	0.057	0.073	0.086	0.098	0.108	0.119	0.129	0.142	0.159	0.334	0.604	0.969
1.80	0.006	0.013	0.024	0.035	0.044	0.053	0.060	0.067	0.075	0.083	0.094	0.105	0.195	0.355	0.628
2.00	0.004	0.008	0.016	0.023	0.029	0.035	0.040	0.045	0.052	0.058	0.067	0.077	0.136	0.238	0.434
2.50	0.002	0.004	0.007	0.010	0.014	0.017	0.020	0.022	0.026	0.031	0.037	0.044	0.080	0.130	0.230
3.00	0.001	0.002	0.004	0.006	0.008	0.010	0.012	0.014	0.017	0.020	0.026	0.031	0.058	0.093	0.158
3.50	0.001	0.001	0.003	0.004	0.006	0.007	0.009	0.010	0.013	0.015	0.019	0.024	0.046	0.073	0.122
4.00	0.001	0.001	0.002	0.003	0.005	0.006	0.007	0.008	0.010	0.012	0.016	0.020	0.038	0.060	0.100
5.00	0.000	0.001	0.001	0.002	0.003	0.004	0.005	0.006	0.007	0.009	0.011	0.014	0.028	0.044	0.073

TABLE D.5.4 *Generalized Three-Parameter (T_r, P_r, ω) Compressibility Factor Tables Fugacity Coefficient Correction, ln (f/P)*[1]

$T_r \backslash P_r$	0.10	0.20	0.40	0.60	0.80	1.00	1.20	1.40	1.70	2.00	2.50	3.00	5.00	7.00	10.00
0.30	−20.221	−20.229	−20.245	−20.261	−20.278	−20.294	−20.310	−20.326	−20.350	−20.374	−20.414	−20.455	−20.615	−20.774	−21.012
0.40	−11.319	−11.329	−11.348	−11.367	−11.386	−11.404	−11.423	−11.442	−11.471	−11.499	−11.546	−11.592	−11.778	−11.961	−12.231
0.50	−6.542	−6.551	−6.569	−6.587	−6.605	−6.623	−6.641	−6.658	−6.685	−6.711	−6.755	−6.799	−6.971	−7.139	−7.386
0.60	−3.790	−3.798	−3.815	−3.831	−3.847	−3.863	−3.879	−3.895	−3.919	−3.943	−3.982	−4.020	−4.172	−4.318	−4.530
0.70	−2.127	−2.134	−2.149	−2.164	−2.178	−2.193	−2.207	−2.221	−2.242	−2.263	−2.298	−2.331	−2.462	−2.587	−2.765
0.75	−0.068	−1.555	−1.569	−1.583	−1.597	−1.611	−1.624	−1.638	−1.658	−1.677	−1.709	−1.741	−1.862	−1.976	−2.138
0.80	−0.046	−0.099	−1.108	−1.121	−1.135	−1.148	−1.161	−1.173	−1.192	−1.210	−1.240	−1.270	−1.381	−1.485	−1.632
0.85	−0.030	−0.064	−0.739	−0.753	−0.765	−0.778	−0.790	−0.802	−0.820	−0.837	−0.865	−0.892	−0.994	−1.088	−1.221
0.90	−0.020	−0.041	−0.090	−0.458	−0.471	−0.483	−0.495	−0.507	−0.523	−0.539	−0.565	−0.590	−0.682	−0.767	−0.885
0.95	−0.012	−0.025	−0.053	−0.085	−0.237	−0.250	−0.261	−0.272	−0.288	−0.303	−0.326	−0.349	−0.431	−0.505	−0.609
1.00	−0.007	−0.014	−0.028	−0.042	−0.057	−0.072	−0.083	−0.093	−0.106	−0.119	−0.139	−0.158	−0.228	−0.293	−0.383
1.05	−0.003	−0.006	−0.011	−0.015	−0.017	−0.015	−0.005	0.014	0.020	0.015	0.003	−0.011	−0.067	−0.120	−0.197
1.10	0.000	0.000	0.002	0.004	0.009	0.016	0.028	0.047	0.077	0.094	0.101	0.097	0.061	0.019	−0.043
1.15	0.002	0.005	0.010	0.017	0.026	0.037	0.052	0.069	0.100	0.129	0.158	0.170	0.160	0.131	0.083
1.20	0.004	0.008	0.017	0.027	0.039	0.052	0.067	0.085	0.115	0.146	0.188	0.213	0.234	0.220	0.186
1.30	0.006	0.012	0.025	0.039	0.054	0.069	0.086	0.104	0.133	0.163	0.211	0.250	0.327	0.344	0.340
1.40	0.007	0.014	0.029	0.045	0.061	0.078	0.095	0.113	0.141	0.170	0.216	0.259	0.370	0.416	0.440
1.50	0.008	0.016	0.032	0.048	0.065	0.082	0.099	0.117	0.144	0.171	0.216	0.258	0.385	0.453	0.502
1.60	0.008	0.016	0.032	0.049	0.066	0.083	0.100	0.117	0.143	0.169	0.212	0.253	0.386	0.469	0.539
1.80	0.008	0.016	0.032	0.049	0.065	0.081	0.097	0.114	0.138	0.162	0.201	0.239	0.371	0.468	0.566
2.00	0.008	0.016	0.031	0.047	0.062	0.077	0.093	0.108	0.130	0.153	0.189	0.224	0.350	0.450	0.562
2.50	0.007	0.014	0.027	0.040	0.054	0.067	0.080	0.093	0.112	0.131	0.161	0.191	0.300	0.395	0.512
3.00	0.006	0.012	0.023	0.035	0.046	0.058	0.069	0.080	0.097	0.113	0.139	0.165	0.262	0.347	0.458
3.50	0.005	0.010	0.021	0.031	0.041	0.051	0.061	0.070	0.085	0.099	0.123	0.146	0.232	0.309	0.412
4.00	0.005	0.009	0.018	0.027	0.036	0.045	0.054	0.063	0.076	0.089	0.110	0.130	0.208	0.279	0.375
5.00	0.004	0.008	0.015	0.022	0.030	0.037	0.044	0.052	0.062	0.073	0.090	0.107	0.173	0.233	0.317

TABLE D.6.1 *Electronic states*

State	n	l	m_l	m_s	No. of states
1s	1	0	0	±1/2	2
2s	2	0	0	±1/2	2
2p	2	1	-1, 0, +1	±1/2	6
3s	3	0	0	±1/2	2
3p	3	1	-1,0,+1	±1/2	6
3d	3	2	-2,-1,0,+1,+2	±1/2	10

n is the main quantum number, **l** is the azimuthal quantum number
$\mathbf{m_l}$ is the magnetic quantum number, $\mathbf{m_s}$ is the spin quantum number

TABLE D.6.2 *Atomic Energy Levels*

Substance	Electron configuration	Term Symbol	Energy (ε/hc) mm^{-1}
H (1 electron)	1s	$^2S_{1/2}$	0.00
	2p	$^2P_{1/2}$	8225.8907
	2s	$^2S_{1/2}$	8225.8942
	2p	$^2P_{3/2}$	8225.9272
	3p	$^2P_{1/2}$	9749.2198
He (2)	$1s^2$	$^1S_0$	0
	1s2s	$^3S_1$	15985.0318
	1s2s	$^1S_0$	16627.170
N (7)	$1s^22s^22p^3$	$^4S_{3/2}$	0
		$^2D_{5/2}$	1922.39
		$^2D_{3/2}$	1923.31
O (8)	$1s^22s^22p^4$	$^3P_2$	0
		$^3P_1$	15.85
		$^3P_0$	22.65
		$^1D_2$	1586.77
F (9)	$1s^22s^22p^5$	$^2P_{3/2}$	0
		$^2P_{1/2}$	40.40
	$1s^22s^22p^43s$	$^4P_{5/2}$	10240.65

TABLE D.7 ***The Error Function***

x	erf(x)	erf(x)	x
0	0	0	0
0.1	0.1125	0.1	0.0889
0.2	0.2227	0.2	0.1791
0.3	0.3286	0.3	0.2725
0.4	0.4284	0.4	0.3708
0.5	0.5205	0.5	0.4769
0.6	0.6039	0.6	0.5951
0.7	0.6778	0.65	0.6609
0.8	0.7421	0.7	0.7329
0.9	0.7969	0.75	0.8134
1.0	0.8427	0.8	0.9062
1.1	0.8802	0.85	1.0179
1.2	0.9103	0.90	1.1631
1.3	0.9340	0.92	1.2379
1.4	0.9523	0.94	1.3299
1.5	0.9661	0.96	1.4522
1.6	0.9764	0.98	1.6450
1.7	0.9838	0.99	1.8214
1.8	0.9891	0.995	1.9849
1.9	0.9928	0.999	2.3268
2.0	0.9953	0.9999	2.7511
2.2	0.9981	0.99999	3.1235
2.8	0.9999		
3.6	1.000000		

$$erf(x) = \frac{2}{\sqrt{\pi}} \int_0^x e^{-z^2}\, dz$$

$$\mathrm{erf}(-x) = -erf(x)$$

$$\mathrm{erfc}(x) = 1 - erf(x)$$

$$\left.\frac{d\mathrm{erf}(x)}{dx}\right|_{x=0} = \frac{2}{\sqrt{\pi}} = 1.12838$$

An approximate closed form expression for the error function is as follows:

$$\mathbf{erf}(x) \cong 1 - A\exp[-B(x+C)^2]$$
$$\mathbf{erfc}(x) \cong A\exp[-B(x+C)^2]$$

with the constants as

$A = 1.5577$ $\quad B = 0.7182$ $\quad C = 0.7856$

giving an accuracy of 0.42 percent.

SECTION E

ENGLISH UNIT TABLES

TABLE E.1 *Critical Constants (English Units)*

Substance	Formula	Molec. Weight	Temp. R	Pressure lbf/in.^{2}	Volume ft^3/lb-mole	Acentric Factor
Ammonia	NH_3	17.031	729.9	1646	1.1613	0.250
Argon	Ar	39.948	271.4	706	1.1998	0.001
Bromine	Br_2	159.808	1058.4	1494	2.0375	0.108
Carbon dioxide	CO_2	44.010	547.4	1070	1.5041	0.239
Carbon monoxide	CO	28.010	239.2	508	1.4929	0.066
Chlorine	Cl_2	70.906	750.4	1157	1.9831	0.090
Deuterium (normal)	D_2	4.032	69.1	241	0.0000	-0.160
Fluorine	F_2	37.997	259.7	757	1.0620	0.054
Helium	He	4.003	9.34	32.9	0.9195	-0.365
Helium^3	He	3.017	5.96	16.5	1.1677	-0.473
Hydrogen (normal)	H_2	2.016	59.76	188.6	1.0428	-0.218
Krypton	Kr	83.800	376.9	798	1.4609	0.005
Neon	Ne	20.183	79.92	400	0.6664	-0.029
Nitric oxide	NO	30.006	324.0	940	0.9243	0.588
Nitrogen	N_2	28.013	227.2	492	1.4385	0.039
Nitrogen dioxide	NO_2	46.006	775.8	1465	2.6879	0.834
Nitrous oxide	N_2O	44.013	557.3	1050	1.5602	0.165
Oxygen	O_2	31.999	278.3	731	1.1758	0.025
Sulfur dioxide	SO_2	64.063	775.4	1143	1.9575	0.256
Water	H_2O	18.015	1165.1	3208	0.9147	0.344
Xenon	Xe	131.300	521.5	847	1.8966	0.008
Acetylene	C_2H_2	26.038	554.9	891	1.8053	0.190
Benzene	C_6H_6	78.114	1012.0	709	4.1488	0.212
n-Butane	C_4H_{10}	58.124	765.4	551	4.0847	0.199
Carbon tetrachloride	CCl_4	153.823	1001.5	661	4.4195	0.193
Chlorodifluoroethane[a] (142b)	CH_3CClF_2	100.495	738.5	616	3.7003	0.250
Chlorodifluoromethane (22)	$CHClF_2$	86.469	664.7	721	2.6527	0.221
Chloroform	$CHCl_3$	119.378	965.5	779	3.8268	0.218
Dichlorodifluoromethane (12)	CCl_2F_2	120.914	693.0	600	3.4712	0.204
Dichlorofluoroethane[a] (141)	CH_3CCl_2F	116.950	866.7	658	4.0367	0.215
Dichlorofluoromethane (21)	$CHCl_2F$	102.923	812.9	751	3.1460	0.210
Dichlorotrifluoroethane[a](123)	$CHCl_2CF_3$	152.930	822.4	532	4.4547	0.282
Difluoroethane[a] (152a)	CHF_2CH_3	66.050	695.5	656	2.8753	0.275
Ethane	C_2H_6	30.070	549.7	708	2.3755	0.099
Ethyl alcohol	C_2H_5OH	46.069	925.0	891	2.6767	0.644
Ethylene	C_2H_4	28.054	508.3	731	2.0888	0.089
n-Heptane	C_7H_{16}	100.205	972.5	397	6.9200	0.349
n-Hexane	C_6H_{14}	86.178	913.5	437	5.9268	0.299
Methane	CH_4	16.043	342.7	667	1.5890	0.011

(Continued)

TABLE E.1 (Continued) *Critical Constants (English Units)*

Substance	Formula	M	Temp. R	Pressure lbf/in.^{2}	Volume ft^3/lb-mole	Acentric Factor
Methyl alcohol	CH_3OH	32.042	922.7	1173	1.8902	0.556
Methyl chloride	CH_3Cl	50.488	749.3	972	2.2250	0.153
n-Octane	C_8H_{18}	114.232	1023.8	361	7.8811	0.398
n-Pentane	C_5H_{16}	72.151	845.5	489	4.8696	0.251
Propane	C_3H_8	44.094	665.6	616	4.2517	0.153
Propene	C_3H_6	42.081	656.8	667	2.8993	0.144
Propyne	C_3H_4	40.065	724.3	817	2.6270	0.215
Tetrafluoroethane[a] (134a)	CF_3CH_2F	102.030	673.6	589	3.1717	0.327

Source: R. C. Reid, J. M. Prausnitz, and B. E. Poling, *The Properties of Gases and Liquids*, fourth edition, McGraw-Hill Book Company, New York, 1987.

[a]Data from M. O. McLinden, NIST Thermophysics Division, 1989.

TABLE E.2 *The Triple Point Constants*

Substance	Formula	T R	P Psia	ρ_f lbm/ft^3
Ammonia	NH_3	351.9	0.8789	45.90
Argon	Ar	150.85	10.00	85.96
Carbon dioxide	CO_2	389.9	75.124	73.60
Copper	Cu	2441	1.146E-5	495.68
Ethane	C_2H_6	162.63	1.639E-4	40.67
Ethylene	C_2H_4	187.17	0.01777	40.89
Helium	He	3.906	0.735	9.133
Hydrogen	H_2	24.91	1.028	4.807
Isobutane	C_4H_{10}	204.39	2.83E-6	46.284
n-butane	C_4H_{10}	242.75	9.77E-5	45.90
Mercury	Hg	421.6	1.885E-8	854.64
Methane	CH_4	163.23	1.696	28.186
Methanol	CH_3OH	316.06	2.661E-5	56.46
Neon	Ne	44.21	6.292	78.03
Nitrogen	N_2	113.67	1.816	54.29
Oxygen	O_2	97.85	0.02122	81.53
Propane	C_3H_8	153.85	2.444E-8	45.79
Silver	Ag	2221	0.00145	582.5
Water	H_2O	491.69	0.08866	62.428
Zinc	Zn	1246	0.7348	414.5

The gas phase is well approximated as an ideal gas for all substances

TABLE E.3 ***Properties of Selected Solids at 77 F***

Substance	ρ lbm/ft³	C_p Btu/lbm-R	k Btu/h-ft-R	$\alpha \times 10^6$ ft²/s
Asphalt	132.3	0.22	1.31	3.85
Bark	21.2	0.30	0.1385	1.86
Brick, common	112.4	0.20	1.31	4.98
Carbon, diamond	202.9	0.122	2527	8767
Carbon, graphite	125-156	0.146	290	1216
Clay	90.5	0.21	2.40	10.76
Coal	75-95	0.30	0.487	1.65
Concrete	137	0.21	2.40	7.12
Cork	9.36	0.449	0.0786	1.60
Earth, coarse	127.4	0.439	1.105	1.69
Glass, plate	156	0.191	2.621	7.54
Glass, wool	12.5	0.158	0.0693	3.01
Granite	172	0.212	5.43	12.76
Ice (32 F)	57.2	0.487	4.212	12.95
Leather	53.7	0.358	0.22-0.28	1.13
Paper	43.7	0.287	0.225	1.54
Plaster	105.5	0.191	1.479	6.29
Plexiglas	73.7	0.344	0.344	1.16
Polystyrene	57.4	0.549	0.655	1.78
Polyvinyl chloride	86.1	0.229	0.281	1.22
Quartz	130-156	0.186	2.62	8.40
Rubber, foam	31.2	0.399	0.168	1.16
Rubber, hard	71.8	0.480	0.30	0.74
Rubber, soft	68.7	0.399	0.243	0.76
Salt, rock	130-156	0.219	13.10	35.6
Sand, dry	93.6	0.191	0.562	2.69
Sandstone	134-144	0.170	3.0-3.9	12.6
Silicon	145.5	0.167	286	1010
Snow, firm	35	0.501	0.861	4.21
Teflon	137	0.248	0.431	1.09
Wood, hard (oak)	44.9	0.301	0.30	1.89
Wood, soft (pine)	31.8	0.33	0.225	1.84
Wool	6.24	0.411	0.0674	2.25

TABLE E.4 *Properties of Metals at 77 F*

Substance	ρ lbm/ft³	C_p Btu/lbm-R	k Btu/h-ft-R	$\alpha \times 10^6$ ft²/s
Aluminum, duralumin	174	0.211	307	717
Copper, commercial	518	0.100	696	1152
Brass, 60-40	524	0.0898	212	385
Gold	1205	0.0308	590	1362
Iron, cast	454	0.100	97	183.3
Iron, steel 0.5% C	489	0.111	101	159.6
Iron. 304 St Steel	488	0.110	25.8	41.3
Lead	708	0.031	65.1	254.1
Magnesium, 2% Mn	111	0.239	213	690
Nickel, 10% Cr	541	0.106	31.8	47.6
Silver, 99.9% Ag	657	0.0564	769	1781
Sodium	60.6	0.288	249	1223
Tin	456	0.0525	125	449
Tungsten	1205	0.032	335	745
Zinc	446	0.0927	226	470

TABLE E.5 *Properties of Some Liquids at 77 F*

Substance	ρ lbm/ft³	C_p Btu/lbm-R	k Btu/h-ft-R	$\alpha \times 10^6$ ft²/s	$\mu \times 10^3$ lbm/s-ft	$\nu \times 10^6$ ft²/s	$\beta \times 10^3$ 1/R	Pr
Ammonia	37.7	1.15	0.297	1.89	0.141	3.81	4.5	2.01
Benzene	54.9	0.41	0.0815	1.00	0.4	7.34	-	7.32
Butane	34.7	0.60	0.0682	0.926	0.110	3.17	3.0	3.43
CCL_4	98.9	0.20	0.0578	0.818	0.61	6.18	--	7.51
CO_2	42.5	0.69	0.0468	0.441	0.048	1.12	25.2	2.54
Engine oil	55.2	0.456	0.0832	0.915	329	5963	1.26	6500
Ethanol	48.9	0.59	0.0971	0.936	0.70	14.32	1.94	15.2
Gasoline	46.8	0.50	0.0670	0.796	0.35	7.46	-	9.32
Glycerine	78.7	0.58	0.165	1.01	538	6835	0.86	6770
Kerosine	50.9	0.48	0.0670	0.764	1.0	19.8	-	24.3
Mercury	848	0.033	4.934	48.65	1.02	1.20	0.326	0.025
Methanol	49.1	0.61	0.110	1.02	0.37	7.52	2.16	7.38
n-octane	43.2	0.53	0.0739	0.893	0.34	7.90	1.87	8.85
Propane	31.8	0.61	0.0549	0.786	0.061	1.92	9.5	2.43
R-12	81.8	0.232	0.0404	0.592	0.168	2.06	4.95	3.46
R-22	74.3	0.30	0.0508	0.635	0.134	1.81	5.85	2.85
R-134a	75.3	0.34	0.0468	0.506	0.134	1.79	5.58	3.53
Water	62.2	1.00	0.347	1.55	0.60	9.61	0.47	6.21

TABLE E.6 ***Properties of Liquid Metals***

Substance	ρ lbm/ft^3	C_p Btu/lbm-R	k Btu/h-ft-R	$\alpha \times 10^6$ ft^2/s	$\nu \times 10^6$ ft^2/s	Pr	T_m F	h_{sf} Btu/lbm
Bismuth, Bi	627	0.033	9.53	126	1.8	0.014	520	23
Lead, Pb	665	0.038	9.42	103	2.6	0.025	620	9.9
Mercury, Hg	855	0.033	4.91	47.4	1.2	0.025	-37.8	4.9
Potassium, K	51.7	0.193	26.6	738	5.6	0.0076	146.1	25.5
Sodium, Na	58	0.33	49.7	722	7.6	0.011	208	48.6
Tin, Sn	434	0.057	18.9	211	2.8	0.013	450	25.5
Zinc, Zn	410	0.12	34.0	179	4.8	0.027	787	48.1
NaK (56/44)	55.4	0.27	14.4	290	7.1	0.026	66	--

TABLE E.7 ***Surface Tension at 77 F, Interphase to Air, Normal Boiling Point and Heat of Evaporation for Some Liquids***

Substance	Formula	$\sigma \times 10^3$ lbf/ft	T_b F	h_{fg} Btu/lbm
Water	H_2O	4.93	212	970
Benzene	C_6H_6	1.93	176	169
Carbon tetrachloride	CCL_4	1.85	171	83
Ethanol	C_2H_5OH	1.51	172	360
Glycerin	$C_2H_5(OH)_3$	4.32	554	419
Hexane	C_6H_{14}	1.26	156	144
Mercury	Hg	33.27	675	127
Methanol	CH_3OH	1.51	149	472
Octane	C_8H_{18}	1.45	259	129
Pentane	C_5H_{12}	1.06	97	153

TABLE E.8 ***Properties of Various Ideal Gases at 77 F, 1 atm* (English Units)***

Gas	Chemical Formula	Molecular Mass	R ft-lbf/lbm-R	$\rho \times 10^3$ lbm/ft^3	C_{po} Btu/lbm-R	C_{vo} Btu/lbm-R	γ C_{po}/C_{vo}
Steam	H_2O	18.015	85.76	1.442	0.447	0.337	1.327
Acetylene	C_2H_2	26.038	59.34	65.55	0.406	0.330	1.231
Air	--	28.97	53.34	72.98	0.240	0.171	1.400
Ammonia	NH_3	17.031	90.72	43.325	0.509	0.392	1.297
Argon	Ar	39.948	38.68	100.7	0.124	0.0745	1.667
Butane	C_4H_{10}	58.124	26.58	150.3	0.410	0.376	1.091
Carbon monoxide	CO	28.01	55.16	70.5	0.249	0.178	1.399
Carbon dioxide	CO_2	44.01	35.10	110.8	0.201	0.156	1.289
Ethane	C_2H_6	30.07	51.38	76.29	0.422	0.356	1.186
Ethanol	C_2H_5OH	46.069	33.54	117.6	0.341	0.298	1.145
Ethylene	C_2H_4	28.054	55.07	71.04	0.370	0.299	1.237
Helium	He	4.003	386.0	10.08	1.240	0.744	1.667
Hydrogen	H_2	2.016	766.5	5.075	3.394	0.241	1.409
Methane	CH_4	16.043	96.35	40.52	0.538	0.415	1.299
Methanol	CH_3OH	32.042	48.22	81.78	0.336	0.274	1.227
Neon	Ne	20.183	76.55	50.81	0.246	0.148	1.667
Nitric oxide	NO	30.006	51.50	75.54	0.237	0.171	1.387
Nitrogen	N_2	28.013	55.15	70.61	0.249	0.178	1.400
Nitrous oxide	N_2O	44.013	35.10	110.8	0.210	0.165	1.274
n-octane	C_8H_{18}	114.23	13.53	5.74	0.409	0.391	1.044
Oxygen	O_2	31.999	48.28	80.66	0.220	0.158	1.393
Propane	C_3H_8	44.094	35.04	112.9	0.401	0.356	1.126
R-12	CCL_2F_2	120.914	12.78	310.9	0.147	0.131	1.126
R-22	$CHCLF_2$	86.469	17.87	221.0	0.157	0.134	1.171
R-134a	CF_3CH_2F	102.03	15.15	262.2	0.203	0.184	1.106
Sulfur dioxide	SO_2	64.059	24.12	163.4	0.149	0.118	1.263
Sulfur trioxide	SO_3	80.053	19.30	204.3	0.152	0.127	1.196

*Or saturation pressure if it is less than 1 atm.

TABLE E.9 ***Transport Properties of Various Ideal Gases at 77 F, 1 atm* (English Units)***

Gas	ρ lbm/ft^3	$k \times 10^3$ Btu/ft-R	$\alpha \times 10^6$ ft^2/s	$\mu \times 10^6$ lbm/s-ft	$\nu \times 10^6$ ft^2/s	Pr	$10^{-6} g\beta/\alpha\nu$ 1/R-ft^3
Steam	1.442	11.3	4876	6.08	4219	0.865	0.00291
Acetylene	65.55	12.36	129.2	7.0	106.6	0.825	4.358
Air	72.98	15.2	241.1	12.4	170	0.704	1.463
Ammonia	43.33	14.4	181.9	6.79	156.6	0.86	2.108
Argon	100.7	10.3	229.6	15.4	152.8	0.667	1.70
Butane	150.3	9.47	42.73	5.0	33.58	0.785	41.77
CO	70.5	14.4	228.7	12.0	169.5	0.741	1.546
CO_2	110.8	9.59	119.6	10	90.95	0.761	5.506
Ethane	76.29	12.1	104.2	6.27	82.2	0.788	7.00
Ethanol	117.6	8.32	57.7	5.8	49.2	0.852	21.13
Ethylene	71.04	11.8	125.2	7.0	98.4	0.785	4.861
Helium	10.08	89.6	1989	13.4	1326	0.667	0.0226
Hydrogen	5.075	108	1743	6.0	1179	0.676	0.0291
Methane	40.52	19.7	250.9	7.46	184.1	0.734	1.30
Methanol	81.78	14.4	146.2	7.4	90.4	0.618	4.53
Neon	50.81	28.6	635.6	21.4	422	0.664	0.223
Nitric oxide	75.54	14.85	230.2	12.8	170.1	0.738	1.53
Nitrogen	70.61	15.0	236.7	12.0	169.4	0.716	1.494
Nitrous oxide	110.8	10.3	122.8	10.0	90.3	0.735	5.411
n-octane	5.74	11	1299	4.9	853.6	0.657	0.0541
Oxygen	80.66	15.4	240.5	13.9	172.2	0.717	1.447
Propane	112.9	10.2	62.4	5.6	49.4	0.792	19.43
R-12	310.9	5.6	34.0	8.5	27.2	0.80	64.72
R-22	221.0	6.36	50.8	8.7	39.5	0.778	29.89
R-134a	262.2	7.74	40.3	~8.4	~32.2	~0.8	~46
Sulfur dioxide	163.4	5.5	63.3	8.6	52.6	0.832	18.0
Sulfur trioxide	204.3	7.5	67.3	~11.0	~53.9	~0.8	~16.5

*Or saturation pressure if it is less than 1 atm. ~ indicates estimated value

TABLE E.10 ***Constant-Pressure Specific Heats of Various Ideal Gases (English Units)***

C_{p0} = Btu/lbmol R $\theta = T$(Rankine)/180

Gas		Range R	Max Error %
N_2	$\bar{C}_{p0} = 39.060 - 512.79\,\theta^{-1.5} + 1072.7\,\theta^{-2} - 820.40\,\theta^{-3}$	300–3500	0.43
O_2	$\bar{C}_{p0} = 37.432 + 0.020\,102\,\theta^{1.5} - 178.57\,\theta^{-1.5} + 236.88\,\theta^{-2}$	300–3500	0.30
H_2	$\bar{C}_{p0} = 56.505 - 702.74\,\theta^{-0.75} + 1165.0\,\theta^{-1} - 560.70\,\theta^{-1.5}$	300–3500	0.60
CO	$\bar{C}_{p0} = 69.145 - 0.704\,63\,\theta^{0.75} - 200.77\,\theta^{-0.5} + 176.76\,\theta^{-0.75}$	300–3500	0.42
OH	$\bar{C}_{p0} = 81.546 - 59.350\,\theta^{0.25} + 17.329\,\theta^{0.75} - 4.2660\,\theta$	300–3500	0.43
NO	$\bar{C}_{p0} = 59.283 - 1.7096\,\theta^{0.5} - 70.613\,\theta^{-0.5} + 74.889\,\theta^{-1.5}$	300–3500	0.34
H_2O	$\bar{C}_{p0} = 143.05 - 183.54\,\theta^{0.25} + 82.751\,\theta^{0.5} - 3.6989\,\theta$	300–3500	0.43
CO_2	$\bar{C}_{p0} = -3.7357 + 30.529\,\theta^{0.5} - 4.1034\,\theta + 0.024\,198\,\theta^2$	300–3500	0.19
NO_2	$\bar{C}_{p0} = 46.045 + 216.10\,\theta^{-0.5} - 363.66\,\theta^{-0.75} + 232.550\,\theta^{-2}$	300–3500	0.26
CH_4	$\bar{C}_{p0} = -672.87 + 439.74\,\theta^{0.25} - 24.875\,\theta^{0.75} + 323.88\,\theta^{-0.5}$	300-2000	0.15
C_2H_4	$\bar{C}_{p0} = -95.395 + 123.15\,\theta^{0.5} - 35.641\,\theta^{0.75} + 182.77\,\theta^{-3}$	300–2000	0.07
C_2H_6	$\bar{C}_{p0} = 6.895 + 17.26\,\theta - 0.6402\,\theta^2 + 0.007\,28\,\theta^3$	300–1500	0.83
C_3H_8	$\bar{C}_{p0} = -4.042 + 30.46\,\theta - 1.571\,\theta^2 + 0.031\,71\,\theta^3$	300–1500	0.40
C_4H_{10}	$\bar{C}_{p0} = 3.954 + 37.12\,\theta - 1.833\,\theta^2 + 0.034\,98\,\theta^3$	300–1500	0.54

Source: From T.C. Scott and R.E. Sonntag. University of Michigan, unpublished 1971, except C_2H_6, C_3H_8, and C_4H_{10} from K.A. Kobe, Petroleum Refiner, 28, No. 2, 113 (1949).

TABLE E.11 ENG ***Thermodynamic Properties of Water***
TABLE E.11.1 ENG ***Saturated Water***

Temp.	Press.	Specific Volume, ft³/lbm			Internal Energy, Btu/lbm		
F T	psia P	Sat. Liquid v_f	Evap. v_{fg}	Sat. Vapor v_g	Sat. Liquid u_f	Evap. u_{fg}	Sat. Vapor u_g
32	0.0887	0.01602	3301.6545	3301.6705	0	1021.21	1021.21
35	0.100	0.01602	2947.5021	2947.5181	2.99	1019.20	1022.19
40	0.122	0.01602	2445.0713	2445.0873	8.01	1015.84	1023.85
45	0.147	0.01602	2036.9527	2036.9687	13.03	1012.47	1025.50
50	0.178	0.01602	1703.9867	1704.0027	18.05	1009.10	1027.15
60	0.256	0.01603	1206.7283	1206.7443	28.08	1002.36	1030.44
70	0.363	0.01605	867.5791	867.5952	38.09	995.64	1033.72
80	0.507	0.01607	632.6739	632.6900	48.08	988.91	1036.99
90	0.699	0.01610	467.5865	467.6026	58.06	982.18	1040.24
100	0.950	0.01613	349.9602	349.9764	68.04	975.43	1043.47
110	1.276	0.01617	265.0548	265.0709	78.01	968.67	1046.68
120	1.695	0.01620	203.0105	203.0267	87.99	961.88	1049.87
130	2.225	0.01625	157.1419	157.1582	97.96	955.07	1053.03
140	2.892	0.01629	122.8567	122.8730	107.95	948.21	1056.16
150	3.722	0.01634	96.9611	96.9774	117.94	941.32	1059.26
160	4.745	0.01639	77.2079	77.2243	127.94	934.39	1062.32
170	5.997	0.01645	61.9983	62.0148	137.94	927.41	1065.35
180	7.515	0.01651	50.1826	50.1991	147.96	920.38	1068.34
190	9.344	0.01657	40.9255	40.9421	157.99	913.29	1071.29
200	11.530	0.01663	33.6146	33.6312	168.03	906.15	1074.18
210	14.126	0.01670	27.7964	27.8131	178.09	898.95	1077.04
212.0	14.696	0.01672	26.7864	26.8032	180.09	897.51	1077.60
220	17.189	0.01677	23.1325	23.1492	188.16	891.68	1079.84
230	20.781	0.01685	19.3677	19.3846	198.25	884.33	1082.58
240	24.968	0.01692	16.3088	16.3257	208.36	876.91	1085.27
250	29.823	0.01700	13.8077	13.8247	218.48	869.41	1087.90
260	35.422	0.01708	11.7503	11.7674	228.64	861.82	1090.46
270	41.848	0.01717	10.0483	10.0655	238.81	854.14	1092.95
280	49.189	0.01726	8.6325	8.6498	249.02	846.35	1095.37
290	57.535	0.01735	7.4486	7.4660	259.25	838.46	1097.71
300	66.985	0.01745	6.4537	6.4712	269.51	830.45	1099.96
310	77.641	0.01755	5.6136	5.6312	279.80	822.32	1102.13
320	89.609	0.01765	4.9010	4.9186	290.13	814.07	1104.20
330	103.00	0.01776	4.2938	4.3115	300.50	805.68	1106.17
340	117.94	0.01787	3.7742	3.7921	310.90	797.14	1108.04
350	134.54	0.01799	3.3279	3.3459	321.35	788.45	1109.80

(Continued)

TABLE E.11.1 ENG (Continued) ***Saturated Water***

Temp.	Press.	Enthalpy, Btu/lbm			Entropy, Btu/lbm R		
F T	psia P	Sat. Liquid h_f	Evap. h_{fg}	Sat. Vapor h_g	Sat. Liquid s_f	Evap. s_{fg}	Sat. Vapor s_g
32	0.0887	0	1075.38	1075.39	0	2.1869	2.1869
35	0.100	2.99	1073.71	1076.70	0.0061	2.1703	2.1764
40	0.122	8.01	1070.89	1078.90	0.0162	2.1430	2.1591
45	0.147	13.03	1068.06	1081.10	0.0262	2.1161	2.1423
50	0.178	18.05	1065.24	1083.29	0.0361	2.0898	2.1259
60	0.256	28.08	1059.59	1087.67	0.0555	2.0388	2.0943
70	0.363	38.09	1053.95	1092.04	0.0746	1.9896	2.0642
80	0.507	48.08	1048.31	1096.39	0.0933	1.9423	2.0356
90	0.699	58.06	1042.65	1100.72	0.1116	1.8966	2.0083
100	0.950	68.04	1036.98	1105.02	0.1296	1.8526	1.9822
110	1.276	78.01	1031.28	1109.29	0.1473	1.8101	1.9574
120	1.695	87.99	1025.55	1113.54	0.1646	1.7690	1.9336
130	2.225	97.97	1019.78	1117.75	0.1817	1.7292	1.9109
140	2.892	107.96	1013.96	1121.92	0.1985	1.6907	1.8892
150	3.722	117.95	1008.10	1126.05	0.2150	1.6533	1.8683
160	4.745	127.95	1002.18	1130.14	0.2313	1.6171	1.8484
170	5.997	137.96	996.21	1134.17	0.2473	1.5819	1.8292
180	7.515	147.98	990.17	1138.15	0.2631	1.5478	1.8109
190	9.344	158.02	984.06	1142.08	0.2786	1.5146	1.7932
200	11.530	168.07	977.87	1145.94	0.2940	1.4822	1.7762
210	14.126	178.13	971.61	1149.74	0.3091	1.4507	1.7599
212.0	14.696	180.13	970.35	1150.49	0.3121	1.4446	1.7567
220	17.189	188.21	965.26	1153.47	0.3240	1.4201	1.7441
230	20.781	198.31	958.81	1157.12	0.3388	1.3901	1.7289
240	24.968	208.43	952.27	1160.70	0.3533	1.3609	1.7142
250	29.823	218.58	945.61	1164.19	0.3677	1.3324	1.7001
260	35.422	228.75	938.84	1167.59	0.3819	1.3044	1.6864
270	41.848	238.95	931.95	1170.90	0.3960	1.2771	1.6731
280	49.189	249.17	924.93	1174.10	0.4098	1.2504	1.6602
290	57.535	259.43	917.76	1177.19	0.4236	1.2241	1.6477
300	66.985	269.73	910.45	1180.18	0.4372	1.1984	1.6356
310	77.641	280.06	902.98	1183.03	0.4507	1.1731	1.6238
320	89.609	290.43	895.34	1185.76	0.4640	1.1483	1.6122
330	103.00	300.84	887.52	1188.36	0.4772	1.1238	1.6010
340	117.94	311.29	879.51	1190.80	0.4903	1.0997	1.5900
350	134.54	321.80	871.30	1193.10	0.5033	1.0760	1.5793

(Continued)

TABLE E.11.1 ENG (Continued) ***Saturated Water***

Temp.	Press.	Specific Volume, ft³/lbm			Internal Energy, Btu/lbm		
F T	psia P	Sat. Liquid v_f	Evap. v_{fg}	Sat. Vapor v_g	Sat. Liquid u_f	Evap. u_{fg}	Sat. Vapor u_g
360	152.93	0.01811	2.9430	2.9611	331.83	779.60	1111.43
370	173.24	0.01823	2.6098	2.6280	342.37	770.57	1112.94
380	195.61	0.01836	2.3203	2.3387	352.95	761.37	1114.31
390	220.17	0.01850	2.0680	2.0865	363.58	751.97	1115.55
400	247.08	0.01864	1.8474	1.8660	374.26	742.37	1116.63
410	276.48	0.01878	1.6537	1.6725	385.00	732.56	1117.56
420	308.52	0.01894	1.4833	1.5023	395.80	722.52	1118.32
430	343.37	0.01909	1.3329	1.3520	406.67	712.24	1118.91
440	381.18	0.01926	1.1998	1.2191	417.61	701.71	1119.32
450	422.13	0.01943	1.0816	1.1011	428.63	690.90	1119.53
460	466.38	0.01961	0.9764	0.9961	439.73	679.82	1119.55
470	514.11	0.01980	0.8826	0.9024	450.92	668.43	1119.35
480	565.50	0.02000	0.7986	0.8186	462.21	656.72	1118.93
490	620.74	0.02021	0.7233	0.7435	473.60	644.67	1118.28
500	680.02	0.02043	0.6556	0.6761	485.11	632.26	1117.37
510	743.53	0.02066	0.5946	0.6153	496.75	619.46	1116.21
520	811.48	0.02091	0.5395	0.5604	508.53	606.23	1114.76
530	884.07	0.02117	0.4896	0.5108	520.46	592.56	1113.02
540	961.51	0.02145	0.4443	0.4658	532.56	578.39	1110.95
550	1044.02	0.02175	0.4031	0.4249	544.85	563.69	1108.54
560	1131.85	0.02207	0.3656	0.3876	557.35	548.42	1105.76
570	1225.21	0.02241	0.3312	0.3536	570.07	532.50	1102.56
580	1324.37	0.02278	0.2997	0.3225	583.05	515.87	1098.91
590	1429.58	0.02318	0.2707	0.2939	596.31	498.44	1094.76
600	1541.13	0.02362	0.2440	0.2676	609.91	480.11	1090.02
610	1659.32	0.02411	0.2193	0.2434	623.87	460.76	1084.63
620	1784.48	0.02465	0.1963	0.2209	638.26	440.20	1078.46
630	1916.96	0.02525	0.1747	0.2000	653.17	418.22	1071.38
640	2057.17	0.02593	0.1545	0.1804	668.68	394.52	1063.20
650	2205.54	0.02673	0.1353	0.1620	684.96	368.66	1053.63
660	2362.59	0.02766	0.1169	0.1446	702.24	340.02	1042.26
670	2528.88	0.02882	0.0990	0.1278	720.91	307.52	1028.43
680	2705.09	0.03031	0.0809	0.1112	741.70	269.26	1010.95
690	2891.99	0.03248	0.0618	0.0943	766.34	220.82	987.16
700	3090.47	0.03665	0.0377	0.0743	801.66	145.92	947.57
705.4	3203.79	0.05053	0	0.0505	872.56	0	872.56

(Continued)

TABLE E.11.1 ENG (Continued) ***Saturated Water***

Temp.	Press.	Enthalpy, Btu/lbm			Entropy, Btu/lbm R		
F T	psia P	Sat. Liquid h_f	Evap. h_{fg}	Sat. Vapor h_g	Sat. Liquid s_f	Evap. s_{fg}	Sat. Vapor s_g
360	152.93	332.35	862.88	1195.23	0.5162	1.0526	1.5688
370	173.24	342.95	854.24	1197.19	0.5289	1.0295	1.5584
380	195.61	353.61	845.36	1198.97	0.5416	1.0067	1.5483
390	220.17	364.33	836.23	1200.56	0.5542	0.9841	1.5383
400	247.08	375.11	826.84	1201.95	0.5667	0.9617	1.5284
410	276.48	385.96	817.17	1203.13	0.5791	0.9395	1.5187
420	308.52	396.89	807.20	1204.09	0.5915	0.9175	1.5090
430	343.37	407.89	796.93	1204.82	0.6038	0.8957	1.4995
440	381.18	418.97	786.34	1205.31	0.6160	0.8740	1.4900
450	422.13	430.15	775.40	1205.54	0.6282	0.8523	1.4805
460	466.38	441.42	764.09	1205.51	0.6404	0.8308	1.4711
470	514.11	452.80	752.40	1205.20	0.6525	0.8093	1.4618
480	565.50	464.30	740.30	1204.60	0.6646	0.7878	1.4524
490	620.74	475.92	727.76	1203.68	0.6767	0.7663	1.4430
500	680.02	487.68	714.76	1202.44	0.6888	0.7447	1.4335
510	743.53	499.59	701.27	1200.86	0.7009	0.7232	1.4240
520	811.48	511.67	687.25	1198.92	0.7130	0.7015	1.4144
530	884.07	523.93	672.66	1196.58	0.7251	0.6796	1.4048
540	961.51	536.38	657.45	1193.83	0.7374	0.6576	1.3950
550	1044.02	549.05	641.58	1190.63	0.7496	0.6354	1.3850
560	1131.85	561.97	624.98	1186.95	0.7620	0.6129	1.3749
570	1225.21	575.15	607.59	1182.74	0.7745	0.5901	1.3646
580	1324.37	588.63	589.32	1177.95	0.7871	0.5668	1.3539
590	1429.58	602.45	570.06	1172.51	0.7999	0.5431	1.3430
600	1541.13	616.64	549.71	1166.35	0.8129	0.5187	1.3317
610	1659.32	631.27	528.08	1159.36	0.8262	0.4937	1.3199
620	1784.48	646.40	505.00	1151.41	0.8397	0.4677	1.3075
630	1916.96	662.12	480.21	1142.33	0.8537	0.4407	1.2943
640	2057.17	678.55	453.33	1131.89	0.8681	0.4122	1.2803
650	2205.54	695.87	423.89	1119.76	0.8831	0.3820	1.2651
660	2362.59	714.34	391.13	1105.47	0.8990	0.3493	1.2483
670	2528.88	734.39	353.83	1088.23	0.9160	0.3132	1.2292
680	2705.09	756.87	309.77	1066.64	0.9350	0.2718	1.2068
690	2891.99	783.72	253.88	1037.60	0.9575	0.2208	1.1783
700	3090.47	822.61	167.47	990.09	0.9901	0.1444	1.1345
705.4	3203.79	902.52	0	902.52	1.0580	0	1.0580

TABLE E.11.2 ENG *Superheated Vapor Water*

Temp. F	v ft³/lbm	u Btu/lbm	h Btu/lbm	s Btu/lbm R	v ft³/lbm	u Btu/lbm	h Btu/lbm	s Btu/lbm R
	1 psia (101.70)				5 psia (162.20)			
Sat.	333.58	1044.02	1105.75	1.9779	73.531	1062.99	1131.03	1.8441
200	392.51	1077.49	1150.12	2.0507	78.147	1076.25	1148.55	1.8715
240	416.42	1091.22	1168.28	2.0775	83.001	1090.25	1167.05	1.8987
280	440.32	1105.02	1186.50	2.1028	87.831	1104.27	1185.53	1.9244
320	464.19	1118.92	1204.82	2.1269	92.645	1118.32	1204.04	1.9487
360	488.05	1132.92	1223.23	2.1499	97.447	1132.42	1222.59	1.9719
400	511.91	1147.02	1241.75	2.1720	102.24	1146.61	1241.21	1.9941
440	535.76	1161.23	1260.37	2.1932	107.03	1160.89	1259.92	2.0154
500	571.53	1182.77	1288.53	2.2235	114.21	1182.50	1288.17	2.0458
600	631.13	1219.30	1336.09	2.2706	126.15	1219.10	1335.82	2.0930
700	690.72	1256.65	1384.47	2.3142	138.08	1256.50	1384.26	2.1367
800	750.30	1294.86	1433.70	2.3549	150.01	1294.73	1433.53	2.1774
900	809.88	1333.94	1483.81	2.3932	161.94	1333.84	1483.68	2.2157
1000	869.45	1373.93	1534.82	2.4294	173.86	1373.85	1534.71	2.2520
1100	929.03	1414.83	1586.75	2.4638	185.78	1414.77	1586.66	2.2864
1200	988.60	1456.67	1639.61	2.4967	197.70	1456.61	1639.53	2.3192
1300	1048.17	1499.43	1693.40	2.5281	209.62	1499.38	1693.33	2.3507
1400	1107.74	1543.13	1748.12	2.5584	221.53	1543.09	1748.06	2.3809
	10 psia (193.19)				14.696 psia (211.99)			
Sat.	38.424	1072.21	1143.32	1.7877	26.803	1077.60	1150.49	1.7567
200	38.848	1074.67	1146.56	1.7927	—	—	—	—
240	41.320	1089.03	1165.50	1.8205	27.999	1087.87	1164.02	1.7764
280	43.768	1103.31	1184.31	1.8467	29.687	1102.40	1183.14	1.8030
320	46.200	1117.56	1203.05	1.8713	31.359	1116.83	1202.11	1.8280
360	48.620	1131.81	1221.78	1.8948	33.018	1131.22	1221.01	1.8516
400	51.032	1146.10	1240.53	1.9171	34.668	1145.62	1239.90	1.8741
440	53.438	1160.46	1259.34	1.9385	36.313	1160.05	1258.80	1.8956
500	57.039	1182.16	1287.71	1.9690	38.772	1181.83	1287.27	1.9262
600	63.027	1218.85	1335.48	2.0164	42.857	1218.61	1335.16	1.9737
700	69.006	1256.30	1384.00	2.0601	46.932	1256.12	1383.75	2.0175
800	74.978	1294.58	1433.32	2.1009	51.001	1294.43	1433.13	2.0584
900	80.946	1333.72	1483.51	2.1392	55.066	1333.60	1483.35	2.0967
1000	86.912	1373.74	1534.57	2.1755	59.128	1373.65	1534.44	2.1330
1100	92.875	1414.68	1586.54	2.2099	63.188	1414.60	1586.44	2.1674
1200	98.837	1456.53	1639.43	2.2428	67.247	1456.47	1639.34	2.2003
1300	104.798	1499.32	1693.25	2.2743	71.304	1499.26	1693.17	2.2318
1400	110.759	1543.03	1747.99	2.3045	75.361	1542.98	1747.92	2.2620
1500	116.718	1587.67	1803.66	2.3337	79.417	1587.63	1803.60	2.2912
1600	122.678	1633.24	1860.25	2.3618	83.473	1633.20	1860.20	2.3194

(Continued)

TABLE E.11.2 ENG (Continued) ***Superheated Water Vapor***

Temp. F	v ft³/lbm	u Btu/lbm	h Btu/lbm	s Btu/lbm R	v ft³/lbm	u Btu/lbm	h Btu/lbm	s Btu/lbm R
	20 psia (227.96)				40 psia (267.26)			
Sat.	20.091	1082.02	1156.38	1.7320	10.501	1092.27	1170.00	1.6767
240	20.475	1086.54	1162.32	1.7405	—	—	—	—
280	21.734	1101.36	1181.80	1.7676	10.711	1097.31	1176.59	1.6857
320	22.976	1116.01	1201.04	1.7929	11.360	1112.81	1196.90	1.7124
360	24.206	1130.55	1220.14	1.8168	11.996	1127.98	1216.77	1.7373
400	25.427	1145.06	1239.17	1.8395	12.623	1142.95	1236.38	1.7606
440	26.642	1159.59	1258.19	1.8611	13.243	1157.82	1255.84	1.7827
500	28.456	1181.46	1286.78	1.8919	14.164	1180.06	1284.91	1.8140
600	31.466	1218.35	1334.80	1.9395	15.685	1217.33	1333.43	1.8621
700	34.466	1255.91	1383.47	1.9834	17.196	1255.14	1382.42	1.9063
800	37.460	1294.27	1432.91	2.0243	18.701	1293.65	1432.08	1.9474
900	40.450	1333.47	1483.17	2.0626	20.202	1332.96	1482.50	1.9859
1000	43.437	1373.54	1534.30	2.0989	21.700	1373.12	1533.74	2.0222
1100	46.422	1414.51	1586.32	2.1334	23.196	1414.16	1585.86	2.0568
1200	49.406	1456.39	1639.24	2.1663	24.690	1456.09	1638.85	2.0897
1300	52.389	1499.19	1693.08	2.1978	26.184	1498.94	1692.75	2.1212
1400	55.371	1542.92	1747.85	2.2280	27.677	1542.70	1747.56	2.1515
1500	58.352	1587.58	1803.54	2.2572	29.169	1587.38	1803.29	2.1807
1600	61.333	1633.15	1860.14	2.2854	30.660	1632.97	1859.92	2.2089
	60 psia (292.73)				80 psia (312.06)			
Sat.	7.177	1098.33	1178.02	1.6444	5.474	1102.56	1183.61	1.6214
320	7.485	1109.46	1192.56	1.6633	5.544	1105.95	1188.02	1.6270
360	7.924	1125.31	1213.29	1.6893	5.886	1122.53	1209.67	1.6541
400	8.353	1140.77	1233.52	1.7134	6.217	1138.53	1230.56	1.6790
440	8.775	1156.01	1253.44	1.7360	6.541	1154.15	1250.98	1.7022
500	9.399	1178.64	1283.00	1.7678	7.017	1177.19	1281.07	1.7346
600	10.425	1216.31	1332.06	1.8165	7.794	1215.28	1330.66	1.7838
700	11.440	1254.35	1381.37	1.8609	8.561	1253.57	1380.31	1.8285
800	12.448	1293.03	1431.24	1.9022	9.322	1292.41	1430.40	1.8700
900	13.452	1332.46	1481.82	1.9408	10.078	1331.95	1481.14	1.9087
1000	14.454	1372.71	1533.19	1.9773	10.831	1372.29	1532.63	1.9453
1100	15.454	1413.81	1585.39	2.0119	11.583	1413.46	1584.93	1.9799
1200	16.452	1455.80	1638.46	2.0448	12.333	1455.51	1638.08	2.0129
1300	17.449	1498.69	1692.42	2.0764	13.082	1498.43	1692.09	2.0445
1400	18.445	1542.48	1747.28	2.1067	13.830	1542.26	1746.99	2.0749
1500	19.441	1587.18	1803.04	2.1359	14.577	1586.99	1802.79	2.1041
1600	20.436	1632.79	1859.70	2.1641	15.324	1632.62	1859.48	2.1323
1800	22.426	1726.69	1975.69	2.2178	16.818	1726.54	1975.50	2.1861
2000	24.415	1824.02	2095.10	2.2685	18.310	1823.88	2094.94	2.2367

(Continued)

TABLE E.11.2 ENG (Continued) *Superheated Water Vapor*

Temp. F	*v* ft³/lbm	*u* Btu/lbm	*h* Btu/lbm	*s* Btu/lbm R	*v* ft³/lbm	*u* Btu/lbm	*h* Btu/lbm	*s* Btu/lbm R
	100 psia (327.85)				120 psia (341.30)			
Sat.	4.4340	1105.76	1187.81	1.6034	3.7302	1108.28	1191.11	1.5886
350	4.5917	1115.39	1200.36	1.6191	3.7835	1112.20	1196.22	1.5950
400	4.9344	1136.21	1227.53	1.6517	4.0785	1133.83	1224.39	1.6287
450	5.2646	1156.20	1253.62	1.6812	4.3600	1154.34	1251.16	1.6590
500	5.5866	1175.72	1279.10	1.7085	4.6330	1174.22	1277.10	1.6868
550	5.9032	1195.02	1304.25	1.7340	4.9002	1193.78	1302.59	1.7127
600	6.2160	1214.23	1329.26	1.7582	5.1636	1213.18	1327.84	1.7371
700	6.8340	1252.78	1379.24	1.8033	5.6825	1251.98	1378.17	1.7825
800	7.4455	1291.78	1429.56	1.8449	6.1948	1291.15	1428.72	1.8243
900	8.0528	1331.45	1480.47	1.8838	6.7029	1330.94	1479.78	1.8633
1000	8.6574	1371.87	1532.08	1.9204	7.2082	1371.46	1531.52	1.9000
1100	9.2599	1413.12	1584.47	1.9551	7.7114	1412.77	1584.01	1.9348
1200	9.8610	1455.21	1637.69	1.9882	8.2132	1454.92	1637.30	1.9679
1300	10.4610	1498.18	1691.76	2.0198	8.7140	1497.93	1691.43	1.9996
1400	11.0602	1542.04	1746.71	2.0502	9.2139	1541.82	1746.42	2.0300
1500	11.6588	1586.79	1802.54	2.0794	9.7133	1586.60	1802.29	2.0592
1600	12.2570	1632.44	1859.25	2.1076	10.2122	1632.26	1859.03	2.0875
1800	13.4525	1726.38	1975.32	2.1614	11.2091	1726.23	1975.14	2.1412
2000	14.6472	1823.74	2094.78	2.2120	12.2052	1823.59	2094.62	2.1919
	140 psia (353.08)				160 psia (363.59)			
Sat.	3.2214	1110.31	1193.77	1.5760	2.8359	1111.99	1195.95	1.5650
400	3.4664	1131.36	1221.16	1.6088	3.0066	1128.81	1217.83	1.5910
450	3.7135	1152.44	1248.64	1.6399	3.2282	1150.49	1246.07	1.6230
500	3.9515	1172.70	1275.07	1.6682	3.4402	1171.15	1273.01	1.6518
550	4.1837	1192.52	1300.90	1.6944	3.6461	1191.25	1299.20	1.6784
600	4.4118	1212.12	1326.41	1.7191	3.8478	1211.05	1324.97	1.7033
700	4.8599	1251.18	1377.09	1.7648	4.2430	1250.38	1376.00	1.7494
800	5.3014	1290.53	1427.87	1.8068	4.6314	1289.89	1427.02	1.7916
900	5.7387	1330.43	1479.10	1.8459	5.0155	1329.92	1478.42	1.8308
1000	6.1730	1371.04	1530.96	1.8827	5.3967	1370.62	1530.40	1.8677
1100	6.6054	1412.42	1583.54	1.9176	5.7759	1412.07	1583.08	1.9026
1200	7.0363	1454.62	1636.91	1.9507	6.1536	1454.32	1636.52	1.9358
1300	7.4661	1497.67	1691.10	1.9824	6.5302	1497.42	1690.77	1.9676
1400	7.8952	1541.60	1746.14	2.0128	6.9061	1541.37	1745.85	1.9980
1500	8.3236	1586.40	1802.04	2.0421	7.2813	1586.20	1801.79	2.0273
1600	8.7516	1632.08	1858.81	2.0704	7.6561	1631.91	1858.59	2.0556
1800	9.6067	1726.08	1974.96	2.1242	8.4049	1725.92	1974.77	2.1094
2000	10.4610	1823.45	2094.46	2.1748	9.1528	1823.31	2094.31	2.1601

(Continued)

TABLE E.11.2 ENG (Continued) *Superheated Water Vapor*

Temp. F	v ft³/lbm	u Btu/lbm	h Btu/lbm	s Btu/lbm R	v ft³/lbm	u Btu/lbm	h Btu/lbm	s Btu/lbm R
	180 psia (373.12)				200 psia (381.86)			
Sat.	2.5333	1113.38	1197.76	1.5553	2.2892	1114.55	1199.28	1.5464
400	2.6482	1126.18	1214.39	1.5749	2.3609	1123.45	1210.83	1.5600
450	2.8504	1148.49	1243.43	1.6078	2.5477	1146.44	1240.73	1.5938
500	3.0424	1169.57	1270.91	1.6372	2.7238	1167.96	1268.77	1.6238
550	3.2279	1189.96	1297.47	1.6641	2.8932	1188.65	1295.72	1.6512
600	3.4091	1209.96	1323.52	1.6893	3.0580	1208.87	1322.05	1.6767
700	3.7631	1249.57	1374.92	1.7357	3.3792	1248.76	1373.82	1.7234
800	4.1102	1289.26	1426.17	1.7781	3.6932	1288.62	1425.31	1.7659
900	4.4529	1329.41	1477.73	1.8175	4.0029	1328.90	1477.04	1.8055
1000	4.7928	1370.20	1529.84	1.8544	4.3097	1369.77	1529.28	1.8425
1100	5.1307	1411.72	1582.61	1.8894	4.6145	1411.36	1582.15	1.8776
1200	5.4670	1454.03	1636.13	1.9227	4.9178	1453.73	1635.74	1.9109
1300	5.8023	1497.16	1690.43	1.9544	5.2200	1496.91	1690.10	1.9427
1400	6.1368	1541.15	1745.56	1.9849	5.5214	1540.93	1745.28	1.9732
1500	6.4707	1586.01	1801.54	2.0142	5.8222	1585.81	1801.29	2.0025
1600	6.8041	1631.73	1858.37	2.0425	6.1225	1631.55	1858.15	2.0308
1800	7.4701	1725.77	1974.59	2.0963	6.7223	1725.62	1974.41	2.0847
2000	8.1353	1823.17	2094.15	2.1470	7.3214	1823.02	2093.99	2.1354
	250 psia (401.03)				300 psia (417.42)			
Sat.	1.8448	1116.73	1202.08	1.5274	1.5441	1118.14	1203.86	1.5115
450	2.0018	1141.09	1233.70	1.5632	1.6361	1135.37	1226.20	1.5365
500	2.1498	1163.81	1263.27	1.5948	1.7662	1159.47	1257.52	1.5701
550	2.2903	1185.30	1291.26	1.6233	1.8878	1181.85	1286.65	1.5997
600	2.4258	1206.09	1318.32	1.6494	2.0041	1203.24	1314.50	1.6266
650	2.5581	1226.49	1344.84	1.6739	2.1168	1224.08	1341.60	1.6516
700	2.6879	1246.71	1371.06	1.6970	2.2269	1244.63	1368.26	1.6751
800	2.9426	1287.02	1423.16	1.7401	2.4421	1285.41	1420.99	1.7187
900	3.1929	1327.61	1475.32	1.7799	2.6528	1326.31	1473.58	1.7589
1000	3.4402	1368.72	1527.87	1.8172	2.8604	1367.65	1526.45	1.7964
1100	3.6854	1410.48	1580.98	1.8524	3.0660	1409.60	1579.80	1.8317
1200	3.9291	1452.98	1634.76	1.8858	3.2700	1452.24	1633.77	1.8653
1300	4.1718	1496.27	1689.27	1.9177	3.4730	1495.63	1688.43	1.8972
1400	4.4136	1540.37	1744.56	1.9483	3.6751	1539.82	1743.84	1.9279
1500	4.6549	1585.32	1800.66	1.9777	3.8767	1584.82	1800.03	1.9573
1600	4.8957	1631.11	1857.59	2.0060	4.0777	1630.66	1857.04	1.9857
1800	5.3763	1725.23	1973.95	2.0599	4.4790	1724.85	1973.50	2.0396
2000	5.8562	1822.67	2093.59	2.1106	4.8794	1822.32	2093.20	2.0904

(Continued)

TABLE E.11.2 ENG (Continued) ***Superheated Vapor Water***

Temp. F	v ft³/lbm	u Btu/lbm	h Btu/lbm	s Btu/lbm R	v ft³/lbm	u Btu/lbm	h Btu/lbm	s Btu/lbm R
	350 psia (431.81)				400 psia (444.69)			
Sat.	1.3267	1119.00	1204.93	1.4978	1.1619	1119.44	1205.45	1.4856
450	1.3733	1129.24	1218.18	1.5125	1.1745	1122.63	1209.57	1.4901
500	1.4913	1154.91	1251.49	1.5481	1.2843	1150.11	1245.17	1.5282
550	1.5998	1178.27	1281.88	1.5790	1.3834	1174.56	1276.95	1.5605
600	1.7025	1200.32	1310.59	1.6068	1.4760	1197.33	1306.58	1.5892
700	1.8975	1242.52	1365.42	1.6562	1.6503	1240.38	1362.54	1.6396
800	2.0846	1283.78	1418.80	1.7004	1.8163	1282.14	1416.59	1.6844
900	2.2670	1325.01	1471.83	1.7409	1.9776	1323.69	1470.07	1.7252
1000	2.4463	1366.58	1525.02	1.7787	2.1357	1365.51	1523.59	1.7632
1100	2.6235	1408.71	1578.63	1.8142	2.2917	1407.81	1577.44	1.7989
1200	2.7993	1451.48	1632.78	1.8478	2.4462	1450.73	1631.79	1.8327
1300	2.9739	1494.99	1687.59	1.8799	2.5995	1494.34	1686.76	1.8648
1400	3.1476	1539.26	1743.12	1.9106	2.7520	1538.70	1742.40	1.8956
1500	3.3208	1584.33	1799.41	1.9401	2.9039	1583.83	1798.78	1.9251
1600	3.4935	1630.22	1856.48	1.9685	3.0553	1629.77	1855.93	1.9535
1700	3.6659	1676.93	1914.36	1.9959	3.2064	1676.52	1913.86	1.9810
1800	3.8380	1724.47	1973.04	2.0225	3.3573	1724.08	1972.59	2.0076
1900	4.0099	1772.82	2032.53	2.0482	3.5080	1772.45	2032.11	2.0333
2000	4.1817	1821.96	2092.80	2.0732	3.6585	1821.61	2092.41	2.0584
	500 psia (467.12)				600 psia (486.33)			
Sat.	0.9283	1119.43	1205.32	1.4645	0.7702	1118.54	1204.06	1.4464
500	0.9924	1139.69	1231.51	1.4922	0.7947	1127.97	1216.21	1.4592
550	1.0792	1166.71	1266.56	1.5279	0.8749	1158.23	1255.36	1.4990
600	1.1583	1191.09	1298.26	1.5585	0.9456	1184.50	1289.49	1.5320
650	1.2327	1213.98	1328.04	1.5860	1.0109	1208.63	1320.87	1.5609
700	1.3040	1236.01	1356.66	1.6112	1.0728	1231.51	1350.62	1.5871
800	1.4407	1278.81	1412.11	1.6571	1.1900	1275.42	1407.55	1.6343
900	1.5723	1321.04	1466.52	1.6986	1.3021	1318.36	1462.92	1.6766
1000	1.7008	1363.34	1520.71	1.7371	1.4108	1361.15	1517.79	1.7155
1100	1.8271	1406.01	1575.06	1.7731	1.5173	1404.20	1572.66	1.7519
1200	1.9518	1449.21	1629.80	1.8071	1.6222	1447.68	1627.80	1.7861
1300	2.0754	1493.05	1685.07	1.8395	1.7260	1491.74	1683.38	1.8186
1400	2.1981	1537.57	1740.96	1.8704	1.8289	1536.44	1739.51	1.8497
1500	2.3203	1582.84	1797.52	1.9000	1.9312	1581.84	1796.26	1.8794
1600	2.4419	1628.88	1854.82	1.9285	2.0330	1627.98	1853.71	1.9080
1700	2.5632	1675.70	1912.87	1.9560	2.1345	1674.88	1911.87	1.9355
1800	2.6843	1723.32	1971.68	1.9826	2.2357	1722.55	1970.78	1.9622
1900	2.8052	1771.72	2031.27	2.0084	2.3367	1771.00	2030.44	1.9880
2000	2.9259	1820.90	2091.62	2.0335	2.4375	1820.20	2090.84	2.0131

(Continued)

TABLE E.11.2 ENG (Continued) *Superheated Vapor Water*

Temp. F	v ft³/lbm	u Btu/lbm	h Btu/lbm	s Btu/lbm R	v ft³/lbm	u Btu/lbm	h Btu/lbm	s Btu/lbm R
	800 psia (518.36)				1000 psia (544.74)			
Sat.	0.5691	1115.02	1199.26	1.4160	0.4459	1109.86	1192.37	1.3903
550	0.6154	1138.83	1229.93	1.4469	0.4534	1114.77	1198.67	1.3965
600	0.6776	1170.10	1270.41	1.4861	0.5140	1153.66	1248.76	1.4450
650	0.7324	1197.22	1305.64	1.5186	0.5637	1184.74	1289.06	1.4822
700	0.7829	1222.08	1337.98	1.5471	0.6080	1212.03	1324.54	1.5135
750	0.8306	1245.65	1368.62	1.5729	0.6490	1237.23	1357.33	1.5412
800	0.8764	1268.45	1398.19	1.5969	0.6878	1261.21	1388.49	1.5664
900	0.9640	1312.88	1455.60	1.6408	0.7610	1307.26	1448.08	1.6120
1000	1.0482	1356.71	1511.88	1.6807	0.8305	1352.17	1505.86	1.6530
1100	1.1300	1400.52	1567.81	1.7178	0.8976	1396.77	1562.88	1.6908
1200	1.2102	1444.60	1623.76	1.7525	0.9630	1441.46	1619.67	1.7260
1300	1.2892	1489.11	1679.97	1.7854	1.0272	1486.45	1676.53	1.7593
1400	1.3674	1534.17	1736.59	1.8167	1.0905	1531.88	1733.67	1.7909
1500	1.4448	1579.85	1793.74	1.8467	1.1531	1577.84	1791.21	1.8210
1600	1.5218	1626.19	1851.49	1.8754	1.2152	1624.40	1849.27	1.8499
1700	1.5985	1673.25	1909.89	1.9031	1.2769	1671.61	1907.91	1.8777
1800	1.6749	1721.03	1968.98	1.9298	1.3384	1719.51	1967.18	1.9046
1900	1.7510	1769.55	2028.77	1.9557	1.3997	1768.11	2027.12	1.9305
2000	1.8271	1818.80	2089.28	1.9808	1.4608	1817.41	2087.74	1.9557
	1250 psia (572.56)				1500 psia (596.38)			
Sat.	0.3454	1101.68	1181.57	1.3619	0.2769	1091.81	1168.67	1.3358
600	0.3786	1129.00	1216.58	1.3954	0.2816	1096.61	1174.78	1.3416
650	0.4267	1167.24	1265.95	1.4409	0.3329	1146.95	1239.34	1.4012
700	0.4670	1198.42	1306.44	1.4766	0.3716	1183.44	1286.60	1.4429
750	0.5030	1226.08	1342.42	1.5070	0.4049	1214.13	1326.52	1.4766
800	0.5364	1251.75	1375.82	1.5341	0.4350	1241.79	1362.53	1.5058
850	0.5680	1276.25	1407.64	1.5589	0.4631	1267.69	1396.23	1.5321
900	0.5984	1300.02	1438.42	1.5819	0.4897	1292.53	1428.46	1.5562
1000	0.6563	1346.37	1498.18	1.6244	0.5400	1340.43	1490.32	1.6001
1100	0.7116	1392.01	1556.62	1.6631	0.5876	1387.16	1550.26	1.6398
1200	0.7652	1437.49	1614.49	1.6990	0.6334	1433.45	1609.25	1.6765
1300	0.8176	1483.08	1672.19	1.7328	0.6778	1479.68	1667.82	1.7108
1400	0.8690	1528.98	1729.98	1.7647	0.7213	1526.06	1726.28	1.7431
1500	0.9197	1575.31	1788.04	1.7952	0.7641	1572.77	1784.86	1.7738
1600	0.9699	1622.15	1846.49	1.8242	0.8064	1619.90	1843.72	1.8031
1700	1.0197	1669.57	1905.44	1.8522	0.8482	1667.53	1902.98	1.8312
1800	1.0693	1717.62	1964.95	1.8791	0.8899	1715.73	1962.73	1.8582
1900	1.1186	1766.32	2025.07	1.9052	0.9313	1764.53	2023.03	1.8843
2000	1.1678	1815.68	2085.82	1.9304	0.9725	1813.97	2083.91	1.9096

(Continued)

TABLE E.11.2 ENG (Continued) ***Superheated Vapor Water***

Temp. F	v ft³/lbm	u Btu/lbm	h Btu/lbm	s Btu/lbm R	v ft³/lbm	u Btu/lbm	h Btu/lbm	s Btu/lbm R
	1750 psia (617.30)				2000 psia (635.99)			
Sat.	0.2268	1080.21	1153.65	1.3109	0.1881	1066.63	1136.25	1.2861
650	0.2627	1122.53	1207.61	1.3603	0.2057	1091.06	1167.18	1.3141
700	0.3022	1166.72	1264.60	1.4106	0.2487	1147.74	1239.79	1.3782
750	0.3341	1201.27	1309.47	1.4485	0.2803	1187.32	1291.07	1.4216
800	0.3622	1231.27	1348.55	1.4801	0.3071	1220.13	1333.80	1.4562
850	0.3878	1258.77	1384.37	1.5080	0.3312	1249.46	1372.03	1.4860
900	0.4119	1284.79	1418.18	1.5334	0.3534	1276.78	1407.58	1.5126
1000	0.4569	1334.34	1482.29	1.5789	0.3945	1328.10	1474.09	1.5598
1100	0.4990	1382.21	1543.79	1.6197	0.4325	1377.17	1537.23	1.6017
1200	0.5392	1429.35	1603.95	1.6570	0.4685	1425.19	1598.58	1.6398
1300	0.5780	1476.23	1663.40	1.6918	0.5031	1472.74	1658.95	1.6751
1400	0.6158	1523.12	1722.55	1.7245	0.5368	1520.15	1718.81	1.7082
1500	0.6530	1570.21	1781.67	1.7555	0.5697	1567.64	1778.48	1.7395
1600	0.6896	1617.64	1840.95	1.7850	0.6020	1615.37	1838.18	1.7692
1700	0.7258	1665.49	1900.53	1.8132	0.6340	1663.45	1898.08	1.7976
1800	0.7617	1713.85	1960.52	1.8404	0.6656	1711.97	1958.32	1.8248
1900	0.7974	1762.76	2021.00	1.8666	0.6971	1760.99	2018.99	1.8511
2000	0.8330	1812.26	2082.02	1.8919	0.7284	1810.56	2080.15	1.8765
	2500 psia (668.30)				3000 psia (695.52)			
Sat.	0.1306	1030.99	1091.41	1.2326	0.0840	968.77	1015.42	1.1575
700	0.1684	1098.70	1176.61	1.3073	0.0977	1003.88	1058.13	1.1944
750	0.2030	1155.21	1249.13	1.3686	0.1483	1114.74	1197.08	1.3122
800	0.2291	1195.69	1301.66	1.4112	0.1757	1167.64	1265.19	1.3675
850	0.2513	1229.54	1345.78	1.4456	0.1973	1207.64	1317.18	1.4080
900	0.2712	1259.90	1385.35	1.4752	0.2160	1241.80	1361.69	1.4413
950	0.2896	1288.20	1422.17	1.5018	0.2328	1272.72	1401.99	1.4705
1000	0.3069	1315.19	1457.19	1.5262	0.2485	1301.69	1439.63	1.4967
1100	0.3393	1366.83	1523.81	1.5704	0.2772	1356.16	1510.05	1.5434
1200	0.3696	1416.71	1587.69	1.6101	0.3036	1408.02	1576.59	1.5847
1300	0.3984	1465.66	1649.95	1.6465	0.3285	1458.45	1640.85	1.6223
1400	0.4261	1514.15	1711.28	1.6804	0.3524	1508.08	1703.69	1.6571
1500	0.4531	1562.47	1772.07	1.7123	0.3754	1557.26	1765.65	1.6895
1600	0.4795	1610.83	1832.64	1.7424	0.3978	1606.27	1827.12	1.7201
1700	0.5055	1659.38	1893.22	1.7711	0.4198	1655.30	1888.38	1.7492
1800	0.5312	1708.23	1953.96	1.7986	0.4416	1704.51	1949.64	1.7769
1900	0.5567	1757.48	2015.01	1.8251	0.4631	1754.00	2011.07	1.8035
2000	0.5820	1807.20	2076.44	1.8506	0.4844	1803.86	2072.78	1.8291

(Continued)

TABLE E.11.2 ENG (Continued) *Superheated Vapor Water*

Temp. F	v ft³/lbm	u Btu/lbm	h Btu/lbm	s Btu/lbm R	v ft³/lbm	u Btu/lbm	h Btu/lbm	s Btu/lbm R
	3500 psia				4000 psia			
650	0.02491	663.52	679.65	0.8629	0.02447	657.71	675.82	0.8574
700	0.03058	759.52	779.32	0.9506	0.02867	742.13	763.35	0.9345
750	0.10460	1058.38	1126.13	1.2440	0.06332	960.69	1007.56	1.1395
800	0.13626	1134.74	1222.99	1.3226	0.10523	1095.04	1172.93	1.2740
850	0.15819	1183.43	1285.88	1.3716	0.12833	1156.47	1251.46	1.3352
900	0.17625	1222.36	1336.51	1.4095	0.14623	1201.47	1309.71	1.3789
950	0.19214	1256.40	1380.85	1.4416	0.16152	1239.20	1358.75	1.4143
1000	0.20663	1287.60	1421.43	1.4699	0.17520	1272.94	1402.62	1.4449
1100	0.23282	1345.17	1495.97	1.5193	0.19954	1333.90	1481.60	1.4973
1200	0.25657	1399.15	1565.32	1.5624	0.22129	1390.11	1553.91	1.5423
1300	0.27871	1451.13	1631.65	1.6012	0.24137	1443.72	1622.38	1.5823
1400	0.29972	1501.93	1696.05	1.6368	0.26029	1495.73	1688.39	1.6188
1500	0.31992	1552.01	1759.22	1.6699	0.27837	1546.73	1752.78	1.6525
1600	0.33953	1601.70	1821.61	1.7010	0.29586	1597.12	1816.11	1.6841
1700	0.35872	1651.24	1883.57	1.7303	0.31291	1647.17	1878.79	1.7138
1800	0.37759	1700.80	1945.36	1.7583	0.32964	1697.11	1941.11	1.7420
1900	0.39624	1750.54	2007.17	1.7851	0.34616	1747.10	2003.32	1.7689
2000	0.41474	1800.56	2069.17	1.8108	0.36251	1797.27	2065.60	1.7948
	6000 psia				8000 psia			
650	0.02322	639.99	665.77	0.8404	0.02239	627.01	660.16	0.8278
700	0.02563	708.08	736.53	0.9028	0.02418	688.59	724.39	0.8844
750	0.02978	788.60	821.66	0.9746	0.02671	755.67	795.21	0.9441
800	0.03942	896.87	940.64	1.0708	0.03061	830.67	875.99	1.0095
850	0.05818	1018.83	1083.42	1.1820	0.03706	915.81	970.67	1.0832
900	0.07588	1102.93	1187.18	1.2599	0.04657	1003.68	1072.63	1.1596
950	0.09009	1162.00	1262.02	1.3140	0.05721	1079.59	1164.28	1.2259
1000	0.10207	1209.11	1322.44	1.3561	0.06722	1141.04	1240.55	1.2791
1100	0.12219	1286.42	1422.08	1.4222	0.08445	1236.84	1361.85	1.3595
1200	0.13928	1352.69	1507.33	1.4752	0.09892	1314.18	1460.62	1.4210
1300	0.15453	1413.30	1584.87	1.5206	0.11161	1382.27	1547.50	1.4718
1400	0.16854	1470.48	1657.61	1.5608	0.12309	1444.85	1627.08	1.5158
1500	0.18169	1525.37	1727.10	1.5972	0.13372	1503.78	1701.74	1.5549
1600	0.19421	1578.68	1794.31	1.6307	0.14373	1560.12	1772.89	1.5904
1700	0.20627	1630.90	1859.93	1.6618	0.15328	1614.58	1841.49	1.6229
1800	0.21801	1682.40	1924.45	1.6910	0.16251	1667.69	1908.27	1.6531
1900	0.22952	1733.45	1988.28	1.7186	0.17151	1719.85	1973.75	1.6815
2000	0.24087	1784.28	2051.72	1.7450	0.18034	1771.38	2038.36	1.7083

TABLE E.11.3 ENG ***Compressed Liquid Water***

Temp. F	v ft³/lbm	u Btu/lbm	h Btu/lbm	s Btu/lbm R	v ft³/lbm	u Btu/lbm	h Btu/lbm	s Btu/lbm R
	250 psia (401.03)				500 psia (467.12)			
Sat.	0.01865	375.37	376.23	0.5680	0.01975	447.69	449.51	0.6490
32	0.0160	0.00	0.74	0.0000	0.01599	0.00	1.48	0.0000
50	0.0160	18.04	18.78	0.0360	0.01599	18.02	19.50	0.0360
100	0.0161	67.95	68.70	0.1295	0.0161	67.87	69.36	0.1293
125	0.0162	92.86	93.61	0.1730	0.0162	92.75	94.24	0.1728
150	0.0163	117.80	118.55	0.2148	0.0163	117.66	119.17	0.2146
175	0.0165	142.79	143.55	0.2550	0.0165	142.62	144.14	0.2547
200	0.0166	167.84	168.61	0.2937	0.0166	167.64	169.18	0.2934
225	0.0168	192.99	193.76	0.3311	0.0168	192.76	194.31	0.3308
250	0.0170	218.25	219.04	0.3674	0.0170	217.99	219.56	0.3670
275	0.0172	243.66	244.46	0.4026	0.0172	243.36	244.95	0.4022
300	0.0174	269.26	270.06	0.4369	0.0174	268.91	270.52	0.4364
325	0.0177	295.07	295.89	0.4703	0.0177	294.68	296.32	0.4698
350	0.0180	321.14	321.97	0.5030	0.0180	320.70	322.36	0.5025
375	0.0183	347.52	348.36	0.5351	0.0183	347.01	348.70	0.5345
400	0.0186	374.25	375.12	0.5667	0.0186	373.68	375.40	0.5660
450	—	—	—	—	0.0194	428.39	430.19	0.6280
	1000 psia (544.74)				1500 psia (596.38)			
Sat.	0.02159	538.37	542.36	0.74318	0.02346	604.95	611.46	0.80821
32	0.01597	0.02	2.98	0.0000	0.01594	0.04	4.47	0.0001
50	0.0160	17.98	20.94	0.0359	0.0159	17.95	22.37	0.0358
100	0.0161	67.70	70.67	0.1290	0.0161	67.53	71.99	0.1287
125	0.0162	92.52	95.51	0.1724	0.0162	92.30	96.78	0.1720
150	0.0163	117.37	120.39	0.2141	0.0163	117.10	121.61	0.2136
175	0.0164	142.28	145.32	0.2542	0.0164	141.95	146.50	0.2536
200	0.0166	167.25	170.32	0.2928	0.0166	166.86	171.46	0.2922
225	0.0168	192.30	195.40	0.3301	0.0167	191.86	196.50	0.3295
250	0.0169	217.46	220.60	0.3663	0.0169	216.95	221.65	0.3655
275	0.0171	242.77	245.94	0.4014	0.0171	242.18	246.93	0.4005
300	0.0174	268.24	271.45	0.4355	0.0173	267.57	272.39	0.4346
325	0.0176	293.91	297.17	0.4688	0.0176	293.16	298.04	0.4678
350	0.0179	319.83	323.14	0.5014	0.0179	318.97	323.93	0.5003
375	0.0182	346.02	349.39	0.5333	0.0182	345.05	350.10	0.5321
400	0.0185	372.55	375.98	0.5647	0.0185	371.45	376.58	0.5634
425	0.0189	399.47	402.97	0.5957	0.0189	398.21	403.45	0.5942
450	0.0193	426.89	430.47	0.6263	0.0193	425.43	430.78	0.6247
500	0.0204	483.77	487.54	0.6874	0.0202	481.76	487.38	0.6852
550	—	—	—	—	0.0216	542.08	548.07	0.7469

(Continued)

TABLE E.11.3 ENG (Continued) *Compressed Liquid Water*								
Temp. F	v ft³/lbm	u Btu/lbm	h Btu/lbm	s Btu/lbm R	v ft³/lbm	u Btu/lbm	h Btu/lbm	s Btu/lbm
	2000 psia (635.99)				4000 psia			
Sat.	0.02565	662.38	671.87	0.8622	—	—	—	—
32	0.0159	0.06	5.95	0.0001	0.0158	0.10	11.79	0.0000
50	0.0159	17.91	23.80	0.0357	0.0158	17.75	29.46	0.0353
100	0.0160	67.36	73.30	0.1284	0.0159	66.71	78.51	0.1271
125	0.0161	92.07	98.04	0.1716	0.0160	91.22	103.09	0.1701
150	0.0162	116.82	122.84	0.2132	0.0161	115.77	127.72	0.2113
175	0.0164	141.62	147.68	0.2531	0.0163	140.36	152.41	0.2510
200	0.0165	166.48	172.60	0.2916	0.0164	165.01	177.17	0.2893
225	0.0167	191.42	197.59	0.3288	0.0166	189.72	202.00	0.3262
250	0.0169	216.45	222.69	0.3648	0.0168	214.51	226.92	0.3620
275	0.0171	241.61	247.93	0.3998	0.0170	239.40	251.96	0.3967
300	0.0173	266.92	273.33	0.4337	0.0172	264.43	277.14	0.4304
325	0.0176	292.42	298.92	0.4669	0.0174	289.61	302.49	0.4632
350	0.0178	318.14	324.74	0.4993	0.0177	314.97	328.04	0.4952
375	0.0181	344.11	350.82	0.5310	0.0179	340.54	353.81	0.5266
400	0.0184	370.38	377.20	0.5621	0.0182	366.34	379.84	0.5573
450	0.0192	424.03	431.13	0.6231	0.0189	418.83	432.83	0.6172
500	0.0201	479.84	487.29	0.6832	0.0198	472.90	487.53	0.6758
550	0.0214	539.24	547.16	0.7440	0.0208	529.44	544.87	0.7340
600	0.0233	605.37	613.99	0.8086	0.0223	590.01	606.51	0.7935
	6000 psia				8000 psia			
50	0.01573	17.57	35.03	0.0348	0.01563	17.38	40.52	0.0342
100	0.01585	66.09	83.69	0.1259	0.01577	65.49	88.83	0.1246
125	0.01595	90.40	108.11	0.1685	0.01586	89.62	113.10	0.1670
150	0.01606	114.76	132.59	0.2096	0.01597	113.81	137.45	0.2078
175	0.01619	139.17	157.14	0.2490	0.01610	138.04	161.87	0.2471
200	0.01633	163.62	181.75	0.2871	0.01623	162.31	186.34	0.2849
225	0.01648	188.12	206.43	0.3238	0.01639	186.61	210.87	0.3214
250	0.01666	212.69	231.18	0.3593	0.01655	210.97	235.47	0.3567
275	0.01684	237.34	256.04	0.3937	0.01673	235.39	260.16	0.3909
300	0.01705	262.10	281.02	0.4271	0.01693	259.91	284.97	0.4241
325	0.01727	286.99	306.16	0.4597	0.01714	284.53	309.91	0.4564
350	0.01751	312.03	331.47	0.4915	0.01737	309.29	335.01	0.4878
375	0.01777	337.25	356.98	0.5225	0.01762	334.19	360.28	0.5186
400	0.01805	362.66	382.70	0.5528	0.01788	359.26	385.73	0.5486
450	0.01869	414.17	434.92	0.6119	0.01848	409.94	437.30	0.6069
500	0.01945	466.89	488.48	0.6692	0.01918	461.56	489.95	0.6633
550	0.02039	521.38	544.02	0.7256	0.02002	514.49	544.13	0.7183
600	0.02159	578.59	602.57	0.7822	0.02106	569.36	600.53	0.7728

TABLE E.11.4 *Saturated Solid-Saturated Vapor Water (English Units)*

		Specific Volume ft^3/lbm		Internal Energy Btu/lbm		
Temp. F T	Press. lbf/in.^{2} P	Sat. Solid v_i	Sat. Vapor $v_g \times 10^{-3}$	Sat. Solid u_i	Evap. u_{ig}	Sat. Vapor u_g
32.02	0.08866	0.017473	3.302	−143.34	1164.5	1021.2
32	0.08859	0.01747	3.305	−143.35	1164.5	1021.2
30	0.08083	0.01747	3.607	−144.35	1164.9	1020.5
25	0.06406	0.01746	4.505	−146.84	1165.7	1018.9
20	0.05051	0.01745	5.655	−149.31	1166.5	1017.2
15	0.03963	0.01745	7.133	−151.75	1167.3	1015.6
10	0.03093	0.01744	9.043	−154.16	1168.1	1013.9
5	0.02402	0.01743	11.522	−156.56	1168.8	1012.2
0	0.01855	0.01742	14.761	−158.93	1169.5	1010.6
−5	0.01424	0.01742	19.019	−161.27	1170.2	1008.9
−10	0.01086	0.01741	24.657	−163.59	1170.8	1007.3
−15	0.00823	0.01740	32.169	−165.89	1171.5	1005.6
−20	0.00620	0.01740	42.238	−168.16	1172.1	1003.9
−25	0.00464	0.01739	55.782	−170.40	1172.7	1002.3
−30	0.00346	0.01738	74.046	−172.63	1173.2	1000.6
−35	0.00256	0.01737	98.890	−174.82	1173.8	998.9
−40	0.00187	0.01737	134.017	−177.00	1174.3	997.3

(Continued)

TABLE E.11.4 (Continued) ***Saturated Solid-Saturated Vapor Water (English Units)***

		Enthalpy Btu/lbm			Entropy Btu/lbm R		
Temp. F T	Press. lbf/in.^{2} P	Sat. Solid h_i	Evap. h_{ig}	Sat. Vapor h_g	Sat. Solid s_i	Evap. s_{ig}	Sat. Vapor s_g
32.02	0.08866	−143.34	1218.7	1075.4	−0.2916	2.4786	2.1869
32	0.08859	−143.35	1218.7	1075.4	−0.2917	2.4787	2.1870
30	0.08083	−144.35	1218.8	1074.5	−0.2938	2.4891	2.1953
25	0.06406	−146.84	1219.1	1072.3	−0.2990	2.5154	2.2164
20	0.05051	−149.31	1219.4	1070.1	−0.3042	2.5422	2.2380
15	0.03963	−151.75	1219.6	1067.9	−0.3093	2.5695	2.2601
10	0.03093	−154.16	1219.8	1065.7	−0.3145	2.5973	2.2827
5	0.02402	−156.56	1220.0	1063.5	−0.3197	2.6256	2.3059
0	0.01855	−158.93	1220.2	1061.2	−0.3248	2.6544	2.3296
−5	0.01424	−161.27	1220.3	1059.0	−0.3300	2.6839	2.3539
−10	0.01086	−163.59	1220.4	1056.8	−0.3351	2.7140	2.3788
−15	0.00823	−165.89	1220.5	1054.6	−0.3403	2.7447	2.4044
−20	0.00620	−168.16	1220.5	1052.4	−0.3455	2.7761	2.4307
−25	0.00464	−170.40	1220.6	1050.2	−0.3506	2.8081	2.4575
−30	0.00346	−172.63	1220.6	1048.0	−0.3557	2.8406	2.4849
−35	0.00256	−174.82	1220.6	1045.7	−0.3608	2.8737	2.5129
−40	0.00187	−177.00	1220.5	1043.5	−0.3659	2.9084	2.5425

TABLE E.12 ENG ***Thermodynamic Properties of Ammonia***
TABLE E.12.1 ENG ***Saturated Ammonia***

Temp.	Press.	Specific Volume, ft³/lbm			Internal Energy, Btu/lbm		
F	psia	Sat. Liquid	Evap.	Sat. Vapor	Sat. Liquid	Evap.	Sat. Vapor
T	P	v_f	v_{fg}	v_g	u_f	u_{fg}	u_g
-60	5.547	0.02277	44.7397	44.7625	-20.92	564.27	543.36
-50	7.663	0.02299	33.0702	33.0932	-10.51	556.84	546.33
-40	10.404	0.02322	24.8464	24.8696	-0.04	549.25	549.20
-30	13.898	0.02345	18.9490	18.9724	10.48	541.50	551.98
-28.0	14.696	0.02350	17.9833	18.0068	12.59	539.93	552.52
-20	18.289	0.02369	14.6510	14.6747	21.07	533.57	554.64
-10	23.737	0.02394	11.4714	11.4953	31.73	525.47	557.20
0	30.415	0.02420	9.0861	9.1103	42.46	517.18	559.64
10	38.508	0.02446	7.2734	7.2979	53.26	508.71	561.96
20	48.218	0.02474	5.8792	5.9039	64.12	500.04	564.16
30	59.756	0.02502	4.7945	4.8195	75.06	491.17	566.23
40	73.346	0.02532	3.9418	3.9671	86.07	482.09	568.15
50	89.226	0.02564	3.2647	3.2903	97.16	472.78	569.94
60	107.641	0.02597	2.7221	2.7481	108.33	463.24	571.56
70	128.849	0.02631	2.2835	2.3098	119.58	453.44	573.02
80	153.116	0.02668	1.9260	1.9526	130.92	443.37	574.30
90	180.721	0.02706	1.6323	1.6594	142.36	433.01	575.37
100	211.949	0.02747	1.3894	1.4168	153.89	422.34	576.23
110	247.098	0.02790	1.1870	1.2149	165.53	411.32	576.85
120	286.473	0.02836	1.0172	1.0456	177.28	399.92	577.20
130	330.392	0.02885	0.8740	0.9028	189.17	388.10	577.27
140	379.181	0.02938	0.7524	0.7818	201.20	375.82	577.02
150	433.181	0.02995	0.6485	0.6785	213.40	363.01	576.41
160	492.742	0.03057	0.5593	0.5899	225.80	349.61	575.41
170	558.231	0.03124	0.4822	0.5135	238.42	335.53	573.95
180	630.029	0.03199	0.4153	0.4472	251.33	320.66	571.99
190	708.538	0.03281	0.3567	0.3895	264.58	304.87	569.45
200	794.183	0.03375	0.3051	0.3388	278.24	287.96	566.20
210	887.424	0.03482	0.2592	0.2941	292.43	269.70	562.13
220	988.761	0.03608	0.2181	0.2542	307.28	249.72	557.00
230	1098.766	0.03759	0.1807	0.2183	323.03	227.47	550.50
240	1218.113	0.03950	0.1460	0.1855	340.05	202.02	542.06
250	1347.668	0.04206	0.1126	0.1547	359.03	171.57	530.60
260	1488.694	0.04599	0.0781	0.1241	381.74	131.74	513.48
270.1	1643.742	0.06816	0	0.0682	446.09	0	446.09

(Continued)

TABLE E.12.1 ENG (Continued) *Saturated Ammonia*							
Temp.	Press.	Enthalpy, Btu/lbm			Entropy, Btu/lbm R		
F T	psia P	Sat. Liquid h_f	Evap. h_{fg}	Sat. Vapor h_g	Sat. Liquid s_f	Evap. s_{fg}	Sat. Vapor s_g
-60	5.547	-20.89	610.19	589.30	-0.0510	1.5267	1.4758
-50	7.663	-10.48	603.73	593.26	-0.0252	1.4737	1.4485
-40	10.404	0	597.08	597.08	0	1.4227	1.4227
-30	13.898	10.54	590.23	600.77	0.0248	1.3737	1.3985
-28.0	14.696	12.65	588.84	601.49	0.0297	1.3641	1.3938
-20	18.289	21.15	583.15	604.31	0.0492	1.3263	1.3755
-10	23.737	31.84	575.85	607.69	0.0731	1.2806	1.3538
0	30.415	42.60	568.32	610.92	0.0967	1.2364	1.3331
10	38.508	53.43	560.54	613.97	0.1200	1.1935	1.3134
20	48.218	64.34	552.50	616.84	0.1429	1.1518	1.2947
30	59.756	75.33	544.18	619.52	0.1654	1.1113	1.2768
40	73.346	86.41	535.59	622.00	0.1877	1.0719	1.2596
50	89.226	97.58	526.68	624.26	0.2097	1.0334	1.2431
60	107.641	108.84	517.46	626.30	0.2314	0.9957	1.2271
70	128.849	120.21	507.89	628.09	0.2529	0.9589	1.2117
80	153.116	131.68	497.94	629.62	0.2741	0.9227	1.1968
90	180.721	143.26	487.60	630.86	0.2951	0.8871	1.1822
100	211.949	154.97	476.83	631.80	0.3159	0.8520	1.1679
110	247.098	166.80	465.59	632.40	0.3366	0.8173	1.1539
120	286.473	178.79	453.84	632.63	0.3571	0.7829	1.1400
130	330.392	190.93	441.54	632.47	0.3774	0.7488	1.1262
140	379.181	203.26	428.61	631.87	0.3977	0.7147	1.1125
150	433.181	215.80	415.00	630.80	0.4180	0.6807	1.0987
160	492.742	228.58	400.61	629.19	0.4382	0.6465	1.0847
170	558.231	241.65	385.35	627.00	0.4586	0.6120	1.0705
180	630.029	255.06	369.08	624.14	0.4790	0.5770	1.0560
190	708.538	268.88	351.63	620.51	0.4997	0.5412	1.0410
200	794.183	283.20	332.80	616.00	0.5208	0.5045	1.0253
210	887.424	298.14	312.27	610.42	0.5424	0.4663	1.0087
220	988.761	313.88	289.63	603.51	0.5647	0.4261	0.9909
230	1098.766	330.67	264.21	594.89	0.5882	0.3831	0.9713
240	1218.113	348.95	234.93	583.87	0.6132	0.3358	0.9490
250	1347.668	369.52	199.65	569.17	0.6410	0.2813	0.9224
260	1488.694	394.41	153.25	547.66	0.6743	0.2129	0.8872
270.1	1643.742	466.83	0	466.83	0.7718	0	0.7718

TABLE E.12.2 ENG *Superheated Ammonia*

Temp. F	v ft³/lbm	h Btu/lbm	s Btu/lbm R	v ft³/lbm	h Btu/lbm	s Btu/lbm R	v ft³/lbm	h Btu/lbm	s Btu/lbm R
	5 psia (-63.09)			10 psia (-41.33)			15 psia (-27.27)		
Sat.	49.32002	588.05	1.4846	25.80648	596.58	1.4261	17.66533	601.75	1.3921
-40	52.3487	599.56	1.5128	25.8962	597.27	1.4277	—	—	—
-20	54.9506	609.53	1.5360	27.2401	607.60	1.4518	17.9999	605.63	1.4010
0	57.5366	619.51	1.5582	28.5674	617.88	1.4746	18.9086	616.22	1.4245
20	60.1099	629.50	1.5795	29.8814	628.12	1.4964	19.8036	626.72	1.4469
40	62.6732	639.52	1.5999	31.1852	638.34	1.5173	20.6880	637.15	1.4682
60	65.2288	649.57	1.6197	32.4809	648.56	1.5374	21.5641	647.54	1.4886
80	67.7782	659.67	1.6387	33.7703	658.80	1.5567	22.4338	657.91	1.5082
100	70.3228	669.84	1.6572	35.0549	669.07	1.5754	23.2985	668.29	1.5271
120	72.8637	680.06	1.6752	36.3356	679.38	1.5935	24.1593	678.70	1.5453
140	75.4015	690.36	1.6926	37.6133	689.75	1.6111	25.0170	689.14	1.5630
160	77.9370	700.74	1.7097	38.8886	700.19	1.6282	25.8723	699.64	1.5803
180	80.4706	711.20	1.7263	40.1620	710.70	1.6449	26.7256	710.21	1.5970
200	83.0026	721.75	1.7425	41.4338	721.30	1.6612	27.5774	720.84	1.6134
220	85.5334	732.39	1.7584	42.7043	731.98	1.6771	28.4278	731.56	1.6294
240	88.0631	743.13	1.7740	43.9737	742.74	1.6928	29.2772	742.36	1.6451
260	90.5918	753.96	1.7892	45.2422	753.61	1.7081	30.1256	753.24	1.6604
280	93.1199	764.90	1.8042	46.5100	764.56	1.7231	30.9733	764.23	1.6755
	20 psia (-16.63)			25 psia (-7.95)			30 psia (-0.57)		
Sat.	13.49628	605.47	1.3680	10.95013	608.37	1.3494	9.22850	610.74	1.3342
0	14.0774	614.54	1.3881	11.1771	612.82	1.3592	9.2423	611.06	1.3349
20	14.7635	625.30	1.4111	11.7383	623.86	1.3827	9.7206	622.39	1.3591
40	15.4385	635.94	1.4328	12.2881	634.72	1.4049	10.1872	633.49	1.3817
60	16.1051	646.51	1.4535	12.8291	645.46	1.4260	10.6447	644.41	1.4032
80	16.7651	657.02	1.4734	13.3634	656.12	1.4461	11.0954	655.21	1.4236
100	17.4200	667.51	1.4925	13.8926	666.73	1.4654	11.5407	665.93	1.4431
120	18.0709	678.01	1.5109	14.4176	677.32	1.4840	11.9820	676.62	1.4618
140	18.7187	688.53	1.5287	14.9395	687.91	1.5020	12.4200	687.29	1.4799
160	19.3640	699.09	1.5461	15.4589	698.54	1.5194	12.8554	697.98	1.4975
180	20.0073	709.71	1.5629	15.9763	709.20	1.5363	13.2888	708.70	1.5145
200	20.6491	720.39	1.5794	16.4920	719.93	1.5528	13.7206	719.47	1.5311
220	21.2895	731.14	1.5954	17.0065	730.72	1.5689	14.1511	730.29	1.5472
240	21.9288	741.97	1.6111	17.5198	741.58	1.5847	14.5804	741.19	1.5630
260	22.5673	752.88	1.6265	18.0322	752.52	1.6001	15.0088	752.16	1.5785
280	23.2049	763.89	1.6416	18.5439	763.55	1.6152	15.4365	763.21	1.5936
300	23.8419	774.99	1.6564	19.0548	774.67	1.6301	15.8634	774.36	1.6085
320	24.4783	786.18	1.6709	19.5652	785.89	1.6446	16.2898	785.59	1.6231

(Continued)

TABLE E.12.2 ENG (Continued) *Superheated Ammonia*

Temp. F	v ft³/lbm	h Btu/lbm	s Btu/lbm R	v ft³/lbm	h Btu/lbm	s Btu/lbm R	v ft³/lbm	h Btu/lbm	s Btu/lbm R
	35 psia (5.89)			40 psia (11.66)			50 psia (21.66)		
Sat.	7.98414	612.73	1.3214	7.04135	614.45	1.3103	5.70491	617.30	1.2917
20	8.2786	620.90	1.3387	7.1964	619.39	1.3206	—	—	—
40	8.6860	632.23	1.3618	7.5596	630.96	1.3443	5.9814	628.37	1.3142
60	9.0841	643.34	1.3836	7.9132	642.26	1.3665	6.2731	640.07	1.3372
80	9.4751	654.29	1.4043	8.2596	653.37	1.3874	6.5573	651.49	1.3588
100	9.8606	665.14	1.4240	8.6004	664.33	1.4074	6.8356	662.70	1.3792
120	10.2420	675.92	1.4430	8.9370	675.21	1.4265	7.1096	673.79	1.3986
140	10.6202	686.67	1.4612	9.2702	686.04	1.4449	7.3800	684.78	1.4173
160	10.9957	697.42	1.4788	9.6008	696.86	1.4626	7.6478	695.73	1.4352
180	11.3692	708.19	1.4959	9.9294	707.69	1.4798	7.9135	706.67	1.4526
200	11.7410	719.01	1.5126	10.2562	718.54	1.4965	8.1775	717.61	1.4695
220	12.1115	729.87	1.5288	10.5817	729.44	1.5128	8.4400	728.59	1.4859
240	12.4808	740.80	1.5447	10.9061	740.40	1.5287	8.7014	739.62	1.5018
260	12.8493	751.80	1.5602	11.2296	751.43	1.5442	8.9619	750.70	1.5175
280	13.2169	762.88	1.5753	11.5522	762.54	1.5594	9.2216	761.86	1.5327
300	13.5838	774.04	1.5902	11.8741	773.72	1.5744	9.4805	773.09	1.5477
320	13.9502	785.29	1.6049	12.1955	785.00	1.5890	9.7389	784.40	1.5624
340	14.3160	796.64	1.6192	12.5163	796.36	1.6034	9.9967	795.80	1.5769
	60 psia (30.19)			70 psia (37.68)			80 psia (44.38)		
Sat.	4.80091	619.57	1.2764	4.14732	621.44	1.2635	3.65200	623.02	1.2523
40	4.9277	625.69	1.2888	4.1738	622.94	1.2665	—	—	—
60	5.1787	637.82	1.3126	4.3961	635.52	1.2912	3.8083	633.16	1.2721
80	5.4217	649.57	1.3348	4.6099	647.62	1.3140	4.0005	645.63	1.2956
100	5.6586	661.05	1.3557	4.8174	659.37	1.3354	4.1861	657.66	1.3175
120	5.8909	672.34	1.3755	5.0201	670.88	1.3556	4.3667	669.39	1.3381
140	6.1197	683.50	1.3944	5.2191	682.21	1.3749	4.5435	680.90	1.3577
160	6.3456	694.59	1.4126	5.4153	693.44	1.3933	4.7174	692.27	1.3763
180	6.5694	705.64	1.4302	5.6093	704.60	1.4110	4.8890	703.55	1.3942
200	6.7915	716.68	1.4472	5.8014	715.73	1.4281	5.0588	714.79	1.4115
220	7.0121	727.73	1.4637	5.9921	726.87	1.4448	5.2270	726.00	1.4283
240	7.2316	738.83	1.4798	6.1816	738.03	1.4610	5.3941	737.23	1.4446
260	7.4501	749.97	1.4955	6.3702	749.23	1.4767	5.5602	748.50	1.4604
280	7.6678	761.17	1.5108	6.5579	760.49	1.4922	5.7254	759.80	1.4759
300	7.8848	772.45	1.5259	6.7449	771.81	1.5073	5.8900	771.17	1.4911
320	8.1011	783.80	1.5406	6.9313	783.21	1.5221	6.0538	782.61	1.5059
340	8.3169	795.24	1.5551	7.1171	794.68	1.5366	6.2172	794.12	1.5205
360	8.5323	806.77	1.5693	7.3025	806.24	1.5509	6.3801	805.71	1.5348

(Continued)

TABLE E.12.2 ENG (Continued) *Superheated Ammonia*

Temp. F	v ft³/lbm	h Btu/lbm	s Btu/lbm R	v ft³/lbm	h Btu/lbm	s Btu/lbm R	v ft³/lbm	h Btu/lbm	s Btu/lbm R
	90 psia (50.45)			100 psia (56.02)			125 psia (68.28)		
Sat.	3.26324	624.36	1.2423	2.94969	625.52	1.2334	2.37866	627.80	1.2143
60	3.3503	630.74	1.2547	2.9831	628.25	1.2387	—	—	—
80	3.5260	643.59	1.2790	3.1459	641.51	1.2637	2.4597	636.11	1.2299
100	3.6947	655.92	1.3014	3.3013	654.16	1.2867	2.5917	649.59	1.2544
120	3.8583	667.88	1.3224	3.4513	666.36	1.3082	2.7177	662.44	1.2770
140	4.0179	679.58	1.3423	3.5972	678.24	1.3283	2.8392	674.83	1.2980
160	4.1745	691.10	1.3612	3.7400	689.91	1.3475	2.9574	686.90	1.3178
180	4.3287	702.50	1.3793	3.8804	701.44	1.3658	3.0730	698.74	1.3366
200	4.4811	713.83	1.3967	4.0188	712.87	1.3834	3.1865	710.44	1.3546
220	4.6319	725.13	1.4136	4.1558	724.25	1.4004	3.2985	722.04	1.3720
240	4.7816	736.43	1.4300	4.2915	735.63	1.4169	3.4091	733.59	1.3887
260	4.9302	747.75	1.4459	4.4261	747.01	1.4329	3.5187	745.13	1.4050
280	5.0779	759.11	1.4615	4.5599	758.42	1.4485	3.6274	756.68	1.4208
300	5.2250	770.53	1.4767	4.6930	769.88	1.4638	3.7353	768.27	1.4362
320	5.3714	782.01	1.4916	4.8254	781.40	1.4788	3.8426	779.89	1.4514
340	5.5173	793.56	1.5063	4.9573	792.99	1.4935	3.9493	791.58	1.4662
360	5.6626	805.18	1.5206	5.0887	804.66	1.5079	4.0555	803.33	1.4807
380	5.8076	816.90	1.5348	5.2196	816.40	1.5220	4.1613	815.15	1.4949
	150 psia (78.79)			175 psia (88.03)			200 psia (96.31)		
Sat.	1.99226	629.45	1.1986	1.71282	630.64	1.1850	1.50102	631.49	1.1731
80	1.9997	630.36	1.2003	—	—	—	—	—	—
100	2.1170	644.81	1.2265	1.7762	639.77	1.2015	1.5190	634.45	1.1785
120	2.2275	658.37	1.2504	1.8762	654.13	1.2267	1.6117	649.71	1.2052
140	2.3331	671.31	1.2723	1.9708	667.67	1.2497	1.6984	663.90	1.2293
160	2.4351	683.80	1.2928	2.0614	680.62	1.2710	1.7807	677.36	1.2514
180	2.5343	695.99	1.3122	2.1491	693.17	1.2909	1.8598	690.30	1.2719
200	2.6313	707.96	1.3306	2.2345	705.44	1.3098	1.9365	702.87	1.2913
220	2.7267	719.79	1.3483	2.3181	717.51	1.3278	2.0114	715.20	1.3097
240	2.8207	731.54	1.3653	2.4002	729.46	1.3451	2.0847	727.35	1.3273
260	2.9136	743.24	1.3818	2.4813	741.33	1.3619	2.1569	739.39	1.3443
280	3.0056	754.93	1.3978	2.5613	753.16	1.3781	2.2280	751.38	1.3607
300	3.0968	766.63	1.4134	2.6406	764.99	1.3939	2.2984	763.33	1.3767
320	3.1873	778.37	1.4287	2.7192	776.84	1.4092	2.3680	775.30	1.3922
340	3.2772	790.15	1.4436	2.7972	788.72	1.4243	2.4370	787.28	1.4074
360	3.3667	801.99	1.4582	2.8746	800.65	1.4390	2.5056	799.30	1.4223
380	3.4557	813.90	1.4726	2.9516	812.64	1.4535	2.5736	811.38	1.4368
400	3.5442	825.88	1.4867	3.0282	824.70	1.4677	2.6412	823.51	1.4511

(Continued)

TABLE E.12.2 ENG (Continued) ***Superheated Ammonia***

Temp. F	v ft³/lbm	h Btu/lbm	s Btu/lbm R	v ft³/lbm	h Btu/lbm	s Btu/lbm R	v ft³/lbm	h Btu/lbm	s Btu/lbm R
	250 psia (110.78)			300 psia (123.20)			350 psia (134.14)		
Sat.	1.20063	632.43	1.1528	0.99733	632.63	1.1356	0.85027	632.28	1.1205
120	1.2384	640.21	1.1663	—	—	—	—	—	—
140	1.3150	655.95	1.1930	1.0568	647.32	1.1605	0.8696	637.87	1.1299
160	1.3863	670.53	1.2170	1.1217	663.27	1.1866	0.9309	655.48	1.1588
180	1.4539	684.34	1.2389	1.1821	678.07	1.2101	0.9868	671.46	1.1842
200	1.5188	697.59	1.2593	1.2394	692.08	1.2317	1.0391	686.34	1.2071
220	1.5815	710.45	1.2785	1.2943	705.55	1.2518	1.0886	700.47	1.2282
240	1.6426	723.05	1.2968	1.3474	718.63	1.2708	1.1362	714.08	1.2479
260	1.7024	735.46	1.3142	1.3991	731.44	1.2888	1.1822	727.32	1.2666
280	1.7612	747.76	1.3311	1.4497	744.07	1.3062	1.2270	740.31	1.2844
300	1.8191	759.98	1.3474	1.4994	756.58	1.3228	1.2708	753.12	1.3015
320	1.8762	772.18	1.3633	1.5482	769.02	1.3390	1.3138	765.82	1.3180
340	1.9328	784.37	1.3787	1.5965	781.43	1.3547	1.3561	778.46	1.3340
360	1.9887	796.59	1.3938	1.6441	793.84	1.3701	1.3979	791.07	1.3496
380	2.0442	808.83	1.4085	1.6913	806.27	1.3850	1.4391	803.67	1.3648
400	2.0993	821.13	1.4230	1.7380	818.72	1.3997	1.4798	816.30	1.3796
420	2.1540	833.48	1.4372	1.7843	831.23	1.4141	1.5202	828.95	1.3942
440	2.2083	845.90	1.4512	1.8302	843.78	1.4282	1.5602	841.65	1.4085
	400 psia (143.97)			600 psia (175.93)			800 psia (200.65)		
Sat.	0.73876	631.50	1.1070	0.47311	625.39	1.0620	0.33575	615.67	1.0242
160	0.7860	647.06	1.1324	—	—	—	—	—	—
180	0.8392	664.44	1.1601	0.4834	630.48	1.0700	—	—	—
200	0.8880	680.32	1.1845	0.5287	652.67	1.1041	—	—	—
220	0.9338	695.21	1.2067	0.5680	671.78	1.1327	0.3769	642.62	1.0645
240	0.9773	709.40	1.2273	0.6035	689.03	1.1577	0.4115	665.08	1.0971
260	1.0192	723.10	1.2466	0.6366	705.06	1.1803	0.4419	684.62	1.1246
280	1.0597	736.47	1.2650	0.6678	720.26	1.2011	0.4694	702.36	1.1489
300	1.0992	749.60	1.2825	0.6976	734.88	1.2206	0.4951	718.93	1.1710
320	1.1379	762.58	1.2993	0.7264	749.09	1.2391	0.5193	734.69	1.1915
340	1.1758	775.45	1.3156	0.7542	763.02	1.2567	0.5425	749.89	1.2108
360	1.2131	788.27	1.3315	0.7814	776.75	1.2737	0.5648	764.68	1.2290
380	1.2499	801.06	1.3469	0.8079	790.34	1.2901	0.5864	779.19	1.2465
400	1.2862	813.85	1.3619	0.8340	803.86	1.3060	0.6074	793.50	1.2634
420	1.3221	826.66	1.3767	0.8595	817.32	1.3215	0.6279	807.68	1.2797
440	1.3576	839.51	1.3911	0.8847	830.76	1.3366	0.6480	821.76	1.2955
460	1.3928	852.39	1.4053	0.9095	844.21	1.3514	0.6677	835.80	1.3109
480	1.4277	865.34	1.4192	0.9340	857.67	1.3658	0.6871	849.80	1.3260

TABLE E.13 ENG *Thermodynamic Properties of R-12*
TABLE E.13.1 ENG *Saturated R-12*

Temp.	Press.	Specific Volume, ft³/lbm			Internal Energy, Btu/lbm		
F	psia	Sat. Liquid	Evap.	Sat. Vapor	Sat. Liquid	Evap.	Sat. Vapor
T	P	v_f	v_{fg}	v_g	u_f	u_{fg}	u_g
-130	0.412	0.00974	70.7205	70.7303	-18.61	76.18	57.57
-120	0.642	0.00982	46.7309	46.7407	-16.57	75.07	58.50
-110	0.970	0.00990	31.7667	31.7766	-14.52	73.96	59.44
-100	1.428	0.00998	22.1537	22.1636	-12.47	72.86	60.39
-90	2.051	0.01007	15.8109	15.8210	-10.41	71.76	61.35
-80	2.881	0.01016	11.5228	11.5329	-8.35	70.67	62.32
-70	3.965	0.01026	8.5584	8.5687	-6.28	69.57	63.29
-60	5.357	0.01036	6.4670	6.4774	-4.20	68.47	64.27
-50	7.117	0.01046	4.9637	4.9742	-2.11	67.37	65.25
-40	9.308	0.01056	3.8645	3.8750	-0.02	66.26	66.24
-30	11.999	0.01067	3.0478	3.0585	2.09	65.13	67.22
-21.6	14.696	0.01077	2.5207	2.5315	3.86	64.19	68.05
-20	15.267	0.01079	2.4322	2.4429	4.21	64.00	68.21
-10	19.189	0.01091	1.9618	1.9727	6.33	62.86	69.19
0	23.849	0.01103	1.5978	1.6089	8.47	61.70	70.17
10	29.335	0.01116	1.3129	1.3241	10.62	60.52	71.15
20	35.736	0.01130	1.0875	1.0988	12.79	59.33	72.12
30	43.148	0.01144	0.9074	0.9188	14.97	58.12	73.08
40	51.667	0.01159	0.7620	0.7736	17.16	56.88	74.04
50	61.394	0.01175	0.6436	0.6554	19.37	55.61	74.99
60	72.433	0.01191	0.5465	0.5584	21.61	54.32	75.92
70	84.888	0.01209	0.4661	0.4782	23.86	52.99	76.85
80	98.870	0.01228	0.3991	0.4114	26.14	51.62	77.76
90	114.491	0.01248	0.3428	0.3553	28.45	50.20	78.65
100	131.864	0.01269	0.2952	0.3079	30.79	48.72	79.52
110	151.110	0.01292	0.2548	0.2677	33.17	47.19	80.36
120	172.349	0.01317	0.2201	0.2333	35.59	45.58	81.17
130	195.708	0.01345	0.1902	0.2036	38.07	43.88	81.95
140	221.315	0.01375	0.1642	0.1780	40.60	42.08	82.68
150	249.307	0.01408	0.1416	0.1556	43.20	40.15	83.35
160	279.821	0.01445	0.1216	0.1360	45.88	38.08	83.96
170	313.003	0.01487	0.1039	0.1187	48.67	35.82	84.48
180	349.003	0.01536	0.0879	0.1033	51.57	33.32	84.89
190	387.978	0.01594	0.0735	0.0894	54.62	30.52	85.14
200	430.092	0.01666	0.0601	0.0767	57.88	27.29	85.17
210	475.517	0.01760	0.0472	0.0648	61.41	23.44	84.85
220	524.432	0.01899	0.0342	0.0531	65.40	18.48	83.88
230	577.027	0.02185	0.0176	0.0394	70.56	10.35	80.91
233.6	597.085	0.02870	0	0.0287	75.69	0	75.69

(Continued)

TABLE E.13.1 ENG (Continued) *Saturated R-12*

Temp.	Press.	Enthalpy, Btu/lbm			Entropy, Btu/lbm R		
F	psia	Sat. Liquid	Evap.	Sat. Vapor	Sat. Liquid	Evap.	Sat. Vapor
T	P	h_f	h_{fg}	h_g	s_f	s_{fg}	s_g
-130	0.412	-18.61	81.58	62.97	-0.0498	0.2474	0.1976
-120	0.642	-16.56	80.62	64.05	-0.0437	0.2373	0.1936
-110	0.970	-14.52	79.66	65.15	-0.0378	0.2278	0.1900
-100	1.428	-12.47	78.71	66.25	-0.0320	0.2188	0.1868
-90	2.051	-10.41	77.76	67.36	-0.0264	0.2103	0.1840
-80	2.881	-8.35	76.81	68.47	-0.0209	0.2023	0.1814
-70	3.965	-6.27	75.85	69.58	-0.0155	0.1946	0.1792
-60	5.357	-4.19	74.89	70.69	-0.0102	0.1874	0.1771
-50	7.117	-2.10	73.91	71.80	-0.0051	0.1804	0.1753
-40	9.308	0	72.91	72.91	0	0.1737	0.1737
-30	11.999	2.11	71.90	74.01	0.0050	0.1673	0.1723
-21.6	14.696	3.89	71.04	74.93	0.0090	0.1622	0.1712
-20	15.267	4.24	70.87	75.11	0.0098	0.1612	0.1710
-10	19.189	6.37	69.82	76.20	0.0146	0.1553	0.1699
0	23.849	8.52	68.75	77.27	0.0193	0.1496	0.1689
10	29.335	10.68	67.65	78.33	0.0240	0.1440	0.1680
20	35.736	12.86	66.52	79.38	0.0285	0.1387	0.1672
30	43.148	15.06	65.36	80.42	0.0330	0.1335	0.1665
40	51.667	17.27	64.16	81.44	0.0375	0.1284	0.1659
50	61.394	19.51	62.93	82.43	0.0418	0.1235	0.1653
60	72.433	21.77	61.64	83.41	0.0462	0.1186	0.1648
70	84.888	24.05	60.31	84.36	0.0505	0.1139	0.1643
80	98.870	26.36	58.92	85.28	0.0548	0.1092	0.1639
90	114.491	28.71	57.46	86.17	0.0590	0.1045	0.1635
100	131.864	31.10	55.93	87.03	0.0632	0.0999	0.1632
110	151.110	33.53	54.31	87.84	0.0675	0.0953	0.1628
120	172.349	36.01	52.60	88.61	0.0717	0.0907	0.1624
130	195.708	38.55	50.77	89.32	0.0759	0.0861	0.1620
140	221.315	41.16	48.81	89.97	0.0802	0.0814	0.1616
150	249.307	43.85	46.68	90.53	0.0845	0.0766	0.1611
160	279.821	46.63	44.37	91.01	0.0889	0.0716	0.1605
170	313.003	49.53	41.83	91.36	0.0934	0.0664	0.1598
180	349.003	52.56	39.00	91.56	0.0980	0.0610	0.1590
190	387.978	55.77	35.79	91.56	0.1028	0.0551	0.1579
200	430.092	59.20	32.08	91.28	0.1079	0.0486	0.1565
210	475.517	62.96	27.60	90.56	0.1133	0.0412	0.1545
220	524.432	67.25	21.79	89.04	0.1194	0.0321	0.1515
230	577.027	72.89	12.23	85.12	0.1274	0.0177	0.1451
233.6	597.085	78.86	0	78.86	0.1359	0	0.1359

TABLE E.13.2 ENG *Superheated R-12*

Temp. F	v ft³/lbm	h Btu/lbm	s Btu/lbm R	v ft³/lbm	h Btu/lbm	s Btu/lbm R	v ft³/lbm	h Btu/lbm	s Btu/lbm R
	5 psia (-62.35)			10 psia (-37.23)			15 psia (-20.75)		
Sat.	6.90689	70.43	0.1776	3.62459	73.22	0.1733	2.48352	75.03	0.1711
-20	7.6938	75.91	0.1907	3.7906	75.53	0.1787	2.4885	75.13	0.1713
0	8.0612	78.58	0.1966	3.9809	78.25	0.1847	2.6201	77.90	0.1775
20	8.4265	81.31	0.2024	4.1691	81.01	0.1906	2.7494	80.71	0.1835
40	8.7903	84.09	0.2081	4.3556	83.83	0.1964	2.8770	83.56	0.1893
60	9.1528	86.92	0.2137	4.5408	86.69	0.2020	3.0031	86.45	0.1950
80	9.5142	89.81	0.2191	4.7248	89.60	0.2075	3.1281	89.38	0.2005
100	9.8747	92.74	0.2245	4.9079	92.55	0.2128	3.2521	92.36	0.2059
120	10.2344	95.72	0.2297	5.0903	95.55	0.2181	3.3754	95.37	0.2112
140	10.5936	98.74	0.2348	5.2720	98.59	0.2232	3.4981	98.43	0.2164
160	10.9523	101.81	0.2398	5.4533	101.67	0.2283	3.6202	101.53	0.2215
180	11.3105	104.92	0.2448	5.6341	104.79	0.2333	3.7419	104.66	0.2265
200	11.6684	108.08	0.2496	5.8145	107.96	0.2381	3.8632	107.83	0.2313
220	12.0260	111.27	0.2544	5.9946	111.16	0.2429	3.9841	111.05	0.2361
240	12.3833	114.50	0.2591	6.1745	114.40	0.2476	4.1049	114.29	0.2408
260	12.7404	117.77	0.2637	6.3541	117.67	0.2522	4.2254	117.57	0.2455
280	13.0973	121.07	0.2682	6.5336	120.98	0.2568	4.3457	120.89	0.2500
	30 psia (11.11)			45 psia (32.29)			60 psia (48.64)		
Sat.	1.29640	78.45	0.1679	0.88274	80.65	0.1663	0.67006	82.30	0.1654
40	1.3969	82.73	0.1767	0.9019	81.85	0.1687	—	—	—
60	1.4644	85.72	0.1826	0.9502	84.94	0.1748	0.6921	84.13	0.1689
80	1.5306	88.73	0.1883	0.9971	88.05	0.1807	0.7296	87.33	0.1750
100	1.5957	91.77	0.1938	1.0429	91.16	0.1863	0.7659	90.53	0.1808
120	1.6600	94.84	0.1992	1.0877	94.30	0.1918	0.8011	93.73	0.1864
140	1.7237	97.95	0.2044	1.1318	97.45	0.1972	0.8355	96.95	0.1919
160	1.7868	101.09	0.2096	1.1753	100.64	0.2024	0.8693	100.18	0.1972
180	1.8494	104.26	0.2146	1.2183	103.85	0.2075	0.9025	103.43	0.2023
200	1.9116	107.46	0.2196	1.2609	107.09	0.2125	0.9353	106.70	0.2074
220	1.9735	110.70	0.2244	1.3031	110.35	0.2174	0.9678	110.00	0.2123
240	2.0351	113.97	0.2291	1.3450	113.65	0.2222	0.9999	113.32	0.2171
260	2.0965	117.28	0.2338	1.3867	116.97	0.2268	1.0318	116.67	0.2218
280	2.1576	120.61	0.2384	1.4282	120.33	0.2314	1.0634	120.04	0.2264
300	2.2186	123.97	0.2429	1.4695	123.71	0.2360	1.0949	123.44	0.2310
320	2.2795	127.37	0.2473	1.5107	127.12	0.2404	1.1262	126.86	0.2354
340	2.3402	130.79	0.2516	1.5517	130.55	0.2447	1.1574	130.31	0.2398

(Continued)

TABLE E.13.2 ENG (Continued) *Superheated R-12*

Temp. F	v ft³/lbm	h Btu/lbm	s Btu/lbm R	v ft³/lbm	h Btu/lbm	s Btu/lbm R	v ft³/lbm	h Btu/lbm	s Btu/lbm R
	90 psia (73.79)			120 psia (93.29)			150 psia (109.45)		
Sat.	0.45144	84.71	0.1642	0.33886	86.46	0.1634	0.26974	87.80	0.1628
100	0.4875	89.17	0.1723	0.3466	87.68	0.1656	—	—	—
120	0.5135	92.54	0.1782	0.3684	91.24	0.1718	0.2801	89.80	0.1663
140	0.5385	95.88	0.1839	0.3890	94.74	0.1778	0.2984	93.50	0.1726
160	0.5627	99.22	0.1894	0.4087	98.20	0.1835	0.3157	97.11	0.1785
180	0.5863	102.56	0.1947	0.4277	101.64	0.1889	0.3320	100.67	0.1841
200	0.6094	105.91	0.1998	0.4461	105.08	0.1942	0.3477	104.21	0.1896
220	0.6321	109.27	0.2049	0.4640	108.51	0.1993	0.3629	107.72	0.1948
240	0.6545	112.64	0.2098	0.4816	111.95	0.2043	0.3776	111.23	0.1999
260	0.6766	116.04	0.2145	0.4989	115.40	0.2092	0.3920	114.73	0.2049
280	0.6985	119.46	0.2192	0.5159	118.86	0.2139	0.4062	118.24	0.2097
300	0.7202	122.89	0.2238	0.5327	122.33	0.2186	0.4201	121.76	0.2144
320	0.7417	126.35	0.2283	0.5493	125.82	0.2231	0.4338	125.29	0.2189
340	0.7630	129.83	0.2327	0.5657	129.33	0.2275	0.4473	128.83	0.2234
360	0.7842	133.33	0.2370	0.5821	132.86	0.2319	0.4607	132.39	0.2278
380	0.8053	136.85	0.2413	0.5983	136.41	0.2362	0.4740	135.96	0.2321
400	0.8264	140.39	0.2454	0.6144	139.98	0.2404	0.4872	139.55	0.2364
	200 psia (131.74)			300 psia (166.18)			600 psia		
Sat.	0.19891	89.44	0.1619	0.12510	91.24	0.1601	—	—	—
180	0.2354	98.92	0.1774	0.1348	94.56	0.1654	—	—	—
200	0.2486	102.65	0.1831	0.1470	98.98	0.1722	—	—	—
220	0.2612	106.33	0.1886	0.1577	103.14	0.1784	—	—	—
240	0.2732	109.96	0.1939	0.1676	107.14	0.1842	0.0475	91.02	0.1534
260	0.2849	113.58	0.1990	0.1769	111.04	0.1897	0.0619	99.74	0.1657
280	0.2962	117.18	0.2039	0.1856	114.88	0.1949	0.0709	105.64	0.1737
300	0.3073	120.78	0.2087	0.1940	118.67	0.2000	0.0781	110.73	0.1805
320	0.3181	124.37	0.2134	0.2021	122.43	0.2049	0.0843	115.42	0.1866
340	0.3288	127.97	0.2179	0.2100	126.17	0.2096	0.0900	119.87	0.1923
360	0.3393	131.58	0.2224	0.2177	129.90	0.2142	0.0953	124.17	0.1976
380	0.3497	135.20	0.2268	0.2252	133.62	0.2187	0.1003	128.35	0.2026
400	0.3599	138.83	0.2310	0.2326	137.35	0.2231	0.1050	132.47	0.2075
420	0.3701	142.48	0.2352	0.2399	141.07	0.2274	0.1095	136.52	0.2121
440	0.3802	146.13	0.2393	0.2470	144.80	0.2316	0.1139	140.54	0.2166
460	0.3901	149.80	0.2434	0.2541	148.54	0.2357	0.1182	144.53	0.2210
480	0.4001	153.49	0.2473	0.2611	152.28	0.2397	0.1223	148.49	0.2253

TABLE E 14 ENG ***Thermodynamic Properties of R-22***
TABLE E.14.1 ENG ***Saturated R-22***

Temp.	Press.	Specific Volume, ft³/lbm			Internal Energy, Btu/lbm		
F	psia	Sat. Liquid	Evap.	Sat. Vapor	Sat. Liquid	Evap.	Sat. Vapor
T	P	v_f	v_{fg}	v_g	u_f	u_{fg}	u_g
-100	2.398	0.01066	18.4219	18.4326	-14.57	99.76	85.19
-90	3.423	0.01077	13.2243	13.2351	-12.22	98.38	86.16
-80	4.782	0.01088	9.6840	9.6949	-9.85	96.98	87.13
-70	6.552	0.01099	7.2208	7.2318	-7.44	95.54	88.10
-60	8.818	0.01111	5.4733	5.4844	-5.01	94.07	89.06
-50	11.674	0.01124	4.2111	4.2224	-2.54	92.56	90.02
-41.4	14.696	0.01135	3.3944	3.4058	-0.37	91.22	90.85
-40	15.222	0.01136	3.2844	3.2957	-0.03	91.01	90.97
-30	19.573	0.01150	2.5934	2.6049	2.51	89.41	91.91
-20	24.845	0.01163	2.0709	2.0826	5.08	87.76	92.84
-10	31.162	0.01178	1.6707	1.6825	7.68	86.07	93.75
0	38.657	0.01193	1.3603	1.3723	10.32	84.33	94.65
10	47.464	0.01209	1.1170	1.1290	13.00	82.53	95.53
20	57.727	0.01226	0.9241	0.9363	15.71	80.67	96.38
30	69.591	0.01243	0.7697	0.7821	18.45	78.76	97.21
40	83.206	0.01262	0.6449	0.6575	21.23	76.79	98.02
50	98.727	0.01282	0.5432	0.5561	24.04	74.75	98.79
60	116.312	0.01303	0.4597	0.4727	26.89	72.65	99.54
70	136.123	0.01325	0.3905	0.4037	29.78	70.46	100.24
80	158.326	0.01349	0.3327	0.3462	32.71	68.19	100.91
90	183.094	0.01375	0.2841	0.2979	35.69	65.83	101.53
100	210.604	0.01404	0.2430	0.2570	38.72	63.37	102.09
110	241.042	0.01435	0.2079	0.2222	41.81	60.78	102.59
120	274.604	0.01469	0.1777	0.1924	44.96	58.05	103.01
130	311.496	0.01508	0.1515	0.1666	48.19	55.14	103.33
140	351.944	0.01552	0.1287	0.1442	51.52	52.02	103.54
150	396.194	0.01603	0.1085	0.1245	54.97	48.63	103.60
160	444.525	0.01663	0.0904	0.1070	58.58	44.88	103.46
170	497.259	0.01737	0.0739	0.0913	62.42	40.62	103.04
180	554.783	0.01833	0.0585	0.0768	66.62	35.57	102.18
190	617.590	0.01973	0.0431	0.0628	71.46	29.10	100.55
200	686.356	0.02244	0.0250	0.0474	78.01	18.81	96.83
204.8	720.698	0.03053	0	0.0305	87.30	0	87.30

(Continued)

TABLE E.14.1 ENG (Continued) *Saturated R-22*

Temp.	Press.	Enthalpy, Btu/lbm			Entropy, Btu/lbm R		
F T	psia P	Sat. Liquid h_f	Evap. h_{fg}	Sat. Vapor h_g	Sat. Liquid s_f	Evap. s_{fg}	Sat. Vapor s_g
-100	2.398	-14.56	107.94	93.37	-0.0373	0.3001	0.2627
-90	3.423	-12.22	106.76	94.54	-0.0309	0.2888	0.2579
-80	4.782	-9.84	105.55	95.71	-0.0246	0.2780	0.2534
-70	6.552	-7.43	104.30	96.87	-0.0183	0.2676	0.2493
-60	8.818	-4.99	103.00	98.01	-0.0121	0.2577	0.2456
-50	11.674	-2.51	101.66	99.14	-0.0060	0.2481	0.2421
-41.4	14.696	-0.34	100.45	100.11	-0.0008	0.2401	0.2393
-40	15.222	0	100.26	100.26	0	0.2389	0.2389
-30	19.573	2.55	98.80	101.35	0.0060	0.2299	0.2359
-20	24.845	5.13	97.28	102.42	0.0119	0.2213	0.2332
-10	31.162	7.75	95.70	103.46	0.0178	0.2128	0.2306
0	38.657	10.41	94.06	104.47	0.0236	0.2046	0.2282
10	47.464	13.10	92.34	105.44	0.0293	0.1966	0.2259
20	57.727	15.84	90.55	106.38	0.0350	0.1888	0.2238
30	69.591	18.61	88.67	107.28	0.0407	0.1811	0.2218
40	83.206	21.42	86.72	108.14	0.0463	0.1735	0.2199
50	98.727	24.27	84.68	108.95	0.0519	0.1661	0.2180
60	116.312	27.17	82.54	109.71	0.0574	0.1588	0.2163
70	136.123	30.12	80.30	110.41	0.0630	0.1516	0.2146
80	158.326	33.11	77.94	111.05	0.0685	0.1444	0.2129
90	183.094	36.16	75.46	111.62	0.0739	0.1373	0.2112
100	210.604	39.27	72.84	112.11	0.0794	0.1301	0.2096
110	241.042	42.45	70.05	112.50	0.0849	0.1230	0.2079
120	274.604	45.71	67.08	112.78	0.0904	0.1157	0.2061
130	311.496	49.06	63.88	112.94	0.0960	0.1083	0.2043
140	351.944	52.53	60.40	112.93	0.1016	0.1007	0.2023
150	396.194	56.14	56.58	112.73	0.1074	0.0928	0.2002
160	444.525	59.95	52.32	112.26	0.1133	0.0844	0.1978
170	497.259	64.02	47.42	111.44	0.1196	0.0753	0.1949
180	554.783	68.50	41.57	110.07	0.1263	0.0650	0.1913
190	617.590	73.71	34.02	107.73	0.1341	0.0524	0.1865
200	686.356	80.86	21.99	102.85	0.1446	0.0333	0.1779
204.8	720.698	91.38	0	91.38	0.1602	0	0.1602

TABLE E.14.2 ENG *Superheated R-22*

Temp. F	v ft³/lbm	h Btu/lbm	s Btu/lbm R	v ft³/lbm	h Btu/lbm	s Btu/lbm R	v ft³/lbm	h Btu/lbm	s Btu/lbm R
	5 psia (-78.62)			10 psia (-55.59)			15 psia (-40.57)		
Sat.	9.30117	95.87	0.2528	4.87779	98.52	0.2440	3.34121	100.19	0.2391
-40	10.2935	101.09	0.2659	5.0838	100.69	0.2493	3.3463	100.28	0.2393
-20	10.8034	103.89	0.2724	5.3460	103.53	0.2559	3.5261	103.16	0.2460
0	11.3114	106.73	0.2787	5.6060	106.41	0.2623	3.7037	106.09	0.2525
20	11.8177	109.64	0.2849	5.8643	109.36	0.2686	3.8794	109.06	0.2588
40	12.3227	112.61	0.2910	6.1212	112.35	0.2747	4.0537	112.09	0.2650
60	12.8265	115.64	0.2969	6.3769	115.40	0.2807	4.2268	115.17	0.2710
80	13.3293	118.72	0.3027	6.6316	118.51	0.2865	4.3989	118.30	0.2769
100	13.8313	121.87	0.3085	6.8855	121.67	0.2923	4.5701	121.48	0.2827
120	14.3327	125.07	0.3141	7.1387	124.89	0.2979	4.7406	124.72	0.2884
140	14.8335	128.33	0.3196	7.3913	128.17	0.3035	4.9105	128.00	0.2940
160	15.3337	131.64	0.3250	7.6434	131.49	0.3089	5.0799	131.34	0.2995
180	15.8336	135.01	0.3304	7.8951	134.87	0.3143	5.2489	134.74	0.3049
200	16.3331	138.44	0.3357	8.1464	138.31	0.3196	5.4174	138.18	0.3102
220	16.8323	141.92	0.3409	8.3974	141.80	0.3248	5.5857	141.68	0.3154
240	17.3312	145.45	0.3460	8.6481	145.34	0.3300	5.7537	145.23	0.3205
260	17.8298	149.03	0.3510	8.8986	148.93	0.3350	5.9215	148.83	0.3256
280	18.3283	152.67	0.3560	9.1489	152.57	0.3400	6.0890	152.48	0.3306
	20 psia (-29.12)			25 psia (-19.73)			30 psia (-11.71)		
Sat.	2.55270	101.44	0.2357	2.07040	102.44	0.2331	1.74388	103.28	0.2310
0	2.7521	105.76	0.2454	2.1808	105.42	0.2397	1.7997	105.08	0.2350
20	2.8867	108.77	0.2518	2.2908	108.47	0.2462	1.8933	108.16	0.2415
40	3.0198	111.83	0.2580	2.3992	111.56	0.2525	1.9853	111.29	0.2479
60	3.1516	114.93	0.2641	2.5063	114.69	0.2586	2.0760	114.45	0.2541
80	3.2823	118.08	0.2701	2.6123	117.86	0.2646	2.1655	117.64	0.2602
100	3.4122	121.28	0.2759	2.7175	121.09	0.2705	2.2542	120.89	0.2661
120	3.5414	124.54	0.2816	2.8219	124.36	0.2762	2.3421	124.18	0.2718
140	3.6700	127.84	0.2872	2.9257	127.68	0.2819	2.4294	127.51	0.2775
160	3.7981	131.19	0.2927	3.0289	131.04	0.2874	2.5162	130.89	0.2830
180	3.9257	134.60	0.2981	3.1318	134.46	0.2928	2.6025	134.32	0.2885
200	4.0529	138.05	0.3034	3.2342	137.93	0.2982	2.6884	137.80	0.2938
220	4.1799	141.56	0.3087	3.3363	141.44	0.3034	2.7739	141.32	0.2991
240	4.3065	145.12	0.3138	3.4382	145.01	0.3086	2.8592	144.90	0.3043
260	4.4329	148.73	0.3189	3.5397	148.62	0.3137	2.9443	148.52	0.3094
280	4.5591	152.38	0.3239	3.6411	152.28	0.3187	3.0291	152.19	0.3144
300	4.6851	156.09	0.3288	3.7423	155.99	0.3236	3.1138	155.90	0.3194
320	4.8109	159.84	0.3337	3.8434	159.75	0.3285	3.1983	159.67	0.3243

(Continued)

TABLE E.14.2 ENG (Continued) ***Superheated R-22***

Temp. F	v ft³/lbm	h Btu/lbm	s Btu/lbm R	v ft³/lbm	h Btu/lbm	s Btu/lbm R	v ft³/lbm	h Btu/lbm	s Btu/lbm R
	40 psia (1.63)			50 psia (12.61)			60 psia (22.03)		
Sat.	1.32853	104.63	0.2278	1.07436	105.69	0.2253	0.90223	106.57	0.2234
20	1.3959	107.54	0.2340	1.0968	106.90	0.2279	—	—	—
40	1.4676	110.73	0.2405	1.1564	110.16	0.2346	0.9486	109.58	0.2295
60	1.5378	113.95	0.2468	1.2145	113.44	0.2410	0.9987	112.92	0.2361
80	1.6068	117.20	0.2530	1.2714	116.74	0.2472	1.0475	116.28	0.2424
100	1.6749	120.48	0.2589	1.3272	120.08	0.2533	1.0952	119.66	0.2486
120	1.7423	123.81	0.2648	1.3822	123.44	0.2592	1.1420	123.06	0.2545
140	1.8090	127.18	0.2705	1.4366	126.84	0.2649	1.1882	126.50	0.2604
160	1.8751	130.59	0.2761	1.4903	130.28	0.2706	1.2338	129.96	0.2660
180	1.9407	134.04	0.2816	1.5436	133.75	0.2761	1.2788	133.47	0.2716
200	2.0060	137.54	0.2869	1.5965	137.28	0.2815	1.3235	137.01	0.2771
220	2.0709	141.08	0.2922	1.6491	140.84	0.2869	1.3678	140.60	0.2824
240	2.1356	144.67	0.2974	1.7013	144.45	0.2921	1.4118	144.22	0.2877
260	2.2000	148.31	0.3026	1.7533	148.10	0.2972	1.4556	147.89	0.2928
280	2.2641	151.99	0.3076	1.8051	151.80	0.3023	1.4991	151.60	0.2979
300	2.3281	155.72	0.3126	1.8567	155.54	0.3073	1.5424	155.35	0.3029
320	2.3919	159.49	0.3175	1.9081	159.32	0.3122	1.5856	159.15	0.3079
340	2.4556	163.31	0.3223	1.9594	163.15	0.3171	1.6286	162.99	0.3127
	70 psia (30.32)			80 psia (37.76)			100 psia (50.77)		
Sat.	0.77766	107.31	0.2217	0.68319	107.95	0.2203	0.54908	109.01	0.2179
40	0.7998	108.97	0.2251	0.6878	108.35	0.2211	—	—	—
60	0.8443	112.39	0.2318	0.7282	111.84	0.2279	0.5650	110.70	0.2212
80	0.8874	115.81	0.2382	0.7671	115.32	0.2345	0.5982	114.32	0.2280
100	0.9293	119.23	0.2445	0.8048	118.80	0.2408	0.6300	117.91	0.2345
120	0.9704	122.68	0.2505	0.8415	122.29	0.2470	0.6608	121.49	0.2408
140	1.0107	126.15	0.2564	0.8775	125.80	0.2529	0.6908	125.08	0.2469
160	1.0504	129.65	0.2621	0.9129	129.33	0.2587	0.7201	128.67	0.2528
180	1.0896	133.18	0.2677	0.9477	132.89	0.2643	0.7488	132.29	0.2586
200	1.1284	136.75	0.2732	0.9821	136.48	0.2699	0.7771	135.93	0.2642
220	1.1669	140.35	0.2786	1.0161	140.10	0.2753	0.8050	139.60	0.2696
240	1.2050	143.99	0.2839	1.0498	143.76	0.2806	0.8326	143.30	0.2750
260	1.2428	147.68	0.2891	1.0833	147.46	0.2858	0.8599	147.03	0.2803
280	1.2805	151.40	0.2942	1.1165	151.20	0.2909	0.8869	150.80	0.2854
300	1.3179	155.17	0.2992	1.1495	154.98	0.2960	0.9137	154.61	0.2905
320	1.3552	158.97	0.3042	1.1823	158.80	0.3009	0.9404	158.45	0.2955
340	1.3923	162.82	0.3090	1.2150	162.66	0.3058	0.9669	162.33	0.3004
360	1.4292	166.72	0.3138	1.2476	166.56	0.3106	0.9932	166.25	0.3052

(Continued)

TABLE E.14.2 ENG (Continued) *Superheated R-22*

Temp. F	v ft³/lbm	h Btu/lbm	s Btu/lbm R	v ft³/lbm	h Btu/lbm	s Btu/lbm R	v ft³/lbm	h Btu/lbm	s Btu/lbm R
	125 psia (64.53)			150 psia (76.38)			175 psia (86.85)		
Sat.	0.43988	110.04	0.2155	0.36587	110.83	0.2135	0.31224	111.45	0.2117
80	0.4622	112.99	0.2210	0.3705	111.56	0.2148	—	—	—
100	0.4896	116.74	0.2279	0.3953	115.50	0.2220	0.3273	114.18	0.2167
120	0.5158	120.46	0.2344	0.4187	119.37	0.2288	0.3488	118.22	0.2238
140	0.5411	124.15	0.2406	0.4410	123.18	0.2353	0.3691	122.17	0.2305
160	0.5656	127.84	0.2467	0.4624	126.97	0.2415	0.3884	126.07	0.2369
180	0.5896	131.53	0.2526	0.4832	130.74	0.2475	0.4070	129.94	0.2430
200	0.6130	135.23	0.2583	0.5034	134.52	0.2533	0.4250	133.79	0.2489
220	0.6360	138.96	0.2638	0.5232	138.31	0.2589	0.4426	137.64	0.2547
240	0.6587	142.71	0.2693	0.5426	142.11	0.2645	0.4597	141.49	0.2603
260	0.6810	146.48	0.2746	0.5618	145.93	0.2698	0.4765	145.36	0.2657
280	0.7032	150.29	0.2798	0.5806	149.78	0.2751	0.4930	149.25	0.2711
300	0.7251	154.13	0.2849	0.5993	153.65	0.2803	0.5094	153.16	0.2763
320	0.7468	158.00	0.2900	0.6177	157.55	0.2854	0.5255	157.10	0.2814
340	0.7683	161.91	0.2949	0.6360	161.49	0.2903	0.5414	161.06	0.2864
360	0.7898	165.85	0.2998	0.6541	165.46	0.2952	0.5572	165.06	0.2913
380	0.8111	169.83	0.3046	0.6721	169.46	0.3001	0.5728	169.08	0.2962
400	0.8322	173.85	0.3093	0.6900	173.50	0.3048	0.5884	173.14	0.3010
	200 psia (96.27)			250 psia (112.76)			300 psia (126.98)		
Sat.	0.27150	111.93	0.2102	0.21352	112.59	0.2074	0.17400	112.90	0.2049
100	0.2755	112.75	0.2116	—	—	—	—	—	—
120	0.2959	117.00	0.2191	0.2204	114.30	0.2104	—	—	—
140	0.3149	121.11	0.2261	0.2379	118.82	0.2180	0.1852	116.20	0.2104
160	0.3327	125.13	0.2327	0.2540	123.14	0.2251	0.2006	120.94	0.2182
180	0.3497	129.10	0.2390	0.2690	127.34	0.2318	0.2146	125.44	0.2253
200	0.3661	133.03	0.2450	0.2833	131.46	0.2381	0.2276	129.78	0.2320
220	0.3820	136.95	0.2509	0.2969	135.53	0.2442	0.2398	134.04	0.2384
240	0.3974	140.87	0.2566	0.3100	139.58	0.2501	0.2515	138.22	0.2445
260	0.4125	144.79	0.2621	0.3228	143.60	0.2558	0.2628	142.38	0.2503
280	0.4273	148.72	0.2675	0.3352	147.63	0.2613	0.2737	146.50	0.2560
300	0.4419	152.67	0.2727	0.3474	151.66	0.2667	0.2843	150.62	0.2615
320	0.4563	156.64	0.2779	0.3593	155.70	0.2719	0.2947	154.74	0.2668
340	0.4705	160.63	0.2830	0.3711	159.76	0.2770	0.3048	158.86	0.2720
360	0.4845	164.65	0.2879	0.3827	163.83	0.2821	0.3148	163.00	0.2771
380	0.4984	168.70	0.2928	0.3942	167.93	0.2870	0.3247	167.15	0.2821
400	0.5122	172.78	0.2976	0.4055	172.05	0.2919	0.3344	171.31	0.2870
420	0.5259	176.89	0.3023	0.4167	176.20	0.2966	0.3440	175.51	0.2919

(Continued)

TABLE E.14.2 ENG (Continued) *Superheated R-22*

Temp. F	v ft³/lbm	h Btu/lbm	s Btu/lbm R	v ft³/lbm	h Btu/lbm	s Btu/lbm R	v ft³/lbm	h Btu/lbm	s Btu/lbm R
	400 psia (150.82)			500 psia (170.50)			600 psia (187.29)		
Sat.	0.12297	112.70	0.2000	0.09053	111.38	0.1947	0.06663	108.51	0.1880
160	0.1305	115.52	0.2046	—	—	—	—	—	—
180	0.1446	121.01	0.2133	0.0987	115.06	0.2005	—	—	—
200	0.1567	126.02	0.2210	0.1122	121.43	0.2103	0.0791	115.12	0.1981
220	0.1677	130.76	0.2281	0.1232	126.95	0.2186	0.0919	122.32	0.2089
240	0.1778	135.31	0.2347	0.1328	132.05	0.2260	0.1020	128.29	0.2175
260	0.1874	139.76	0.2410	0.1416	136.89	0.2328	0.1106	133.70	0.2252
280	0.1965	144.13	0.2470	0.1498	141.57	0.2392	0.1184	138.79	0.2321
300	0.2052	148.45	0.2527	0.1576	146.14	0.2453	0.1256	143.67	0.2386
320	0.2137	152.74	0.2583	0.1650	150.63	0.2512	0.1323	148.41	0.2448
340	0.2219	157.01	0.2637	0.1721	155.08	0.2568	0.1387	153.05	0.2507
360	0.2299	161.27	0.2690	0.1789	159.49	0.2622	0.1449	157.63	0.2563
380	0.2378	165.54	0.2741	0.1856	163.88	0.2675	0.1508	162.16	0.2618
400	0.2455	169.81	0.2791	0.1921	168.26	0.2727	0.1566	166.66	0.2671
420	0.2530	174.09	0.2841	0.1985	172.63	0.2777	0.1622	171.15	0.2722
440	0.2605	178.38	0.2889	0.2048	177.01	0.2826	0.1677	175.62	0.2773
460	0.2679	182.69	0.2936	0.2109	181.40	0.2875	0.1730	180.09	0.2822
480	0.2752	187.02	0.2983	0.2170	185.80	0.2922	0.1783	184.57	0.2870
	700 psia (201.88)			800 psia			900 psia		
Sat.	0.04365	101.02	0.1750	—	—	—	—	—	—
220	0.0671	116.05	0.1975	0.0422	104.39	0.1788	0.0305	95.43	0.1648
240	0.0788	123.79	0.2087	0.0600	118.02	0.1986	0.0434	109.91	0.1857
260	0.0879	130.09	0.2176	0.0702	125.88	0.2097	0.0557	120.85	0.2011
280	0.0956	135.73	0.2253	0.0782	132.34	0.2186	0.0643	128.53	0.2117
300	0.1025	141.01	0.2324	0.0851	138.13	0.2263	0.0713	135.01	0.2203
320	0.1089	146.05	0.2389	0.0913	143.54	0.2333	0.0774	140.87	0.2279
340	0.1149	150.93	0.2451	0.0970	148.70	0.2399	0.0830	146.36	0.2349
360	0.1206	155.69	0.2510	0.1023	153.69	0.2460	0.0881	151.60	0.2413
380	0.1260	160.39	0.2566	0.1074	158.56	0.2519	0.0929	156.67	0.2475
400	0.1312	165.02	0.2621	0.1122	163.34	0.2575	0.0975	161.62	0.2533
420	0.1363	169.62	0.2674	0.1169	168.07	0.2630	0.1018	166.49	0.2589
440	0.1412	174.20	0.2725	0.1214	172.76	0.2682	0.1060	171.29	0.2643
460	0.1460	178.76	0.2775	0.1258	177.41	0.2734	0.1101	176.05	0.2695
480	0.1507	183.31	0.2824	0.1301	182.05	0.2783	0.1141	180.77	0.2746
500	0.1553	187.87	0.2872	0.1343	186.67	0.2832	0.1179	185.47	0.2795
520	0.1599	192.42	0.2919	0.1384	191.30	0.2880	0.1217	190.17	0.2844

TABLE E.15 ENG ***Thermodynamic Properties of R-134a***
TABLE E.15.1 ENG ***Saturated R-134a***

Temp.	Press.	Specific Volume, ft³/lbm			Internal Energy, Btu/lbm		
F T	psia P	Sat. Liquid v_f	Evap. v_{fg}	Sat. Vapor v_g	Sat. Liquid u_f	Evap. u_{fg}	Sat. Vapor u_g
-100	0.951	0.01077	39.5032	39.5139	50.47	94.15	144.62
-90	1.410	0.01083	27.3236	27.3345	52.03	93.89	145.92
-80	2.047	0.01091	19.2731	19.2840	53.96	93.27	147.24
-70	2.913	0.01101	13.8538	13.8648	56.19	92.38	148.57
-60	4.067	0.01111	10.1389	10.1501	58.64	91.26	149.91
-50	5.575	0.01122	7.5468	7.5580	61.27	89.99	151.26
-40	7.511	0.01134	5.7066	5.7179	64.04	88.58	152.62
-30	9.959	0.01146	4.3785	4.3900	66.90	87.09	153.99
-20	13.009	0.01159	3.4049	3.4165	69.83	85.53	155.36
-15.3	14.696	0.01166	3.0350	3.0466	71.25	84.76	156.02
-10	16.760	0.01173	2.6805	2.6922	72.83	83.91	156.74
0	21.315	0.01187	2.1340	2.1458	75.88	82.24	158.12
10	26.787	0.01202	1.7162	1.7282	78.96	80.53	159.50
20	33.294	0.01218	1.3928	1.4050	82.09	78.78	160.87
30	40.962	0.01235	1.1398	1.1521	85.25	76.99	162.24
40	49.922	0.01253	0.9395	0.9520	88.45	75.16	163.60
50	60.311	0.01271	0.7794	0.7921	91.68	73.27	164.95
60	72.271	0.01291	0.6503	0.6632	94.95	71.32	166.28
70	85.954	0.01313	0.5451	0.5582	98.27	69.31	167.58
80	101.515	0.01335	0.4588	0.4721	101.63	67.22	168.85
90	119.115	0.01360	0.3873	0.4009	105.04	65.04	170.09
100	138.926	0.01387	0.3278	0.3416	108.51	62.77	171.28
110	161.122	0.01416	0.2777	0.2919	112.03	60.38	172.41
120	185.890	0.01448	0.2354	0.2499	115.62	57.85	173.48
130	213.425	0.01483	0.1993	0.2142	119.29	55.17	174.46
140	243.932	0.01523	0.1684	0.1836	123.04	52.30	175.34
150	277.630	0.01568	0.1415	0.1572	126.89	49.21	176.11
160	314.758	0.01620	0.1181	0.1343	130.86	45.85	176.71
170	355.578	0.01683	0.0974	0.1142	134.99	42.12	177.11
180	400.392	0.01760	0.0787	0.0963	139.32	37.91	177.23
190	449.572	0.01862	0.0614	0.0801	143.97	32.94	176.90
200	503.624	0.02013	0.0444	0.0645	149.19	26.59	175.79
210	563.438	0.02334	0.0238	0.0471	156.18	16.17	172.34
214.1	589.953	0.03153	0	0.0315	164.65	0	164.65

(Continued)

TABLE E.15.1 ENG (Continued) ***Saturated R-134a***

Temp.	Press.	Enthalpy, Btu/lbm			Entropy, Btu/lbm R		
F	psia	Sat. Liquid	Evap.	Sat. Vapor	Sat. Liquid	Evap.	Sat. Vapor
T	P	h_f	h_{fg}	h_g	s_f	s_{fg}	s_g
-100	0.951	50.47	101.10	151.57	0.1563	0.2811	0.4373
-90	1.410	52.04	101.02	153.05	0.1605	0.2733	0.4338
-80	2.047	53.97	100.58	154.54	0.1657	0.2649	0.4306
-70	2.913	56.19	99.85	156.04	0.1715	0.2562	0.4277
-60	4.067	58.65	98.90	157.55	0.1777	0.2474	0.4251
-50	5.575	61.29	97.77	159.06	0.1842	0.2387	0.4229
-40	7.511	64.05	96.52	160.57	0.1909	0.2300	0.4208
-30	9.959	66.92	95.16	162.08	0.1976	0.2215	0.4191
-20	13.009	69.86	93.72	163.59	0.2044	0.2132	0.4175
-15.3	14.696	71.28	93.02	164.30	0.2076	0.2093	0.4169
-10	16.760	72.87	92.22	165.09	0.2111	0.2051	0.4162
0	21.315	75.92	90.66	166.58	0.2178	0.1972	0.4150
10	26.787	79.02	89.04	168.06	0.2244	0.1896	0.4140
20	33.294	82.16	87.36	169.53	0.2310	0.1821	0.4132
30	40.962	85.34	85.63	170.98	0.2375	0.1749	0.4124
40	49.922	88.56	83.83	172.40	0.2440	0.1678	0.4118
50	60.311	91.82	81.97	173.79	0.2504	0.1608	0.4112
60	72.271	95.13	80.02	175.14	0.2568	0.1540	0.4108
70	85.954	98.48	77.98	176.46	0.2631	0.1472	0.4103
80	101.515	101.88	75.84	177.72	0.2694	0.1405	0.4099
90	119.115	105.34	73.58	178.92	0.2757	0.1339	0.4095
100	138.926	108.86	71.19	180.06	0.2819	0.1272	0.4091
110	161.122	112.46	68.66	181.11	0.2882	0.1205	0.4087
120	185.890	116.12	65.95	182.07	0.2945	0.1138	0.4082
130	213.425	119.88	63.04	182.92	0.3008	0.1069	0.4077
140	243.932	123.73	59.90	183.63	0.3071	0.0999	0.4070
150	277.630	127.70	56.49	184.18	0.3135	0.0926	0.4061
160	314.758	131.81	52.73	184.53	0.3200	0.0851	0.4051
170	355.578	136.09	48.53	184.63	0.3267	0.0771	0.4037
180	400.392	140.62	43.74	184.36	0.3336	0.0684	0.4020
190	449.572	145.52	38.05	183.56	0.3409	0.0586	0.3995
200	503.624	151.07	30.73	181.80	0.3491	0.0466	0.3957
210	563.438	158.61	18.65	177.26	0.3601	0.0278	0.3879
214.1	589.953	168.09	0	168.09	0.3740	0	0.3740

TABLE E.15.2 ENG ***Superheated R-134a***

Temp. F	v ft³/lbm	h Btu/lbm	s Btu/lbm R	v ft³/lbm	h Btu/lbm	s Btu/lbm R	v ft³/lbm	h Btu/lbm	s Btu/lbm R
	5 psia (-53.51)			10 psia (-29.85)			15 psia (-14.44)		
Sat.	8.3676	158.53	0.4236	4.3732	162.10	0.4190	2.9885	164.42	0.4168
-20	9.1149	164.47	0.4377	4.4879	163.92	0.4232	—	—	—
0	9.5533	168.11	0.4458	4.7168	167.66	0.4315	3.1033	167.19	0.4229
20	9.9881	171.83	0.4537	4.9417	171.45	0.4396	3.2586	171.06	0.4311
40	10.4202	175.63	0.4615	5.1637	175.31	0.4475	3.4109	174.97	0.4391
60	10.8502	179.52	0.4691	5.3836	179.24	0.4552	3.5610	178.95	0.4469
80	11.2786	183.50	0.4766	5.6019	183.25	0.4628	3.7093	183.00	0.4545
100	11.7059	187.56	0.4840	5.8189	187.34	0.4702	3.8563	187.12	0.4620
120	12.1322	191.71	0.4913	6.0350	191.51	0.4775	4.0024	191.31	0.4694
140	12.5578	195.95	0.4985	6.2503	195.77	0.4848	4.1476	195.59	0.4767
160	12.9828	200.28	0.5056	6.4650	200.11	0.4919	4.2922	199.95	0.4838
180	13.4073	204.69	0.5126	6.6791	204.54	0.4989	4.4364	204.39	0.4909
200	13.8314	209.19	0.5195	6.8929	209.05	0.5059	4.5801	208.91	0.4978
220	14.2551	213.77	0.5263	7.1064	213.64	0.5127	4.7234	213.51	0.5047
240	14.6786	218.44	0.5331	7.3195	218.32	0.5195	4.8665	218.19	0.5115
260	15.1019	223.19	0.5398	7.5324	223.07	0.5262	5.0093	222.96	0.5182
280	15.5250	228.02	0.5464	7.7452	227.91	0.5328	5.1519	227.80	0.5248
300	15.9478	232.93	0.5530	7.9577	232.83	0.5394	5.2943	232.72	0.5314
320	16.3706	237.92	0.5595	8.1701	237.82	0.5459	5.4365	237.73	0.5379
	30 psia (15.15)			40 psia (28.83)			50 psia (40.08)		
Sat.	1.5517	168.82	0.4136	1.1787	170.81	0.4125	0.9506	172.41	0.4118
20	1.5725	169.82	0.4157	—	—	—	—	—	—
40	1.6559	173.93	0.4240	1.2157	173.18	0.4173	—	—	—
60	1.7367	178.05	0.4321	1.2796	177.42	0.4256	1.0045	176.76	0.4203
80	1.8155	182.21	0.4400	1.3413	181.67	0.4336	1.0563	181.10	0.4285
100	1.8929	186.43	0.4477	1.4015	185.95	0.4414	1.1062	185.45	0.4364
120	1.9691	190.70	0.4552	1.4604	190.27	0.4490	1.1549	189.83	0.4441
140	2.0445	195.03	0.4625	1.5184	194.65	0.4565	1.2026	194.26	0.4516
160	2.1192	199.44	0.4697	1.5757	199.09	0.4637	1.2495	198.74	0.4590
180	2.1933	203.92	0.4769	1.6324	203.60	0.4709	1.2957	203.28	0.4662
200	2.2670	208.48	0.4839	1.6886	208.18	0.4780	1.3415	207.89	0.4733
220	2.3403	213.11	0.4908	1.7444	212.84	0.4849	1.3869	212.56	0.4803
240	2.4133	217.82	0.4976	1.7999	217.57	0.4918	1.4319	217.31	0.4872
260	2.4860	222.61	0.5044	1.8552	222.37	0.4985	1.4766	222.13	0.4939
280	2.5585	227.47	0.5110	1.9102	227.25	0.5052	1.5211	227.02	0.5007
300	2.6309	232.41	0.5176	1.9650	232.20	0.5118	1.5655	231.99	0.5073
320	2.7030	237.43	0.5241	2.0196	237.23	0.5184	1.6096	237.03	0.5138
340	2.7750	242.53	0.5306	2.0741	242.34	0.5248	1.6536	242.15	0.5203
360	2.8469	247.70	0.5370	2.1285	247.52	0.5312	1.6974	247.34	0.5267

(Continued)

TABLE E.15.2 ENG (Continued) ***Superheated R-134a***

Temp. F	v ft³/lbm	h Btu/lbm	s Btu/lbm R	v ft³/lbm	h Btu/lbm	s Btu/lbm R	v ft³/lbm	h Btu/lbm	s Btu/lbm R
	60 psia (49.72)			70 psia (58.20)			80 psia (65.81)		
Sat.	0.7961	173.75	0.4113	0.6844	174.90	0.4108	0.5996	175.91	0.4105
60	0.8204	176.06	0.4157	0.6882	175.32	0.4116	—	—	—
80	0.8657	180.51	0.4241	0.7291	179.89	0.4203	0.6262	179.24	0.4168
100	0.9091	184.94	0.4322	0.7679	184.41	0.4285	0.6617	183.86	0.4252
120	0.9510	189.38	0.4400	0.8051	188.92	0.4364	0.6954	188.44	0.4332
140	0.9918	193.86	0.4476	0.8411	193.45	0.4441	0.7279	193.03	0.4410
160	1.0318	198.38	0.4550	0.8763	198.01	0.4516	0.7595	197.64	0.4485
180	1.0712	202.95	0.4623	0.9107	202.62	0.4589	0.7903	202.28	0.4559
200	1.1100	207.59	0.4694	0.9446	207.28	0.4661	0.8205	206.98	0.4632
220	1.1484	212.29	0.4764	0.9781	212.01	0.4731	0.8503	211.72	0.4702
240	1.1865	217.05	0.4833	1.0112	216.79	0.4801	0.8796	216.53	0.4772
260	1.2243	221.89	0.4902	1.0440	221.65	0.4869	0.9087	221.41	0.4841
280	1.2618	226.80	0.4969	1.0765	226.57	0.4937	0.9375	226.34	0.4909
300	1.2991	231.78	0.5035	1.1088	231.57	0.5003	0.9661	231.35	0.4975
320	1.3362	236.83	0.5101	1.1410	236.63	0.5069	0.9945	236.43	0.5041
340	1.3732	241.96	0.5166	1.1729	241.77	0.5134	1.0227	241.58	0.5107
360	1.4100	247.16	0.5230	1.2048	246.98	0.5199	1.0508	246.80	0.5171
380	1.4468	252.43	0.5294	1.2365	252.26	0.5262	1.0788	252.09	0.5235
400	1.4834	257.78	0.5357	1.2681	257.62	0.5325	1.1066	257.46	0.5298
	90 psia (72.72)			100 psia (79.08)			125 psia (93.09)		
Sat.	0.5331	176.81	0.4102	0.4794	177.61	0.4100	0.3814	179.28	0.4094
80	0.5457	178.55	0.4135	0.4809	177.83	0.4104	—	—	—
100	0.5788	183.28	0.4221	0.5122	182.68	0.4192	0.3910	181.06	0.4126
120	0.6100	187.95	0.4303	0.5414	187.44	0.4276	0.4171	186.08	0.4214
140	0.6398	192.60	0.4382	0.5691	192.15	0.4356	0.4413	190.98	0.4297
160	0.6685	197.25	0.4458	0.5957	196.86	0.4433	0.4642	195.84	0.4377
180	0.6965	201.94	0.4532	0.6215	201.58	0.4508	0.4861	200.68	0.4454
200	0.7239	206.66	0.4605	0.6466	206.34	0.4581	0.5073	205.52	0.4529
220	0.7508	211.44	0.4676	0.6712	211.15	0.4653	0.5278	210.40	0.4601
240	0.7773	216.27	0.4746	0.6954	216.00	0.4723	0.5480	215.32	0.4673
260	0.8035	221.16	0.4815	0.7193	220.91	0.4792	0.5677	220.28	0.4743
280	0.8294	226.12	0.4883	0.7429	225.88	0.4861	0.5872	225.30	0.4811
300	0.8551	231.14	0.4950	0.7663	230.92	0.4928	0.6064	230.38	0.4879
320	0.8806	236.23	0.5017	0.7895	236.03	0.4994	0.6254	235.51	0.4946
340	0.9059	241.39	0.5082	0.8125	241.20	0.5060	0.6442	240.71	0.5012
360	0.9311	246.62	0.5146	0.8353	246.44	0.5124	0.6629	245.98	0.5077
380	0.9561	251.92	0.5210	0.8580	251.75	0.5188	0.6814	251.32	0.5141
400	0.9811	257.29	0.5274	0.8806	257.13	0.5252	0.6998	256.72	0.5205

(Continued)

TABLE E.15.2 ENG (Continued) ***Superheated R-134a***

Temp. F	v ft³/lbm	h Btu/lbm	s Btu/lbm R	v ft³/lbm	h Btu/lbm	s Btu/lbm R	v ft³/lbm	h Btu/lbm	s Btu/lbm R
	150 psia (105.13)			175 psia (115.73)			200 psia (125.25)		
Sat.	0.3150	180.61	0.4089	0.2669	181.68	0.4085	0.2304	182.53	0.4080
120	0.3332	184.57	0.4159	0.2719	182.88	0.4105	—	—	—
140	0.3554	189.72	0.4246	0.2933	188.34	0.4198	0.2459	186.82	0.4152
160	0.3761	194.75	0.4328	0.3126	193.58	0.4284	0.2645	192.33	0.4242
180	0.3955	199.72	0.4407	0.3305	198.71	0.4365	0.2814	197.64	0.4327
200	0.4141	204.67	0.4484	0.3474	203.78	0.4444	0.2971	202.85	0.4407
220	0.4321	209.63	0.4558	0.3636	208.84	0.4519	0.3120	208.01	0.4484
240	0.4496	214.62	0.4630	0.3791	213.90	0.4592	0.3262	213.15	0.4559
260	0.4666	219.64	0.4701	0.3943	218.98	0.4664	0.3400	218.31	0.4631
280	0.4833	224.70	0.4770	0.4091	224.10	0.4734	0.3534	223.48	0.4702
300	0.4998	229.82	0.4838	0.4236	229.26	0.4803	0.3664	228.69	0.4772
320	0.5160	235.00	0.4906	0.4379	234.47	0.4871	0.3792	233.94	0.4840
340	0.5320	240.23	0.4972	0.4519	239.74	0.4937	0.3918	239.24	0.4907
360	0.5479	245.52	0.5037	0.4658	245.06	0.5003	0.4042	244.60	0.4973
380	0.5636	250.88	0.5102	0.4795	250.45	0.5068	0.4165	250.01	0.5038
400	0.5792	256.31	0.5166	0.4931	255.90	0.5132	0.4286	255.48	0.5103
	250 psia (141.87)			300 psia (156.14)			350 psia (168.69)		
Sat.	0.1783	183.75	0.4068	0.1428	184.43	0.4055	0.1167	184.63	0.4039
160	0.1955	189.46	0.4162	0.1467	185.84	0.4078	—	—	—
180	0.2117	195.28	0.4255	0.1637	192.53	0.4184	0.1275	189.13	0.4110
200	0.2261	200.84	0.4340	0.1779	198.59	0.4278	0.1425	196.01	0.4216
220	0.2394	206.26	0.4421	0.1905	204.35	0.4364	0.1550	202.24	0.4309
240	0.2519	211.60	0.4498	0.2020	209.93	0.4445	0.1660	208.14	0.4395
260	0.2638	216.90	0.4573	0.2128	215.43	0.4522	0.1761	213.86	0.4476
280	0.2752	222.21	0.4646	0.2230	220.88	0.4597	0.1856	219.49	0.4553
300	0.2863	227.52	0.4717	0.2328	226.31	0.4669	0.1945	225.06	0.4627
320	0.2971	232.86	0.4786	0.2423	231.75	0.4740	0.2031	230.61	0.4699
340	0.3076	238.24	0.4854	0.2515	237.21	0.4809	0.2114	236.16	0.4769
360	0.3180	243.66	0.4921	0.2605	242.70	0.4877	0.2194	241.74	0.4838
380	0.3282	249.13	0.4987	0.2693	248.24	0.4944	0.2273	247.33	0.4906
400	0.3382	254.65	0.5052	0.2779	253.81	0.5009	0.2349	252.97	0.4972

(Continued)

TABLE E.15.2 ENG (Continued) *Superheated R-134a*

Temp. F	v ft^3/lbm	h Btu/lbm	s Btu/lbm R	v ft^3/lbm	h Btu/lbm	s Btu/lbm R	v ft^3/lbm	h Btu/lbm	s Btu/lbm R
	400 psia (179.92)			500 psia (199.36)			600 psia		
Sat.	0.0965	184.37	0.4020	0.0655	181.96	0.3960	—	—	—
190	0.1066	189.00	0.4092	—	—	—	—	—	—
200	0.1146	192.92	0.4152	0.0666	182.54	0.3969	—	—	—
210	0.1215	196.50	0.4205	0.0786	189.06	0.4067	0.0211	155.64	0.3554
220	0.1277	199.86	0.4255	0.0867	193.80	0.4137	0.0507	182.23	0.3948
230	0.1333	203.08	0.4302	0.0933	197.89	0.4197	0.0627	190.23	0.4065
240	0.1386	206.19	0.4347	0.0990	201.62	0.4251	0.0703	195.56	0.4142
250	0.1436	209.22	0.4390	0.1042	205.13	0.4300	0.0763	200.02	0.4205
260	0.1484	212.20	0.4432	0.1089	208.47	0.4347	0.0815	204.02	0.4261
270	0.1530	215.13	0.4472	0.1133	211.71	0.4392	0.0861	207.74	0.4312
280	0.1573	218.03	0.4512	0.1174	214.86	0.4435	0.0903	211.27	0.4360
290	0.1616	220.90	0.4550	0.1214	217.95	0.4476	0.0942	214.67	0.4406
300	0.1657	223.76	0.4588	0.1252	220.99	0.4517	0.0978	217.96	0.4450
310	0.1697	226.60	0.4625	0.1288	224.00	0.4556	0.1013	221.18	0.4492
320	0.1737	229.44	0.4662	0.1323	226.98	0.4594	0.1046	224.34	0.4533
330	0.1775	232.27	0.4698	0.1357	229.93	0.4632	0.1077	227.46	0.4572
340	0.1813	235.09	0.4733	0.1390	232.87	0.4669	0.1108	230.54	0.4611
350	0.1850	237.92	0.4769	0.1423	235.80	0.4705	0.1138	233.59	0.4649
360	0.1886	240.75	0.4803	0.1454	238.73	0.4741	0.1166	236.62	0.4686
370	0.1922	243.58	0.4838	0.1485	241.64	0.4777	0.1194	239.64	0.4723
380	0.1957	246.42	0.4872	0.1516	244.56	0.4812	0.1222	242.64	0.4759
390	0.1992	249.27	0.4905	0.1546	247.48	0.4846	0.1249	245.64	0.4794
400	0.2027	252.12	0.4939	0.1575	250.39	0.4880	0.1275	248.63	0.4829

TABLE E.16 *Ideal-Gas Properties of Air, English Units Standard Entropy at 1 atm = 101.325 kPa = 14.696 lbf/in.²*

T R	u Btu/lbm	h Btu/lbm	s^0 Btu/lbm R	P_r	v_r
400	68.212	95.634	1.56788	0.39046	379.523
440	75.047	105.212	1.59071	0.54470	299.264
480	81.887	114.794	1.61155	0.73825	240.877
520	88.733	124.383	1.63074	0.97670	197.244
536.67	91.589	128.381	1.63831	1.09071	182.288
540	92.160	129.180	1.63979	1.11458	179.491
560	95.589	133.980	1.64852	1.26592	163.885
600	102.457	143.590	1.66510	1.61217	137.880
640	109.340	153.216	1.68063	2.02204	117.260
680	116.242	162.860	1.69524	2.50257	100.666
720	123.167	172.528	1.70906	3.06119	87.1367
760	130.118	182.221	1.72216	3.70585	75.9775
800	137.099	191.944	1.73463	4.44496	66.6778
840	144.114	201.701	1.74653	5.28751	58.8555
880	151.165	211.494	1.75791	6.24303	52.2211
920	158.255	221.327	1.76884	7.32166	46.5519
960	165.388	231.202	1.77935	8.53415	41.6744
1000	172.564	241.121	1.78947	9.89193	37.4523
1040	179.787	251.086	1.79924	11.40706	33.7768
1080	187.058	261.099	1.80868	13.09232	30.5609
1120	194.378	271.161	1.81783	14.96119	27.7339
1160	201.748	281.273	1.82670	17.02788	25.2381
1200	209.168	291.436	1.83532	19.30735	23.0259
1240	216.640	301.650	1.84369	21.81531	21.0581
1280	224.163	311.915	1.85184	24.56826	19.3017
1320	231.737	322.231	1.85977	27.58348	17.7290
1360	239.362	332.598	1.86751	30.87907	16.3168
1400	247.037	343.016	1.87506	34.47392	15.0451
1440	254.762	353.483	1.88243	38.38777	13.8972
1480	262.537	364.000	1.88964	42.64121	12.8585
1520	270.359	374.565	1.89668	47.25567	11.9165
1560	278.230	385.177	1.90357	52.25344	11.0603
1600	286.146	395.837	1.91032	57.65771	10.2807
1650	296.106	409.224	1.91856	65.02144	9.40126
1700	306.136	422.681	1.92659	73.10700	8.61487
1750	316.232	436.205	1.93444	81.96560	7.90980
1800	326.393	449.794	1.94209	91.65077	7.27604
1850	336.616	463.445	1.94957	102.2183	6.70505

(Continued)

TABLE E.16 (Continued) ***Ideal-Gas Properties of Air, English Units Standard Entropy at 1 atm = 101.325 kPa = 14.696 lbf/in.²***

T R	u Btu/lbm	h Btu/lbm	s^0 Btu/lbm R	P_r	v_r
1900	346.901	477.158	1.95689	113.7264	6.18944
1950	357.243	490.928	1.96404	126.2356	5.72284
2000	367.642	504.755	1.97104	139.8090	5.29973
2050	378.096	518.636	1.97790	154.5119	4.91531
2100	388.602	532.570	1.98461	170.4125	4.56538
2150	399.158	546.554	1.99119	187.5812	4.24627
2200	409.764	560.588	1.99765	206.0915	3.95477
2300	431.114	588.793	2.01018	247.4432	3.44359
2400	452.640	617.175	2.02226	295.1096	3.01292
2500	474.330	645.721	2.03391	349.7802	2.64791
2600	496.175	674.421	2.04517	412.1964	2.33683
2700	518.165	703.267	2.05606	483.1554	2.07031
2800	540.286	732.244	2.06659	563.4304	1.84110
2900	562.532	761.345	2.07681	653.9284	1.64296
3000	584.895	790.564	2.08671	755.5802	1.47096
3100	607.369	819.894	2.09633	869.3694	1.32104
3200	629.948	849.328	2.10567	996.3336	1.18988
3300	652.625	878.861	2.11476	1137.566	1.07472
3400	675.396	908.488	2.12361	1294.214	0.97327
3500	698.257	938.204	2.13222	1467.483	0.88360
3600	721.203	968.005	2.14062	1658.635	0.80410
3700	744.230	997.888	2.14880	1869.020	0.73341
3800	767.334	1027.848	2.15679	2100.030	0.67037
3900	790.513	1057.882	2.16459	2353.126	0.61401
4000	813.763	1087.988	2.17221	2629.834	0.56350
4100	837.081	1118.162	2.17967	2931.747	0.51810
4200	860.466	1148.402	2.18695	3260.527	0.47722
4300	883.913	1178.705	2.19408	3617.908	0.44032
4400	907.422	1209.069	2.20106	4005.693	0.40694
4500	930.989	1239.492	2.20790	4425.759	0.37669
4600	954.613	1269.972	2.21460	4880.058	0.34921
4700	978.292	1300.506	2.22117	5370.617	0.32421
4800	1002.023	1331.093	2.22761	5899.541	0.30143
4900	1025.806	1361.732	2.23392	6469.012	0.28062
5000	1049.638	1392.419	2.24012	7081.293	0.26159
5100	1073.518	1423.155	2.24621	7738.728	0.24415
5200	1097.444	1453.936	2.25219	8443.744	0.22815
5300	1121.414	1484.762	2.25806	9198.851	0.21345
5400	1145.428	1515.632	2.26383	10006.645	0.19992

TABLE E.17 ***Ideal-Gas Properties of Various Substances (English Units), Entropies at 1 atm Pressure***

	Nitrogen, Diatomic (N_2) $\bar{h}^o_{f,537}$ = 0 Btu/lb mol M = 28.013		Nitrogen, Monatomic (N) $\bar{h}^o_{f,537}$ = 203 216 Btu/lb mol M = 14.007	
T ***R***	$\bar{h}^o - \bar{h}^o_{f,537}$ **Btu/lb mol**	$\bar{s}^o$ **Btu/lbmol/R**	$\bar{h}^o - \bar{h}^o_{f,537}$ **Btu/lb mol**	$\bar{s}^o$ **Btu/lbmol/R**
0	−3727	0	−2664	0
200	−2341	38.877	−1671	31.689
400	−950	43.695	−679	35.130
537	0	45.739	0	36.589
600	441	46.515	314	37.143
800	1837	48.524	1307	38.571
1000	3251	50.100	2300	39.679
1200	4693	51.414	3293	40.584
1400	6169	52.552	4286	41.349
1600	7681	53.561	5279	42.012
1800	9227	54.472	6272	42.597
2000	10804	55.302	7265	43.120
2200	12407	56.066	8258	43.593
2400	14034	56.774	9251	44.025
2600	15681	57.433	10244	44.423
2800	17345	58.049	11237	44.791
3000	19025	58.629	12230	45.133
3200	20717	59.175	13223	45.454
3400	22421	59.691	14216	45.755
3600	24135	60.181	15209	46.038
3800	25857	60.647	16202	46.307
4000	27587	61.090	17195	46.562
4200	29324	61.514	18189	46.804
4400	31068	61.920	19183	47.035
4600	32817	62.308	20178	47.256
4800	34571	62.682	21174	47.468
5000	36330	63.041	22171	47.672
5500	40745	63.882	24670	48.148
6000	45182	64.654	27186	48.586
6500	49638	65.368	29724	48.992
7000	54109	66.030	32294	49.373
7500	58595	66.649	34903	49.733
8000	63093	67.230	37559	50.076
8500	67603	67.777	40270	50.405
9000	72125	68.294	43040	50.721
9500	76658	68.784	45875	51.028
10000	81203	69.250	48777	51.325

(Continued)

TABLE E.17 (Continued) ***Ideal-Gas Properties of Various Substances (English Units), Entropies at 1 atm Pressure***

	Oxygen, Diatomic (O_2) $\bar{h}^o_{f,537} = 0$ Btu/lb mol $M = 31.999$		Oxygen, Monatomic (O) $\bar{h}^o_{f,537} = 107\ 124$ Btu/lb mol $M = 16.00$	
T R	**$\bar{h}^o - \bar{h}^o_{f,537}$ Btu/lb mol**	**$\bar{s}^o$ Btu/lbmol/R**	**$\bar{h}^o - \bar{h}^o_{f,537}$ Btu/lb mol**	**$\bar{s}^o$ Btu/lbmol/R**
0	−3733	0	−2891	0
200	−2345	42.100	−1829	33.041
400	−955	46.920	−724	36.884
537	0	48.973	0	38.442
600	446	49.758	330	39.023
800	1881	51.819	1358	40.503
1000	3366	53.475	2374	41.636
1200	4903	54.876	3383	42.556
1400	6487	56.096	4387	43.330
1600	8108	57.179	5389	43.999
1800	9761	58.152	6389	44.588
2000	11438	59.035	7387	45.114
2200	13136	59.844	8385	45.589
2400	14852	60.591	9381	46.023
2600	16584	61.284	10378	46.422
2800	18329	61.930	11373	46.791
3000	20088	62.537	12369	47.134
3200	21860	63.109	13364	47.455
3400	23644	63.650	14359	47.757
3600	25441	64.163	15354	48.041
3800	27250	64.652	16349	48.310
4000	29071	65.119	17344	48.565
4200	30904	65.566	18339	48.808
4400	32748	65.995	19334	49.039
4600	34605	66.408	20330	49.261
4800	36472	66.805	21327	49.473
5000	38350	67.189	22325	49.677
5500	43091	68.092	24823	50.153
6000	47894	68.928	27329	50.589
6500	52751	69.705	29847	50.992
7000	57657	70.433	32378	51.367
7500	62608	71.116	34924	51.718
8000	67600	71.760	37485	52.049
8500	72633	72.370	40063	52.362
9000	77708	72.950	42658	52.658
9500	82828	73.504	45270	52.941
10000	87997	74.034	47897	53.210

(Continued)

TABLE E.17 (Continued) ***Ideal-Gas Properties of Various Substances (English Units), Entropies at 1 atm Pressure***

	Carbon Dioxide (CO_2) $\bar{h}^o_{f,537} = -169\,184$ Btu/lb mol $M = 44.01$		Carbon Monoxide (CO) $\bar{h}^o_{f,537} = -47\,518$ Btu/lb mol $M = 28.01$	
T R	$\bar{h}^o - \bar{h}^o_{f,537}$ **Btu/lb mol**	$\bar{s}^o$ **Btu/lbmol/R**	$\bar{h}^o - \bar{h}^o_{f,537}$ **Btu/lb mol**	$\bar{s}^o$ **Btu/lbmol/R**
0	−4026	0	−3728	0
200	−2636	43.466	−2343	40.319
400	−1153	48.565	−951	45.137
537	0	51.038	0	47.182
600	573	52.047	441	47.959
800	2525	54.848	1842	49.974
1000	4655	57.222	3266	51.562
1200	6927	59.291	4723	52.891
1400	9315	61.131	6220	54.044
1600	11798	62.788	7754	55.068
1800	14358	64.295	9323	55.992
2000	16982	65.677	10923	56.835
2200	19659	66.952	12549	57.609
2400	22380	68.136	14197	58.326
2600	25138	69.239	15864	58.993
2800	27926	70.273	17547	59.616
3000	30741	71.244	19243	60.201
3200	33579	72.160	20951	60.752
3400	36437	73.026	22669	61.273
3600	39312	73.847	24395	61.767
3800	42202	74.629	26128	62.236
4000	45105	75.373	27869	62.683
4200	48021	76.084	29614	63.108
4400	50948	76.765	31366	63.515
4600	53885	77.418	33122	63.905
4800	56830	78.045	34883	64.280
5000	59784	78.648	36650	64.641
5500	55739	68.649	39393	61.477
6000	74660	81.360	45548	66.263
6500	82155	82.560	50023	66.979
7000	89682	83.675	54514	67.645
7500	97239	84.718	59020	68.267
8000	104823	85.697	63539	68.850
8500	112434	86.620	68069	69.399
9000	120071	87.493	72610	69.918
9500	127734	88.321	77161	70.410
10000	135426	89.110	81721	70.878

(Continued)

TABLE E.17 (Continued) ***Ideal-Gas Properties of Various Substances (English Units), Entropies at 1 atm Pressure***

	Water (H_2O) $\bar{h}^o_{f,537} = -103\,966$ Btu/lb mol, $M = 18.015$		Hydroxyl (OH) $\bar{h}^o_{f,537} = 16\,761$ Btu/lb mol, $M = 17.007$	
T R	$\bar{h}^o - \bar{h}^o_{f,537}$ Btu/lb mol	$\bar{s}^o$ Btu/lbmol/R	$\bar{h}^o - \bar{h}^o_{f,537}$ Btu/lb mol	$\bar{s}^o$ Btu/lbmol/R
0	–4258	0	–3943	0
200	–2686	37.209	–2484	36.521
400	–1092	42.728	–986	41.729
537	0	45.076	0	43.852
600	509	45.973	452	44.649
800	2142	48.320	1870	46.689
1000	3824	50.197	3280	48.263
1200	5566	51.784	4692	49.549
1400	7371	53.174	6112	50.643
1600	9241	54.422	7547	51.601
1800	11178	55.563	9001	52.457
2000	13183	56.619	10477	53.235
2200	15254	57.605	11978	53.950
2400	17388	58.533	13504	54.614
2600	19582	59.411	15054	55.235
2800	21832	60.245	16627	55.817
3000	24132	61.038	18220	56.367
3200	26479	61.796	19834	56.887
3400	28867	62.520	21466	57.382
3600	31293	63.213	23114	57.853
3800	33756	63.878	24777	58.303
4000	36251	64.518	26455	58.733
4200	38774	65.134	28145	59.145
4400	41325	65.727	29849	59.542
4600	43899	66.299	31563	59.922
4800	46496	66.852	33287	60.289
5000	49114	67.386	35021	60.643
5500	55739	68.649	39393	61.477
6000	62463	69.819	43812	62.246
6500	69270	70.908	48272	62.959
7000	76146	71.927	52767	63.626
7500	83081	72.884	57294	64.250
8000	90069	73.786	61851	64.838
8500	97101	74.639	66434	65.394
9000	104176	75.448	71043	65.921
9500	111289	76.217	75677	66.422
10000	118440	76.950	80335	66.900

(Continued)

TABLE E.17 (Continued) ***Ideal-Gas Properties of Various Substances (English Units), Entropies at 1 atm Pressure***

	Hydrogen (H_2) $\bar{h}^o_{f,537}$ = 0 Btu/lb mol M = 2.016		Hydrogen, Monatomic (H) $\bar{h}^o_{f,537}$ = 93 723 Btu/lb mol M = 1.008	
T R	$\bar{h}^o-\bar{h}^o_{f,537}$ Btu/lb mol	$\bar{s}^o$ Btu/lbmol/R	$\bar{h}^o-\bar{h}^o_{f,537}$ Btu/lb mol	$\bar{s}^o$ Btu/lbmol/R
0	−3640	0	−2664	0
200	−2224	24.703	−1672	22.473
400	−927	29.193	−679	25.914
537	0	31.186	0	27.373
600	438	31.957	314	27.927
800	1831	33.960	1307	29.355
1000	3225	35.519	2300	30.463
1200	4622	36.797	3293	31.368
1400	6029	37.883	4286	32.134
1600	7448	38.831	5279	32.797
1800	8884	39.676	6272	33.381
2000	10337	40.441	7265	33.905
2200	11812	41.143	8258	34.378
2400	13309	41.794	9251	34.810
2600	14829	42.401	10244	35.207
2800	16372	42.973	11237	35.575
3000	17938	43.512	12230	35.917
3200	19525	44.024	13223	36.238
3400	21133	44.512	14215	36.539
3600	22761	44.977	15208	36.823
3800	24407	45.422	16201	37.091
4000	26071	45.849	17194	37.346
4200	27752	46.260	18187	37.588
4400	29449	46.655	19180	37.819
4600	31161	47.035	20173	38.040
4800	32887	47.403	21166	38.251
5000	34627	47.758	22159	38.454
5500	39032	48.598	24641	38.927
6000	43513	49.378	27124	39.359
6500	48062	50.105	29606	39.756
7000	52678	50.789	32088	40.124
7500	57356	51.434	34571	40.467
8000	62094	52.045	37053	40.787
8500	66889	52.627	39535	41.088
9000	71738	53.182	42018	41.372
9500	76638	53.712	44500	41.640
10000	81581	54.220	46982	41.895

(Continued)

TABLE E.17 (Continued) ***Ideal-Gas Properties of Various Substances (English Units), Entropies at 1 atm Pressure***

	Nitric Oxide (NO) $\bar{h}^{o}_{f,537}$ = 38 818 Btu/lb mol M = 30.006		Nitrogen Dioxide (NO_2) $\bar{h}^{o}_{f,537}$ = 14 230 Btu/lb mol M = 46.005	
T R	**$\bar{h}^{o}-\bar{h}^{o}_{f,537}$ Btu/lb mol**	**$\bar{s}^{o}$ Btu/lbmol/R**	**$\bar{h}^{o}-\bar{h}^{o}_{f,537}$ Btu/lb mol**	**$\bar{s}^{o}$ Btu/lbmol/R**
0	–3952	0	–4379	0
200	–2224	24.703	–1672	22.473
400	–927	29.193	–679	25.914
537	0	31.186	0	27.373
600	438	31.957	314	27.927
800	1831	33.960	1307	29.355
1000	3225	35.519	2300	30.463
1200	4622	36.797	3293	31.368
1400	6029	37.883	4286	32.134
1600	7448	38.831	5279	32.797
1800	8884	39.676	6272	33.381
2000	10337	40.441	7265	33.905
2200	11812	41.143	8258	34.378
2400	13309	41.794	9251	34.810
2600	14829	42.401	10244	35.207
2800	16372	42.973	11237	35.575
3000	17938	43.512	12230	35.917
3200	19525	44.024	13223	36.238
3400	21133	44.512	14215	36.539
3600	22761	44.977	15208	36.823
3800	24407	45.422	16201	37.091
4000	26071	45.849	17194	37.346
4200	27752	46.260	18187	37.588
4400	29449	46.655	19180	37.819
4600	31161	47.035	20173	38.040
4800	32887	47.403	21166	38.251
5000	34627	47.758	22159	38.454
5500	41726	68.965	63395	84.990
6000	43513	49.378	27124	39.359
6500	48062	50.105	29606	39.756
7000	52678	50.789	32088	40.124
7500	57356	51.434	34571	40.467
8000	62094	52.045	37053	40.787
8500	66889	52.627	39535	41.088
9000	71738	53.182	42018	41.372
9500	76638	53.712	44500	41.640
10000	81581	54.220	46982	41.895

TABLE E.18 ***Enthalpy of Formation, Gibbs Function of Formation, and Absolute Entropy of Various Substances at 77 F, 1 atm Pressure***

Substance	Formula	M	State	$\bar{h}_f^o$ Btu/lbmol	$\bar{g}_f^o$ Btu/lbmol	$\bar{s}_f^o$ Btu/lbmol R
Water	H_2O	18.015	gas	−103 966	−98 279	45.076
Water	H_2O	18.015	liq	−122 885	−101 973	16.707
Hydrogen peroxide	H_2O_2	34.015	gas	−58 515	−45 347	55.623
Ozone	O_3	47.998	gas	+61 339	+70 150	57.042
Carbon (graphite)	C	12.011	solid	0	0	1.371
Carbon monoxide	CO	28.011	gas	−47 518	−58 962	47.182
Carbon dioxide	CO_2	44.010	gas	−169 184	−169 556	51.038
Methane	CH_4	16.043	gas	−32 190	−21 841	44.459
Acetylene	C_2H_2	26.038	gas	+97 477	+ 91 412	47.972
Ethene	C_2H_4	28.054	gas	+22.557	+29 402	52.360
Ethane	C_2H_6	30.070	gas	−36 432	−14 167	54.812
Propene	C_3H_6	42.081	gas	+8 783	+26 981	63.761
Propane	C_3H_8	44.094	gas	−44 669	−10 099	64.442
Butane	C_4H_{10}	58.124	gas	−54 256	−6 922	73.215
Pentane	C_5H_{12}	72.151	gas	−62 984	−3 600	83.318
Benzene	C_6H_6	78.114	gas	+35 675	+55 760	64.358
Hexane	C_6H_{14}	86.178	gas	−71 926	−73	92.641
Heptane	C_7H_{16}	100.205	gas	−80 782	+3 439	102.153
n-Octane	C_8H_{18}	114.232	gas	−89 682	+7 049	111.399
n-Octane	C_8H_{18}	114.232	liq	−107 526	+2 770	86.122
Methanol	CH_3OH	32.042	gas	−86 543	−69 905	57.227
Ethanol	C_2H_5OH	46.069	gas	−101 032	−72 399	67.434
Ammonia	NH_3	17.031	gas	−19 656	−6 948	45.969
T-T-Diesel	$C_{14.4}H_{24.9}$	198.06	liq	−74 807	+76 748	125.609
Sulfur	S	32.06	solid	0	0	7.656
Sulfur dioxide	SO_2	64.059	gas	−127 619	−129 030	59.258
Sulfur trioxide	SO_3	80.058	gas	−170 148	−159 515	61.302
Nitrogen oxide	N_2O	44.013	gas	+35 275	+44 782	52.510
Nitromethane	CH_3NO_2	61.04	liq	−48 624	−6 249	41.034

SOME SELECTED REFERENCES

R. E. Sonntag and G. J. Van Wylen, *Introduction to Thermodynamics, Classical and Statistical, Third Edition,* John Wiley & Sons, New York, 1991.

Water: J. H. Keenan, F. G. Keyes, P. C. Hill and J. G. Moore, *Steam Tables,* John Wiley and Sons, Inc., New York, 1969.

Ammonia: *Thermodynamic Properties of Ammonia*, Haar, L. and Gallagher, J.S. Jour. Phys. Chem. Ref. Data, Vol. 7, No. 30, p. 635, 1978.

Argon: *Thermodynamic properties of argon from the triple point to 1200 K with pressures to 1000 MPa,* R.B. Stewart and R.T. Jacobsen, J. Phys. Chem. Ref. Data, vol. 18, no. 2, 1989.

Butane: *Thermophysical properties of fluids. II. Methane, Ethane, Propane, Isobutane and normal butane,* B.A. Younglove and J.F. Ely, J. Phys. Chem. Ref. Data, vol. 16, no 4, pp. 577-798, 1987.

Ethane: *Thermophysical properties of Ethane,* D.G. Friend, H. Ingham and J.F. Ely, J. Phys. Chem. Ref. Data, vol 20, no 2, pp. 275-340, 1991.

Ethylene: *Thermodynamic properties of ethylene from the freezing line to 450 K at pressures to 260 MPa,* M. Jahangirl, R.T. Jacobsen, R.B. Stewart and R.D. McCarty, J. Phys. Chem. Ref. Data, vol. 15, no. 2, pp. 593-645, 1986.

Methane: *Thermophysical properties of Methane,* D.G. Friend, J.F. Ely and H. Ingham, J. Phys. Chem. Ref. Data, vol 18, no 2, pp. 583-632, 1989.

Nitrogen: *Thermodynamic properties of nitrogen from the freezing line to 2000 K at pressures to 1000 MPa,* R.T. Jacobsen, R.B. Stewart and M. Jahangirl, J. Phys. Chem. Ref. Data, vol. 15, no. 2, 1986.

Oxgen: *Thermodynamic properties of oxygen from the triple point to 300 K with pressures to 80 MPa,* R.B. Stewart, R.T. Jacobsen and W. Wagner, J. Phys. Chem. Ref. Data, vol. 20, no. 5, 1991.

Propane: *Thermophysical properties of fluids. II. Methane, Ethane, Propane, Isobutane and normal butane,* B.A. Younglove and J.F. Ely, J. Phys. Chem. Ref. Data, vol. 16, no 4, pp. 577-798, 1987.

R-11, R-12, R-13, R-14, R-21, R-22
R-23, R-113, R-114, R-500, R-502
R-C318:

Refrigerant Equations, Downing, R.C., ASHRAE Paper 2313, Trans. ASHRAE, Vol. 80, part II, p. 158, 1979.

R-22: *Application of Nonlinear Regression in the Development of a Wide Range Formulation for HCFC-22*, Kamei, A., Beyerlein, S.W. and Jacobsen, R.T. Int. Journal of Thermophysics, Vol. 16, No. 5, p. 1155, 1995.

R-123: *Thermodynamic Properties of Refrigerant 123 (2,2-Dichloro-1,1,1-Trifluoroethane),* B.A. Younglove and M.O. McLinden, J. Chem. Phys. Ref. Data Vol 23, no 5, pp. 731-779, 1994.

R-134a: *Thermodynamic charts, tables, and equations for refrigerant HFC-134a*, C. Piao, H. Sato and K. Watanabe, ASHRAE Transactions Vol 97 (2), pp. 268-284, 1991.

An International Standard Formulation for the Thermodynamic Properties of 1,1,1,2-Tetrafluoroethane (HFC-134a) for Temperatures From 170 K to 455 K and Pressures up to 70 MPa, Tillner-Roth, R. and Baehr, H.D. Jour. Phys. Chem. Ref. Data, Vol. 23, No. 5, p. 657, 1994.

R-152a: Yungil Kim, Ph.D. thesis, University of Michigan, 1993.

A Fundamental Equation of State for 1,1-Difluoroethane (HFC-152a) Tillner-Roth, R., Int. Jour. Thermophysics, Vol. 16, No. 1, 1995.

Transport properties:

The properties of gases & liquids, Fourth edition, by Robert C. Reid, John M. Prausnitz and Bruce E. Poling, Mc Graw Hill, 1987.

CRC Handbook of Chemistry and Physics, 74th edition, David R. Lide editor, CRC Press, 1993.

Heat Transfer, Adrian Bejan, John Wiley & Sons, 1993.

Introduction to Heat Transfer, Frank P. Incropera and David P. DeWitt,2nd edition, John Wiley & Sons, 1990.